Advances in HEAT TRANSFER

Edited by

James P. Hartnett

Energy Resources Center
University of Illinois at Chicago
Chicago, Illinois

Thomas F. Irvine, Jr.

Department of Mechanical Engineering
State University of New York at Stony Brook
Stony Brook, New York

Volume 19

ACADEMIC PRESS, INC.
Harcourt Brace Jovanovich, Publishers
San Diego New York Berkeley Boston
London Sydney Tokyo Toronto

ACADEMIC PRESS, INC.
San Diego, California 92101

United Kingdom Edition published by
ACADEMIC PRESS LIMITED
24-28 Oval Road, London NW1 7DX

Library of Congress Catalog Card Number: 63-22329

ISBN 0-12-020019-8 (alk. paper)

PRINTED IN THE UNITED STATES OF AMERICA
89 90 91 92 9 8 7 6 5 4 3 2 1

ADVANCES IN HEAT TRANSFER

Volume 19

CONTENTS

Heat Transfer to Newtonian and Non-Newtonian Fluids in Rectangular Ducts

JAMES P. HARTNETT AND MILIVOJE KOSTIC

Melting and Freezing

L. S. YAO

Department of Mechanical and Aerospace Engineering,
Arizona State University,
Tempe, Arizona 85287

J. PRUSA

Department of Mechanical Engineering,
Iowa State University,
Ames, Iowa 50011

I. Introduction

Freezing and melting are important processes in our world. Undoubtedly, such processes played a decisive role in the formation of the Earth. Since near-homogeneous accretion was almost certainly the way that the Earth was formed, melting and freezing were key processes at work in the initial period of planetary growth and differentiation of the interior. Melting and freezing processes have also provided the necessary continuous segregation mechanism for the tectonic activity underlying magma migration and volcanism. Melting and freezing of the ice caps near the poles are critical in stabilizing the world's climate. Snow, which accumulates on mountains in the winter, is a means provided by nature to store water. The melting of this snow is important to the distribution of water for human needs.

Unwelcome freezing and melting processes do occur and can cause hazards. Icy roads are slippery and often dangerous for moving vehicles, and iced wings can ground aircraft. Melting of frozen soils can destroy the foundations of man-made structures in permafrost regions. An interesting description of many serious problems in cold regions has been presented by Lunardini [1]. Numerous examples can be used to demonstrate the importance of melting and freezing processes in nature. Understanding the mechanisms of those phenomena provides the means to cope with nature and to improve our living standards.

At the present time, phase-change phenomena have become an indispensable part of many processes, for example, manufacturing of glass, crystals, and metal alloys; continuous casting of ingots, fabrics, and wires; food preservation and cryopreservation of biological cells; and desalination of seawater and freezing-out separation processes. One of the most important inventions that improves our life-style is refrigeration. Phase-change is the key mechanism that allows heat to be transferred from cold to hot and provides a shield to unbearable natural cycles. Less important, but more intimate, are products of refrigeration, such as ice cream and popsicles, which certainly sweeten our lives and are an unforgettable part of our childhood.

It is difficult to prepare a complete list of phase-change phenomena, let alone to review the literature about phase-change studies. Since there are unlimited incentives to study phase-change processes, several books, monographs, and review articles are available and provide good sources of information. Even though we will concentrate our survey on more recent efforts, it is still difficult to complete the task since the literature concerned with this problem is continuously increasing. We hope that our survey presents a reasonable picture of current knowledge in this problem area. The chosen bibliography is restricted to the scope of our survey and sometimes depends on availability; unavoidably, it is biased somewhat to our preferences. We believe other important contributions can be easily identified through the references cited in this article. An effort has been made to translate particular results into a consistent system of nomenclature. It is our hope that this will enhance physical understanding of the subject. Finally, we want to point out that our survey is restricted to phase-change problems from solid to liquid or to gas. The physics of liquid/gas phase-changes are somewhat different and therefore are not reviewed in this article.

A rather complete survey of the available literature before 1964 in heat conduction with change of phase has been provided by Bankoff [2] who describes integral–equation methods, variational methods, boundary-layer methods, and perturbation methods for small time. A more complete review of integral methods has been made available by Goodman [3]. A good collection of classical solutions and the mathematical background of phase-change problems and other moving-boundary problems whose formulations can be reduced to phase-change problems are included in the translated monograph entitled "The Stefan Problem" [4]. Two volumes, which contain the proceedings of workshops held at Oxford University in 1974 [5] and at Oak Ridge National Laboratory in 1977 [6], contain information on the application of phase-change processes, the development of mathematical theories, and methods of solutions. A recent book by Crank [7] provides a thorough review of most available methods.

Because of their simplicity, one-dimensional, phase-change problems involving conduction heat transfer are the most frequently used model problems for the development of solution methods. This survey starts with the consideration of such one-dimensional methods, either exact or approximate, which have been widely used in the literature. The substantial information about the physics of phase-change processes that can, in fact, be learned from studies of these simple models is discussed in depth in Section II. The extension of these methods to multidimensional conduction phase-change problems as well as the additional physics attributable to geometrical complexities are also reviewed there.

The initial concern about the influence of convection on the phase-change processes can be traced back to more than 20 years ago. The significance of convection was confirmed only after 1977. The convection phase-change models used before 1977 are speculative. Usually an empirical factor was introduced to adjust the heat-transfer coefficient to account for convection effects. This factor was typically determined by matching with available experimental data. This kind of approach can be considered as a global correction of the inadequacy of the conduction models, and it cannot properly model the details of the local phase-change mechanism.

Convincing flow visualization of melting around a hot horizontal cylinder revealed that convection is likely the dominant heat-transfer mode in phase-change processes. Gone were the simple one-dimensional models; irregular geometry was recognized as an inherent property of phase-change problems. Substantial efforts have since been devoted to the clarification of the importance of convection effects. Successful solution methods must be able to handle complex and moving geometries; consequently, geometry complicates experimental techniques and degrades the accuracy of experimental results. Several solutions have been obtained by adopting methods related to body-fitted coordinates, which are an effective way to treat complex and moving geometries. At present, the effects of moving boundaries can be efficiently analyzed by these methods. For extremely complex geometries, numerical methods are currently limited by the *finite* storage and speed of the current generation of computers. In the forseeable future, such a restriction can be expected to lessen, but it is doubtful that it can be completely removed. Of course, no guarantee can be made that further limitations of numerical methods will not surface. Most available analytical solutions have been obtained for problems associated with natural convection. A survey of the methods that produced these solutions and comparisons with experimental data form the major part of Section III. A collection of literature by Viskanta [8, 9] is particularly related to thermal-storage applications.

Relatively little analytical effort has been devoted to the effect of forced convection on phase-change processes. The phenomena are complicated because of possible flow separations, instabilities, and turbulence. Interesting phase-change problems influenced by forced convection have been comprehensively described by Epstein and Cheung [10], and presented in an expanded form [11]. Most of these forced-convection, phase-change problems could probably be analyzed by using the same methods developed for natural convection *if the flow is laminar.* At present, our knowledge about forced-convection effects on phase-change problems has accumulated through scaled experiments, and the reliability of predictions relies heavily on empirical correlations. A brief review of the physics of these problems and some speculations about their solutions make up the remaining part of Section III.

In the last section, a few additional problem areas in phase-change are discussed. Frequently, model simplifications, sometimes oversimplified, are necessary for more complicated phenomena. A common simplification is the neglect of moving boundaries. Sometimes substantial discrepancies between the model studies and the experimental observations occur. In light of the progress of our knowledge in phase change in the past decade, we suggest that moving-boundary effects could be one of the leading candidates for future study to remedy these discrepancies. Important physical ideas, which have so far been ignored in model studies, are pointed out, and possible approaches to tackle these ideas are also discussed. Such discussions have to be speculative in nature and are subject to modification as more is learned in the future.

II. Conduction-Dominated Melting–Freezing Problems

A. Historical Development

In 1831, Lame and Clapeyron [4] published the first analytical work on phase-change problems. They considered the freezing of a liquid held at the freezing temperature by constant-temperature cooling. They found that the formation of the crust was proportional to the square root of time but did not determine the constant of proportionality.

During the 1860s, Neumann solved exactly the one-dimensional phase-change problem with isothermal boundary conditions in a series of unpublished lectures [12, 13]. He considered the case of freezing an initially superheated liquid (the liquid is initially above the freezing temperature; in a melting problem, the analogous condition is the initial subcooling of a solid below the melting temperature). Since the temperatures of both phases are to be determined, this is known as a two-phase problem. Neumann assumed trial forms for the temperatures [2] in terms of the error function, a known solution of the equation of linear heat flow [12]. He then determined that these solutions satisfied the governing equations and all boundary conditions provided that the position of the solid–liquid interface varied as the square root of time. Three partial differential equations were required to be solved simultaneously: the heat diffusion equations for the solid and liquid phase-change material (PCM) and an equation derived from an energy balance at the solid–liquid interface. Neumann's results were not published until 1912 [12, 13].

In 1889, Stefan published articles on two-phase problems in infinite and semi-infinite planes [4]. In another paper published in 1891, he solved a one-phase problem motivated by observations of the formation of polar ice [4, 12–14]. Stefan determined that, for the case of isothermal cooling, the position of the solid–liquid interface varied as the square root of time. By assuming that the position of the solid–liquid interface varies linearly with time [12, 14], he also found a solution for a one-phase freezing problem in which the temperature has a simple exponential character. However, this solution has no physical significance since it requires that the temperature at the cooled surface decrease exponentially fast in time. More general open-form solutions were also developed by Stefan for problems in which the temperature at the cooled surface was an arbitrary function of time [4, 14]. Because Neumann's work was still some 20 years from publication, it was not realized that Stefan's work, in part, repeated Neumann's earlier work. Consequently, it was Stefan's contributions that were first recognized and first developed a general awareness of phase-change problems, particularly among geophysicists. In light of Stefan's seminal role, phase-change problems have since been named Stefan problems in his honor. Since moving boundaries are the outstanding feature of phase-change problems, various other moving-boundary problems are frequently also referred to as Stefan problems.

Boltzmann recognized that the one-phase problem with isothermal boundary conditions could be solved through the use of the similarity transformation

$$\eta = \bar{x}/(\alpha\bar{t})^{1/2} \tag{1}$$

which reduced the heat diffusion equation to an ordinary differential equation [15]. This transformation is now known as the Boltzmann transformation [2].

Since these pioneering efforts of the previous century, a tremendous amount of work has occurred on Stefan problems. A large number of techniques have been developed in order to attack a wide variety of problems that are variations and generalizations of the elementary problems analyzed by these first investigators. Unfortunately, no single method appears to work best in all classes of problems. It is the purpose of the remainder of this chapter to present the major methods for solving phase-change problems and to interpret fundamentally important results obtained by their use. The goal is to develop a coherent understanding of the basic physics of the phase-change process in order to better decide which method (or combination of methods) may be most appropriate in a given situation.

B. SIMILARITY TRANSFORMATION

It is a matter of record that in almost all cases where an exact closed form solution can be determined the problem admits a similarity solution [12, 16]. Thus, we begin our discussion with this method and explore the considerable information that it yields.

Consider the melting of a solid in the semi-infinite plane $0 \leq \bar{x} < \infty$, at a subcooled temperature T_c for time $\bar{t} < 0$. At time $\bar{t} = 0$, a sudden heating is applied to the left face at $\bar{x} = 0$. The left face is maintained at a temperature $T_h > T_0 > T_c$, where T_0 is the phase-change temperature. The melt region grows with time, its thickness being denoted by R (see Fig. 1). Finally, assume that both the solid and liquid phases are homogeneous, but that the values of thermophysical properties for each phase may differ. In particular, suppose that there is a change in density of the PCM upon melting. As a result of mass conservation, this density change will cause a fluid motion to occur toward the solid–liquid interface if the liquid is more dense, and away from the solid–liquid interface if the solid is more dense. This motion of the melt may be interpreted as a suction or blowing effect along the solid–liquid interface. In order to accommodate this fluid motion, let us assume that the heated surface is permeable to the melt. Thus, only one-dimensional fluid motion results.

The formulation of the problem is

$$\frac{\partial \bar{T}_l}{\partial \bar{t}} + (1 - \rho_s/\rho_l)\dot{R}\frac{\partial \bar{T}_l}{\partial \bar{x}} = \alpha_l \frac{\partial^2 \bar{T}_l}{\partial \bar{x}^2} \tag{2a}$$

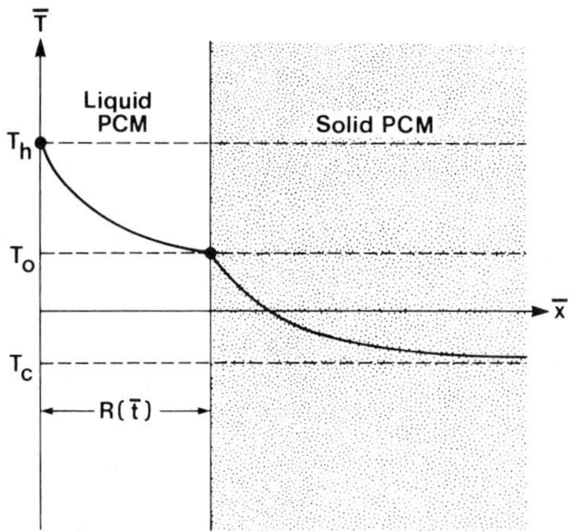

FIG. 1. Geometry of the one-dimensional, two-phase Stefan problem.

$$\frac{\partial \bar{T}_s}{\partial \bar{t}} = \alpha_s \frac{\partial^2 \bar{T}_s}{\partial \bar{t}^2} \tag{2b}$$

$$\rho_s h_{sl} \dot{R} = -\left[k_l \frac{\partial \bar{T}_l}{\partial \bar{x}} - k_s \frac{\partial \bar{T}_s}{\partial \bar{x}} \right]_{\bar{x}=R} \tag{2c}$$

subject to

$$\bar{R} = 0, \bar{T}_s = T_c \qquad \text{for all} \qquad \bar{x} \geq 0$$

and

$$\bar{T}_l \qquad \text{does not exist for} \qquad \bar{t} < 0 \tag{2d}$$

$$\bar{T}_l = T_h \qquad \text{at} \qquad \bar{x} = 0, \qquad \bar{T}_l = T_0 = \bar{T}_s \qquad \text{at} \qquad \bar{x} = R$$

and

$$\bar{T}_s \rightarrow T_c \qquad \text{as} \qquad \bar{x} \rightarrow \infty \qquad \text{for} \qquad \bar{t} > 0 \tag{2e}$$

Here $\dot{R} = \partial R/\partial \bar{t}$. The second term in Eq. (2a) represents the convection effect due to density change. The magnitude of the melt velocity, $-(1 - \rho_s/\rho_l)\dot{R}$, is determined using a mass balance on an infinitesimal control volume along the interface. Equation (2c) is also known as the Stefan condition. It predicts the position of the solid–liquid interface and is based upon an energy balance on an infinitesimal control volume along the interface.

A similarity solution is obtained by rescaling the spatial dimension using R as the characteristic length. By assuming that $R\dot{R}/\alpha_l = 2\mu^2$, where μ is a yet to be determined constant, use of the similarity transformations $\eta_l = \bar{x}/R + \rho_s/\rho_l - 1$ and $\eta_s = \bar{x}/R$ will reduce the system of partial differential equations [Eqs. (2a)–(2c)] into the following coupled system of ordinary differential equations:

$$\frac{d^2 T_l}{d\eta_l^2} + 2\mu^2 \eta_l \frac{dT_l}{d\eta_l} = 0 \tag{3a}$$

$$\frac{d^2 T_s}{d\eta_s^2} + 2\mu^2 \lambda \eta_s \frac{dT_s}{d\eta_s} = 0 \tag{3b}$$

$$2\mu^2 = -\text{Ste}\left[\left(\frac{dT_l}{d\eta_l} \right)_\Delta - \text{Sb}\left(\frac{dT_s}{d\eta_s} \right)_1 \right] \tag{3c}$$

where

$$T_l = (\bar{T}_l - T_0)/(T_h - T_0), \qquad T_s = (\bar{T}_s - T_c)/(T_0 - T_c)$$

$$\text{Ste} = \frac{\rho_l}{\rho_s} \frac{c_l(T_h - T_0)}{h_{sl}} \qquad \text{(Stefan number)}$$

$$\mathrm{Sb} = \frac{k_s}{k_l}\frac{T_0 - T_c}{T_h - T_0} \qquad \text{(subcooling number)}$$

$$\lambda = \alpha_l/\alpha_s, \qquad \Delta = \rho_s/\rho_l$$

subject to

$$T_l = 1 \quad \text{at} \quad \eta_l = \varepsilon = \Delta - 1, \qquad T_l = 0 \quad \text{at} \quad \eta_l = \Delta$$

$$T_s = 1 \quad \text{at} \quad \eta_s = 1 \quad \text{and} \quad T_s \to 0 \quad \text{as} \quad \eta_s \to \infty \tag{3d}$$

Equations (3a) and (3b) may be solved independently of Eq. (3c). The resulting distributions for T_l and T_s may then be substituted into Eq. (3c), with these results:

$$T_l = \frac{\operatorname{erf}(\mu\Delta) - \operatorname{erf}(\mu\eta_l)}{\operatorname{erf}(\mu\Delta) - \operatorname{erf}(\mu\varepsilon)} \tag{4a}$$

$$T_s = \frac{\operatorname{erfc}(\mu\lambda^{1/2}\eta_s)}{\operatorname{erfc}(\mu\lambda^{1/2})} \tag{4b}$$

$$\mu = \frac{\mathrm{Ste}}{\sqrt{\pi}}\left\{\frac{e^{-\mu^2\Delta^2}}{\operatorname{erf}(\mu\Delta) - \operatorname{erf}(\mu\varepsilon)} - \frac{\mathrm{Sb}\,\lambda^{1/2}e^{-\mu^2\lambda}}{\operatorname{erfc}(\mu\lambda^{1/2})}\right\} \tag{4c}$$

Given values of the dimensionless parameters Ste, Sb, λ, and Δ, one solves the transcendental equation [Eq. (4c)] to determine μ. Given μ, the temperature distributions become known. The solid–liquid interface may be determined from

$$R = 2\mu(\alpha_l\bar{t})^{1/2} \tag{4d}$$

thus completing the solution. Henceforth, we will refer to the solution [Eq. (4)], for the case $\Delta = 1$, as the Stefan–Neumann solution. The Stefan number may be thought of as the ratio of available sensible heat to latent heat, whereas the subcooling number may be interpreted as the ratio of sensible cooling of solid PCM to sensible heating of the liquid PCM.

Several outstanding characteristics of Stefan problems may now be pointed out. Let us consider the mathematical nature of phase-change problems. The fact that the position of the solid–liquid interface is unknown *a priori* and must be determined as part of the solution makes the problem nonlinear. The nonlinearity appears in the formulation in the Stefan condition [Eq. (2c)]. Carslaw and Jaeger [12] demonstrate its existence explicitly. Since the nonlinearity generally excludes the possibility of superposition of solutions, the number of analytical techniques available for solving Stefan problems is considerably reduced.

Next, observe that temperature gradients in the melt and solid near the solid–liquid interface and the rate of growth of the melt thickness all become unbounded as $\bar{t} \to 0$. In fact, Eq. (2) becomes singular as the melt degenerates. The presence of the initial singularity complicates the character of the problem greatly; general questions such as existence and uniqueness of solutions can only be answered away from the singularity, for $\bar{t} > 0$ [4, 16]. The presence of the singularity is due to the sudden appearance of melt at $\bar{t} = 0^+$ when none existed the moment before, at $\bar{t} = 0^-$. This interpretation makes it clear that other singularities will occur whenever one of the phases degenerates.

The nonlinear character of Stefan problems coupled with their singular behavior as a phase degenerates combine to make these problems intrinsically very difficult to solve. Sophisticated techniques are generally required to obtain approximate analytical solutions. Even numerical methods that work well on many nonlinear problems often perform quite poorly on Stefan problems. They simply cannot handle the extraordinarily large gradients that occur in the neighborhood of a singularity or along the solid–liquid interface [17]. Of course, it is possible to "diffuse" the effects of these large gradients by using sufficiently coarse increments in time and space or by explicitly adding an artificial viscosity term into a numerical method. Generally, however, a successful and accurate numerical method requires that some very careful analytical modeling be built into it. Most often the best method for solving a Stefan problem is a combination of analytical and numerical methods [17].

Let us now consider the physical nature of phase-change problems. Since μ is constant, it is clear from Eq. (4d) that the melt region thickness $R(\bar{t})$ grows like $\bar{t}^{1/2}$. This is a general characteristic of all one-dimensional phase-change problems driven by constant temperature boundary conditions. Corresponding to this growth rate is the fact that the initial singularity is a square root singularity, that is, various terms in the dimensional formulation become unbounded like $1/\bar{t}^{1/2}$ as $\bar{t} \to 0$.

Of the four independent dimensionless parameters appearing in Eqs. (4a)–(4c), Ste is by far the most important in determining the melting rate. A straightforward Taylor series expansion of Eq. (4c) indicates that

$$\lim_{\mathrm{Ste} \to 0} \mu = (\mathrm{Ste}/2)^{1/2} \tag{5a}$$

Thus, for sufficiently small values of Ste, the effects of Sb, λ, and Δ all vanish. Remarkably, if there is a negligible subcooling effect and density change ($\mathrm{Sb} \to 0$ and $\Delta \to 1$), Eq. (5a) predicts the value of μ accurately even for values of Ste that are not small. For example, the result predicted by Eq. (5a) is in error by at most 10.3% for $\mathrm{Ste} \leq 0.7$. Table I summarizes a more detailed comparison between the exact μ value for $\mathrm{Sb} = 0$, $\Delta = 1$, and the approximate prediction given by Eq. (5a). If Eq. (5a) is substituted into Eq. (4d), an

TABLE I

COMPARISON OF PREDICTED μ VALUES

Ste	Exact μ value[a]	Approximate μ value[b]	Percentage error
.01	.070594	.070711	+0.17
.02	.099670	.10000	+0.33
.05	.15682	.15811	+0.83
.10	.22002	.22361	+1.6
.20	.30643	.31623	+3.2
.50	.46479	.50000	+7.6
1.00	.62006	.70711	+14.0

[a] Determined by Eq. (4c).
[b] Determined by Eq. (5a).

extremely simple equation for the thickness of the melt region is obtained

$$R/(\alpha_l \bar{t})^{1/2} = (2\ \mathrm{Ste})^{1/2} \tag{5b}$$

Moderate amounts of subcooling will, however, cause significant changes in the value of μ if the value of Ste is not small. If subcooling is present, thermal energy is required to heat the PCM in order to raise its temperature to the melting point. Less energy is then left for melting. Figure 2 shows the percent decrease in the melting rate that subcooling causes. For example, if Ste = 1,

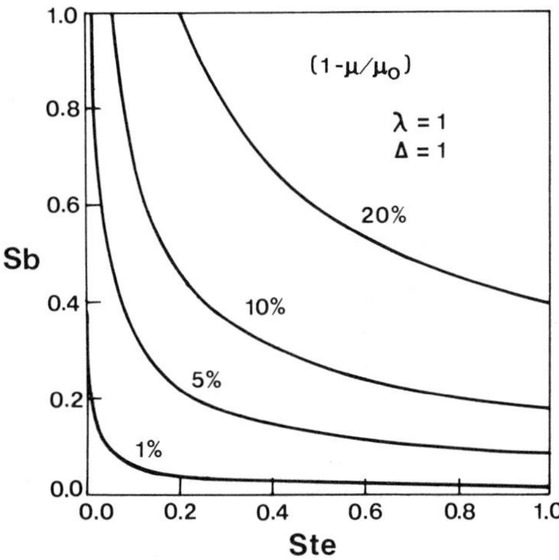

FIG. 2. Reduction in melting rate caused by uniform subcooling of the solid phase.

$\lambda = 1$, and $\Delta = 1$, then a subcooling number of only Sb = 0.174 will cause a 10% decrease in the melting rate as compared to a rate with Sb = 0.

The effect of the ratio of thermal diffusivities λ is to enhance or diminish the effect of subcooling according to whether $\lambda < 1$ or $\lambda > 1$, respectively. If $\lambda < 1$, then the solid PCM has an enhanced value of diffusivity, and it warms up more easily for a given input of heat [18]. Less energy is needed for sensible heating and more is available for phase change. If $\lambda > 1$, then the reverse argument applies. For example, if Ste = 1, Sb = 1, and $\Delta = 1$, then a value of $\lambda = 0.5$ enhances the melting rate by 16.7%, whereas a value of $\lambda = 2.0$ decreases the melting rate by 18.7%, when compared to the melting rate that occurs when $\lambda = 1$.

The effect of the density ratio Δ is to enhance or diminish the melting rate. If $\Delta > 1$, then the blowing effect at the solid–liquid interface diminishes the melting rate since thermal energy is being convected away from the interface. The suction effect occurring along the solid–liquid interface when $\Delta < 1$ has the opposite effect. However, note that when $\Delta = 1.1$ and 0.9, the decrease and increase are only 1.6% and 1.7%, respectively, when Ste = 1 and Sb = 0. These results clearly show that the effects of density change generally are surprisingly small, since a 10% change in density is representative of a great number of PCMs. Even if more extreme changes in density are considered, for example, $\Delta = 2.0$ and $\Delta = 0.5$, the increase and decrease in melting rates are still only 14.0% and 9.7%, respectively, for Ste = 1 and Sb = 0. It is interesting to observe that the melt velocity induced by density changes becomes unbounded in magnitude like $1/\bar{t}^{1/2}$ as $\bar{t} \to 0$. The magnitude of this convection effect is $0[\text{Ste}(1 - \Delta)]$ compared with that of conduction. All effects of density change decrease as Ste $\to 0$ and/as Sb increases.

In summary, the similarity solution has revealed that subcooling effects can be quite important while those due to density change are negligible. A word of caution is in order, however. Effects due to density change are negligible only if the Stefan number is defined as in the present discussion, following Eq. (3c). It is common to see the Stefan number defined in the literature as $c_l(T_h - T_0)/h_{sl}$. This definition, missing a factor of ρ_l/ρ_s, will not correctly model the effects of density change. In effect, the influence of density change may be modeled by first- and second-order approximations. To first order, the effects are well approximated simply by using the correct definition of Ste (in a freezing process, it is $c_s(T_0 - T_c)/h_{sl}$). The second-order approximation appears in the Stefan condition, Eqs. (2c) and (3c); these are the effects that are termed negligible. If the incorrect definition of Ste is used, then the predicted melting rate will be in error by a factor of $0(1/\Delta^{1/2})$.

Very few additional Stefan problems are known to have exact close-form solutions. Cases in which the domain is extended to the entire plane, the PCM has a melting range (limit of vanishing heat of fusion), and multiphases exist simultaneously (e.g., three or more distinct phases of PCM) are all solvable in closed form using the method of similarity transformation [12]. Additional exact solutions are available for Stefan problems in one-dimensional cylindrical and spherical systems [2, 12] and in an infinite wedge [19]. Generally, key requirements for a successful similarity transformation are an unbounded domain, constant thermophysical properties, and constant-temperature boundary conditions driving the phase-change process.

C. Power Series Expansions

It is natural to consider the effects of different types of boundary conditions on the solution of Stefan problems, in particular, those of constant heat flux and convection. These types of boundary conditions do not admit similarity solutions. By assuming the following power series for R and $\bar{T}$,

$$R = a_0 + a_1\bar{t} + \cdots \tag{6a}$$

$$\bar{T} = (b_{00} + b_{01}\bar{x} + \cdots) + \bar{t}(b_{10} + b_{11}\bar{x} + \cdots) + \cdots \tag{6b}$$

it is possible to substitute the series into Eq. (2) and determine the unknown coefficients to obtain a solution. Consider one-phase freezing problems in a semi-infinite domain, $\bar{x} \geq 0$, driven by constant heat flux or convection boundary conditions at the cooled surface, $\bar{x} = 0$. Then the solutions for the position of the solid–liquid interface may, to leading order, in both cases be written in the form [12]

$$R/(\alpha_1\bar{t}/l) = \mathrm{Ste} + 0(\mathrm{Ste}^2\,\bar{t}/(l^2/\alpha_1)) \tag{6c}$$

where $\mathrm{Ste} = 1$ and $l = \rho_s\alpha_s h_{sl}/q$ [19] for the constant heat flux case; $\mathrm{Ste} = c_s(T_0 - T_\infty)/h_{s1}$, $l = k_s/h$, h is the heat transfer coefficient, and T_∞ is the bulk temperature of the heat transfer fluid [20, 21] for the convection case. If the characteristic length l is allowed to be some other explicit scale, then the constant-flux Stefan number is defined as $\mathrm{Ste} = c_s ql/h_{sl}k_s$ [22]. For the convection case, Ste in Eq. (6c) must be replaced with $\mathrm{Ste} \cdot \mathrm{Bi}$, where $\mathrm{Bi} = hl/k_s$.

The most fundamental difference between one-phase Stefan problems with flux or convection boundary conditions and those with isothermal boundary conditions is apparent in Eq. (6c). The initial growth rate of the new phase is linear in time, $R \sim \bar{t}$, whereas $R \sim \bar{t}^{1/2}$ for the case of isothermal boundary conditions. Tao presents the first four terms of the series solutions for R for the cases of convection and constant-flux boundary conditions [23–25] (the coefficient of the second term in the constant heat flux solution [25] is incorrectly given as -3, its value should be $-\frac{1}{2}$). Tao's expansions are in terms of three different types of polynomials of complimentary error functions rather than the more naive series of Eqs. (6a) and (6b). These polynomials allow Tao to determine analytical solutions for one-dimensional Stefan problems with arbitrary initial conditions as well as variable boundary conditions. Questions on existence as well as convergence of the resulting series are satisfied. Tao has further determined that in two-phase problems the initial growth rate of R is the same order for all three types of boundary conditions, that is, $R \sim \bar{t}^{1/2}$. This contrasts sharply with the behavior of the one-phase solution [25]. Two-phase problems with flux and convection-driven melting also differ fundamentally from the case of isothermally driven melting in that a premelt solution occurs. Unlike the isothermal case, the other two boundary conditions result in an initial period of sensible heating whereby the subcooled solid is brought up to the melting temperature. Tao further determines that similarity solutions can occur only if the heat flux at $\bar{x} = 0$ varies like $1/\bar{t}^{1/2}$ [24]. This corresponds to the case of isothermal heating.

An earlier work by Evans *et al.* [26] determined solutions for one- and two-phase melting problems with a variable surface heat flux. They give the first five terms of the power series solution for R for the case of the one-phase problem. For a constant heat flux, their result agrees with Eq. (6c). For the two-phase problem, if the heat flux is assumed to be continuous in time throughout the premelt and melting processes, their solution indicates a zero initial rate of growth for the melt. This behavior is not consistent with Tao's results.

Although Eq. (6c) gives only the leading term of the solution for the position of the solid–liquid interface, it can give a very reasonable estimate of the growth rate even for moderate values of time and the Stefan number. Since no exact closed form solution exists for the case of constant flux and convection boundary conditions, a straightforward comparison cannot be made to test the validity as was done for Eq. (5b). The full power-series solutions as given in [24–26] are not very helpful in this endeavor because their radii of convergence are too small to be helpful in establishing quantitative limits, but quantitative estimates can be made using the approximate results of Cho and Sunderland [20]. For the constant flux case with Ste = 1, Eq. (6c) overpredicts the melting rate roughly by 10% and 20% at $\bar{t}/(l^2/\alpha_1)$ equal to 0.3 and 0.8, respectively. For the convection case with Ste = 1 and Bi = 1, Eq. (6c) overpredicts the melting rate by the same amount roughly at $\bar{t}/(l^2/\alpha_1)$ equal to 0.1 and 0.4. Of course, Eq. (6c) becomes exact as $\bar{t} \to 0$. Note that the initial intervals of time during which Eq. (6c) is a valid approximation will lengthen considerably as the Stefan number decreases in value.

Power-series expansions find their greatest use in the determination of short-time solutions to the Stefan problem. For small times, series convergence is generally very robust and only a few terms are needed to provide accurate solutions. Also, the range of validity of a short-time solution is often considerable [27]. Finally, power-series methods can give very accurate solutions in the neighborhood of any singularities that may occur as a phase degenerates. For later times, more terms are required to provide an accurate answer and power-series methods begin to lose their attractiveness.

D. Integrodifferential Equations

Reduction of the Stefan problem [Eq. (2)] to a system of integrodifferential equations provides another avenue of attack. It has been found that questions on existence and uniqueness of solutions can often be more conveniently addressed if the formulation can be rewritten in terms of integrodifferential equations [4, 28]. Such formulations may also serve as the basis for determining various asymptotic solutions [4, 29]. A wide variety of methods of reduction exist.

Lightfoot [2, 12] considered the solid–liquid interface to be a moving heat source of strength $h_{sl}\rho_s\dot{R}$. Assuming that $R(\bar{t})$ was an arbitrary, though specified, function of time, he determined the temperature distribution resulting from this source by using the Green's function:

$$G(\bar{x},\bar{t}\,|\,R(\tau),\tau) = \frac{\exp\left\{-\dfrac{[\bar{x}-R(\tau)]^2}{4\alpha(\bar{t}-\tau)}\right\} - \exp\left\{-\dfrac{[\bar{x}+R(\tau)]^2}{4\alpha(\bar{t}-\tau)}\right\}}{2\sqrt{\pi\alpha(\bar{t}-\tau)}} \tag{7a}$$

for a unit instantaneous plane source at $\bar{x} = R(\bar{t})$ [2]. This solution was then superposed with a second solution satisfying the initial and boundary conditions in order to obtain the desired PCM temperature for a two-phase freezing problem. An integrodifferential equation for $R(\bar{t})$ was then determined by evaluating the superposed temperature distribution at the solid–liquid interface, thus completing the solution. In dimensionless form, the results were

$$\frac{\bar{T}-T_c}{T_0-T_c} = (1+\mathrm{Su})\,\mathrm{erf}\left\{\frac{\bar{x}}{2\sqrt{\alpha\bar{t}}}\right\} + \frac{1}{\mathrm{Ste}}\int_0^{\bar{t}} \dot{R}(\tau)G(\bar{x},\bar{t}\,|\,R(\tau),\tau)\,d\tau \tag{7b}$$

$$1 = (1+\mathrm{Su})\,\mathrm{erf}\left\{\frac{R(\bar{t})}{2\sqrt{\alpha\bar{t}}}\right\} + \frac{1}{\mathrm{Ste}}\int_0^{\bar{t}} \dot{R}(\tau)G(R(\bar{t}),\bar{t}\,|\,R(\tau),\tau)\,d\tau \tag{7c}$$

Here, $\mathrm{Su} = k_l(T_h - T_0)/k_s(T_0 - T_c)$ is the superheating number, and it is the ratio of sensible heating of the liquid PCM to sensible cooling of the solid PCM. The superheating number is the counterpart of the subcooling number appearing in melting problems. For the case of an isothermally driven phase change, Lightfoot assumed that $R \sim \bar{t}^{1/2}$. As a result, the integrodifferential equation [Eq. (7c)] for R degenerated into a transcendental equation for the constant of proportionality, and he was able to recover the exact Stefan–Neumann solution with his method. An interesting feature of the solution is that it does not differentiate between solid and liquid PCM temperature distributions. Only one equation for the temperature results [Eq. (7b)]; whether the phase is liquid or solid depends upon which side of the solid–liquid interface the PCM is located. An important constraint tied to this feature is that the thermophysical properties of the solid and liquid PCM must be identical. Since the Stefan problem is nonlinear, it is curious to see Lightfoot's method—based upon linear techniques—correctly recover the Stefan–Neumann solution. What occurs is that the *a priori* assumption that $R(\bar{t})$ is specified allows the nonlinearity to be concentrated in Eq. (7c) [2]. Budhia and Kreith used Lightfoot's method to solve the phase-change problem in an infinite wedge [19]. They assumed an approximate hyperbolic form for the shape of the solid–liquid interface. In order to predict phase-change results when the solid and liquid PCM properties differed from each other, they introduced a semianalytical correction term. While Budhia and Kreith's method required a sudden increase in surface temperature, the two surfaces of the wedge could be at different isothermal temperatures.

Other methods based upon the use of Green's functions have also been developed. Kolodner [2] developed one method which is free of the constant-property constraint of Lightfoot's method. To begin, Kolodner extends the finite domain of solution through all space and uses jump conditions in the temperature and its gradient at the solid–liquid interface. The jump conditions are achieved through suitable use of single or double layers of sources at the solid–liquid interface. The temperature distribution extending beyond the original domain is called the complementary solution, and it is the continuous function that satisfies the heat diffusion equation in such a way that the boundary condition at the fixed surface of the finite domain is satisfied. The final results are a coupled set of integrodifferential equations for the temperature and the position of solid–liquid interface. Techniques based upon integrations of $T(G_{xx} + G_t) - G(T_{xx} - T_t)$ in both finite and infinite domains, where G is an appropriate Green's function, also exist [4, 29]. Such methods again result in systems of coupled integrodifferential equations.

Kolodner's method is an example of a class of techniques called embedding methods. The basic idea of such methods is to consider each phase with a time variant domain to be part of, or "embedded" in, a larger domain with fixed boundaries. The boundary conditions along the larger domain are not directly determined by the physics of the Stefan problem. Instead, they are determined so that the boundary conditions along the solid–liquid interface and any other physically real boundaries are satisfied. An embedding method developed by Boley [27] introduces an unknown heat flux $q(\bar{t})$ at the nonphysical fixed boundary of the time-invariant domain. A system of integrodifferential equations for $q(\bar{t})$, $R(\bar{t})$, and the temperature are then determined through the use of DuHamel's superposition integral [2, 17]. Boley's embedding method was extended to certain types of multidimensional Stefan problems by Sikarskie and Boley [30]. Without internal heat generation, Boley distinguishes three classes of melting problems.

Class 1: melting starting simultaneously at all points of a surface
Class 2: melting starting simultaneously at all points of a portion of a surface
Class 3: melting starting at a point of a surface

All one-dimensional problems are Class 1 problems. Solutions to a few Class 3 problems indicate that initially the solid–liquid interface propagates much faster along the surface ($R \sim \bar{t}^{1/2}$) than into the interior ($R \sim \bar{t}^{3/2}$) of the PCM [27].

Transform methods can also be used to solve one-dimensional Stefan problems [28, 29]. Over infinite intervals, Fourier transforms may be used to reduce Stefan problems to initial-value problems. For finite intervals, the Laplace transform is more useful and leads to a boundary-value problem. Inversion of the solutions to the subsidiary equations using either method then leads to an integrodifferential equation, often of the Volterra or Fredholm type. Difficulties arise when using the Laplace transform if $R(\bar{t})$ is not monotone. As in Lightfoot's method, it is interesting to observe that with transform methods, linear techniques again seem to be useful despite the nonlinearity of Stefan problems.

Additional methods exist for reducing the formulation of the Stefan problem [Eq. (2)] into a system of integrodifferential equations [17, 29]. Although all of the methods for producing integrodifferential equations offer the considerable advantage of providing exact solutions in integral form, there are some distinct disadvantages. Generally, the solid and liquid phases must be homogeneous, since in large part these methods depend upon linear techniques (with the exception of Lightfoot's method, the solid and liquid properties may differ from each other). Also, it is usually difficult to evaluate the resulting integral equations. For short-time solutions, power-series methods may have to be employed, while numerical methods may be used to evaluate the integrodifferential equation for later times [17, 27].

E. COORDINATE TRANSFORMATIONS

Using an embedding method is one way to replace the moving-boundary problem characteristic of Stefan problems with another problem characterized by a fixed domain. An alternate method, which has the advantage of not requiring the determination of an additional unknown function, is to transform or map the time-variant domain into an invariant domain. Thus, the moving-boundary problem is transformed into a fixed-boundary problem. While a number of ways exist to do this, perhaps the simplest is an elementary stretching and translation of the spatial coordinate according to

$$x = \frac{\bar{x} - S}{R - S} + V \tag{8a}$$

where $R(\bar{t})$ gives the position of the solid–liquid interface, S gives the position of the second boundary of the PCM domain (R gives the first), and V is the distance the coordinate is to be translated. Either S or V may be constants or functions of time. In multidimensional problems, R, S, and V may be functions of spatial position also.

For an example of how the transformation method works, let us again consider the one-dimensional melting of a semi-infinite solid. For simplicity, assume that $\mathrm{Sb} = 0$, so that this is a one-phase problem and $\Delta = 1$, so that there is no fluid motion. Let $S = 0$ be the position of the isothermally heated surface, and set $V = 0$. Then the mapping $(\bar{x}, \bar{t}) \rightarrow (x, t)$ is defined by

$$x = \bar{x}/(lB), \qquad t = \bar{t}/(l^2/\alpha_1) \tag{8b}$$

When these dimensionless coordinates are substituted into the dimensional formulation, Eq. (2), the following dimensionless formulation results

$$B^2 \frac{\partial T_1}{\partial t} - xB\dot{B}\frac{\partial T_1}{\partial x} = \frac{\partial^2 T_1}{\partial x^2} \tag{8c}$$

$$B\dot{B} = -\,\mathrm{Ste}\left.\frac{\partial T_1}{\partial x}\right|_1 \tag{8d}$$

subject to

$$B = 0, \quad T_1 \quad \text{does not exist for} \quad t < 0 \tag{8e}$$

$$T_1 = 1 \quad \text{at} \quad x = 0 \quad \text{and}$$

$$T_1 = 0 \quad \text{at} \quad x = 1 \quad \text{for} \quad t > 0 \tag{8f}$$

Here, $\dot{B} = \partial B/\partial t$, $B = R/l$ is the dimensionless position of the solid–liquid interface, and l is an arbitrary characteristic length. The nonlinear convective term on the left-hand side of the heat diffusion equation [Eq. (8c)] for the melt represents the motion of the solid–liquid interface. In the physical space, the solid–liquid interface does move with time. This information cannot simply be lost from the problem and it reappears in the governing equation itself. Equation (8d) is the Stefan condition. Observe that by using the coordinate transformation, the nonlinear character of the Stefan problem is plainly brought out. Also, it can be observed that some of the nonlinearity is transferred from the Stefan condition into the governing equation. In multidimensions, use of the transformation [Eq. (8a)] would also allow an irregular shaped solid–liquid interface to be mapped into a simpler geometry, such as a plane, cylinder, or sphere. Additional terms with coefficients containing first- and second-order derivatives of B with respect to spatial variables would appear in Eq. (8c). These terms would introduce the variable curvature of the solid–liquid interface in physical space directly into the transformed governing equations.

Use of the transformation does not by itself solve a Stefan problem. It only allows one the considerable convenience of working with a fixed domain. Once the transformation has been employed, any one of a large number of techniques, exact or approximate, may be used to complete the solution procedure. In the recent literature, the transformation method has become especially popular for numerical methods, but we wish to emphasize that it may also be very effective for a number of analytical and semianalytical methods.

The earliest use of the transformation appears to have been made in 1938 by Prandtl in his transposition theorem [31]. Prandtl used the coordinate mapping

$$y = \bar{y} + f(\bar{x}) \tag{9a}$$

to transform a two-dimensional boundary-layer problem along a flat plate into one along the curved plate $\bar{y} = -f(\bar{x})$. Here, $\bar{y}$ is the coordinate normal to the plate while $\bar{x}$ is tangential to the plate. Equation (9a) can be obtained from Eq. (8a) by setting $V = 0$, $S = -f$, and $R = 1 - f$. Glauert generalized the theorem to three-dimensional flows in 1957.

In a paper published in 1949, Zener [32] considered the use of the transformation $x = x/R$, given by Eq. (8a) with $S = 0 = V$ to estimate an approximate solution to the growth rate of a spherical particle. With $R - S = l$ and $V = (S - R)/l$ in Eq. (8a), the transformation can also take on the form $x = (\bar{x} - R)/l$. Redozubov used this form in a 1951 paper on a two-phase problem in a semi-infinite solid [13]. The final solution was obtained using Mellin transforms.

In 1949, Landau published a work on ablation [33] in which he suggested the use of the transformation with $V = 0$ for finite domains. In the ablation problem, melt is removed from the neighborhood of the solid–liquid interface as soon as it is formed. For the case of a semi-infinite solid, an exact steady-state solution exists if constant-flux heating is used [12, 33], which in dimensionless form is

$$T_s = e^{-\dot{B}x} \qquad \text{and} \qquad \dot{B} = \mathrm{Ste}/(1 + \mathrm{Ste} \cdot \mathrm{Sb}) \tag{9b}$$

where $\mathrm{Sb} = (T_0 - T_c)/(ql/k_s)$. The differential equations used to determine this solution are easily generated by using the coordinate transformation in the form attributable to Redozubov. Clearly, the melting rate increases with the Stefan number and decreases with subcooling. For the case of a finite domain, a simple exact solution such as Eq. (9b) does not exist.

Landau examined the ablation problem for the semi-infinite solid after dropping the steady-state assumption. Using standard linear methods, he determined the required premelt solution for a time variant heat flux $q(\bar{t})$ at $\bar{x} = 0$. Solutions for the ablation process were generated only for the case of constant q. After immobilizing the solid–liquid interface using an appropriate form of Eq. (8a), Landau analytically determined asymptotic solutions to the ablation problem for the limits of negligible heat capacity ($\mathrm{Ste} \to 0$) and negligible latent heat ($\mathrm{Ste} \to \infty$). For cases with $\mathrm{Ste} \sim 1$, Landau used a finite-difference method to determine the solution. An interesting result from Landau's solutions is that the initial melting rate (at the moment melting begins) is zero if $\mathrm{Ste} < \infty$, and it is proportional to $1/\mathrm{Sb}$ if the Stefan number is infinite. Landau justified this result physically in terms of a required continuity of the temperature gradient at the solid–liquid interface. This initial behavior does not appear to be consistent with that predicted by Tao [25] for constant heat-flux, phase change. Whether or not the difference is due to melt removal in the ablation process is not clear, since the thermal conditions at the solid–liquid interface are similar with or without ablation. The initial behavior predicted by Landau does seem to be consistent with the behavior predicted by Evans *et al.* [26]. In all cases, as $t \to \infty$, the melting rate approached the asymptotic steady-state value given by Eq. (9b).

Independently, Yang [34] used the coordinate transformation $x = \bar{x} - R$ ($S = R - 1$ and $V = -1$) to analytically determine short- and long-time solutions for freezing in plane stagnation flow. Although the fluid mechanics and heat transfer in the fluid were considered to be two dimensional, the phase-change process itself was modeled as being one dimensional. More recently, Duda *et al.* [35] generalized the one-dimensional form of the transformation to solve a two-dimensional, diffusion-controlled, phase-change problem in a cylindrical geometry. This consisted of replacing $\bar{x}$ by $\bar{r}$, where $\bar{r}$ is the radial position, in Eq. (8a) and recognizing that $R = R(\bar{r}, \bar{t})$. The transformation is suitable for two-dimensional, simply connected geometries. Saitoh [36] further generalized the transformation by letting $S = S(\bar{\xi})$ in Eq. (8a), where $\bar{\xi}$ is an independent position vector to specify the shape of the container. Duda *et al.* [35] and Saitoh [36] used numerical methods to solve the transformed equations. All curvature terms and moving-boundary-induced convective terms are properly retained in these transformations.

Sparrow *et al.* [37] applied a coordinate transformation to determine the effects of subcooling in a one-dimensional melting problem around an isothermally heated cylinder. Two independent transformations were used, one for each phase of the PCM. An implicit finite-difference method was used to solve the resulting fixed-domain formulation. Poor performance by the numerical method near the initial singularity was avoided by starting it some time after the beginning of melting (a melt thickness equal to one percent of the cylinder radius was the criterion used). For the small initial interval of time preceding the numerical solution, the Stefan–Neumann solution was assumed to be accurate. Sparrow *et al.* found that the subcooling effect could be quite substantial in reducing the melting rate when $\mathrm{Sb} \geq 2$. Reductions to a factor of five were obtained. This supported the results of an earlier work by Tien and Churchill [38] on freezing around an isothermally cooled cylinder. Both these results are foreshadowed by the Stefan–Neumann solution [Eqs. (4a)–(4c), see Fig. 2]. Interestingly, subcooling also acted to increase the heat transfer at the cylinder surface, despite the decrease in melting rate. This again agrees with Eq. (4). Initially, the heat flux at the cylinder surface varied like $1/\bar{t}^{1/2}$, in accordance with the Stefan–Neumann solution, but, as time advanced and the curvature effects of the geometry gained importance, this gradually changed so that q increased more rapidly. These curvature effects were observed to be enhanced by larger values of Sb. Finally, Sparrow *et al.* observed that the effects of Ste on the melting rate could almost be eliminated simply by using a dimensionless time:

$$\tau = t\,\mathrm{Ste} \tag{9c}$$

where t is the usual Fourier number, dimensionless time, as defined in Eq. (8b). With this definition, melting curves corresponding to different values of Ste very nearly collapse onto one another. This similitude is easily explained by reference to Eq. (5b), where the first-order dependence of the melting rate on the value of the Stefan number is given. Equation (5b) is correct for any geometry with finite curvature.

A power-series solution for the one-phase freezing problem inside a sphere was developed by Hill and Kucera [39]. Their formulation used a coordinate transformation to immobilize the solid–liquid interface, which in physical space moved inward from the sphere's surface. A convective boundary condition was applied along the surface, and, to leading order, the result for interface motion agrees with Eq. (6c). By setting the radius of the remaining liquid PCM equal to zero, Hill and Kucera obtained a result for the time required to freeze the entire sphere. They provided the following upper and lower bounds for this final time t_f:

$$\frac{2 + \mathrm{Bi}}{6\mathrm{Bi} \cdot \mathrm{Ste}} \leq t_\mathrm{f} \leq \frac{(2 + \mathrm{Bi})(1 + \mathrm{Ste})}{6\mathrm{Bi} \cdot \mathrm{Ste}} \tag{9d}$$

Note that as Bi $\to$ 0 and Ste $\to$ 0, this result indicates that the freezing time is inversely proportional to Bi $\cdot$ Ste. Beyond being intuitively correct, this result suggests that convection-driven phase change shows similitude with respect to the product of Stefan and Biot numbers. The theoretical underpinnings for this are in our discussion of Eq. (6c). Hill and Kucera use tSte as their dimensionless time for presenting data. If instead the time

$$\tau = t\,\mathrm{Bi} \cdot \mathrm{Ste} \tag{9e}$$

is used, then their data for differing values of Bi will be found to very nearly collapse onto a single curve. For example, they determine the freezing times for three different cases of Bi, all with the same value of Ste = 0.10. For values of Bi = 1.0, 0.5, and 0.25, Hill and Kucera find $t_f = 0.53$, 0.88, and 1.56, respectively (from Fig. 2 of Ref. [39]). Using Eq. (9e), the rescaled times become 0.53, 0.44, and 0.39. Even more striking is that Eq. (6c) can actually be used to predict the position of the solid–liquid interface in a spherical geometry. The data generated by Hill and Kucera [39] agree to within 10% of the predictions made by Eq. (6c) up to $t/t_f = 0.60$, and are within 20% up to $t/t_f = 0.75$ for all three cases of the Biot number. As $t \to 0$, Eq. (6c) becomes exact.

Although each of these works used the coordinate transformation [Eq. (8a)] in one form or another, none of them indicated that use of the transformation was advantageous over other methods. Although the evidence is not unequivocal, various works in the literature do indicate that the mapping of a Stefan problem into a fixed domain can and often is a decided advantage. This advantage may exist in a wide variety of techniques of solution. For example, if the approximate technique of a heat balance integral is used to solve a one-dimensional Stefan problem, use of the transformation may improve the accuracy by up to a factor of three, all other considerations being the same [3, 40] (see Section II,H for details).

Furzeland [41] made a comparative study between four different types of numerical methods for solving one-dimensional Stefan problems. Two of these methods were based upon use of the coordinate transformation. Since the nodes move in step with the solid–liquid interface, these types of numerical methods are also known as moving-grid methods. The other two methods employed fixed grids. Furzeland found that the fixed-grid methods gave inferior numerical results, especially for larger values of Ste. Another interesting observation was that the Crank–Nicolson finite-difference method (second-order time truncation error) does not always give better results than an implicit forward-time (first-order time truncation error) method. In fact, the latter type of finite differencing was found to be slightly more accurate near the initial singularity.

As has been demonstrated by Yao [42], the precise application of the boundary conditions on the *actual* boundary is the key reason why the coordinate transformation method provides more accurate results. This fact cannot be solely explained by using the Stokes linearization process to estimate the error. There are two errors to consider in this process. First, the application of the boundary conditions at a distance ε, slightly away from the actual boundary, introduces a "boundary condition" error. An additional error is present because of the approximate form of the governing equations in the limit $\varepsilon \to 0$. Even though both errors are of order ε, the boundary condition error can be substantial for small but finite ε. This behavior was demonstrated by comparison of the perturbation solution in transformed coordinates with the Stokes linearized solution.

Murray and Landis [43] developed a moving-grid method using a coordinate transformation that was much more accurate than the fixed-grid methods then available. The latter methods, termed conventional by Murray and Landis, produced solutions that exhibited pathological behavior, for example, the solid–liquid interface moved in a series of irregular spurts rather than smoothly. This shortcoming was because of the poor implementation of the boundary conditions along the solid–liquid interface. Using linear interpolation between the nodes, Murray and Landis developed a special set of *ad hoc* numerical boundary conditions to improve the fixed-grid method. This improved method, which carefully accounted for the position of the solid–liquid interface, indeed gave results almost as good as the moving-grid method. However, the resulting numerical algorithm was considerably more complicated. Despite Murray and Landis's success, the *ad hoc* boundary conditions at the solid–liquid interface generally do not completely neutralize pathological behavior in the predicted solution from fixed-grid methods. An example of this behavior is given by Pujado *et al.* [44] who, despite careful efforts to develop the finite-difference equations and an involved computational strategy, found that oscillations in the predicted position of the solid–liquid interface resulted. The usefulness of the moving grid approach has also been demonstrated for finite-element methods. Lynch and O'Neill [45] use finite elements to determine solutions to several one-dimensional Stefan problems. They used separate moving meshes for each phase in the two-phase problem and found that accurate solutions could be obtained even when relatively large time steps were used.

Although the work cited here clearly indicates that using a coordinate transformation is advantageous, there are exceptions to this viewpoint. For example, Duda and Vrentas [46] solved a one-dimensional diffusion problem equivalent to a one-phase Stefan problem with density change. They offered two types of perturbation solutions. One of them uses the coordinate transformation, and the other does not. The perturbation solution for the time-variant domain outperforms the one based upon use of the coordinate transformation. In predicting the position of the solid–liquid interface, on average, n terms of the former are about as accurate as $n + 1$ terms of the latter. This directly contradicts the data presented by Yao [42].

If the behavior of a solution in the neighborhood of a singularity is known either from an exact analytical solution or from scale analysis, then it becomes possible to modify the transformation in such a way as to make the dimensionless formulation nonsingular [18]. This tremendously enhances the power of the transformation method. We will refer to this modification of the transformation as the extended form. For example, in the case of isothermal melting, it is known that near the initial singularity $B \sim t^{1/2}$. Thus, we define

$$B = X\sqrt{2t} \tag{10a}$$

where, in general, the gap function, X, is a slowly varying function of time. With this decomposition, the dimensionless formulation Eq. (8) becomes

$$2tX^2\frac{\partial T_l}{\partial t} - x(X^2 + 2tX\dot{X})\frac{\partial T_l}{\partial x} = \frac{\partial^2 T_l}{\partial x^2} \tag{10b}$$

$$X^2 + 2tX\dot{X} = -\mathrm{Ste}\left.\frac{\partial T_l}{\partial x}\right|_1 \tag{10c}$$

subject to

$$T_l = \mathrm{erfc}(xX/\sqrt{2})/\mathrm{erf}(X/\sqrt{2})$$

$$X = \mathrm{Ste}\sqrt{\frac{2}{\pi}}e^{-X^2/2}/\mathrm{erf}(X/\sqrt{2}) \qquad \text{at} \qquad t = 0^+ \tag{10d}$$

and

$$T_l = 1 \quad \text{at} \quad x = 0 \quad \text{and} \quad T_l = 0 \quad \text{at} \quad x = 1 \tag{10e}$$

Now the temperature has a well-defined initial condition [Eq. (10d)] found by determining the solution of the limiting forms of Eqs. (10b) and (10c) as $t \to 0^+$. Unlike Eq. (2) or even Eqs. (8c)–(8f), Eqs. (10b)–(10d) can be solved with great accuracy using a numerical method starting right at the moment that phase change begins. Since X is at most only a slowly varying function of time, there are no terms in Eqs. (10b) or (10c) that will ever change rapidly with time. If X is constant, as in the case of the melting or freezing in a semi-infinite solid, then the extended form of the transformation leads directly to a similarity solution. For different types of boundary conditions, Eq. (10a) should be modified to reflect the different orders of the singularity. For one-phase problems with flux or convective type boundary conditions, one would employ

$$B = Xt \qquad \text{and} \qquad T = \frac{\bar{T}_1 - T_0}{B\,\Delta T} \tag{10f}$$

since the growth rate is initially linear in time as indicated by Eq. (6c). Here, $\Delta T = ql/k$ for the constant heat flux case and $T_\infty - T_0$ for the convective case. If, in addition, one used the dimensionless time [Eqs. (9c) and (9e)], the resulting solutions would show a degree of similitude that would become perfect as Ste $\rightarrow 0$, in accord with Eq. (6c). Landau actually used the extended form of the transformation in his work on the ablation of a semi-infinite solid [33]. Crank also used an extended form of the transformation in a work solving a one-phase Stefan problem [47]. Crank pointed out how to solve a two-phase problem using two transformations but did not actually solve such a problem.

F. Weak Formulations and the Enthalpy Method

The formulation [Eq. (2)] and solution [Eq. (4)] of the one-dimensional, two-phase Stefan problem are also known as the classical formulation and solution. In classical formulations, one solves for the temperature and solid–liquid interface position. The moving boundary appears explicitly in the form of a Stefan condition, such as Eq. (2c). The classical solution exhibits a high degree of continuity and consists of separate temperature functions, one for each PCM phase, such as Eq. (4a) and (4b).

Equation (2) can be reformulated to obtain what is known as a weak solution to the Stefan problem. For the one-dimensional, two-phase problem, a weak solution is defined to be a pair of bounded functions $(\bar{T}, \bar{h})$ such that

$$\bar{h} = \begin{cases} \rho c \bar{T}, & \text{for} \quad \bar{T} < T_0 \\ \rho(c\bar{T} + h_{\mathrm{sl}}), & \text{for} \quad \bar{T} > T_0 \end{cases} \tag{11a}$$

which satisfy the integral equation

$$\int_0^\infty \int_{-\infty}^\infty \left(\bar{h}\frac{\partial \phi}{\partial \bar{t}} + k\bar{T}\frac{\partial^2 \phi}{\partial \bar{x}^2} \right) d\bar{x}\, d\bar{t} = -\int_{-\infty}^\infty \phi(\bar{x}, 0)\bar{h}(\bar{x}, 0)\, d\bar{x} \tag{11b}$$

for all suitable test functions $\phi(\bar{x}, \bar{t})$ [28]. This integrodifferential equation is formulated by substituting enthalpy into the unsteady term of the heat diffusion equation, multiplying by the test function ϕ, and integrating the equation throughout the half-plane $(\bar{x}, \bar{t})$. The test function is required to decay sufficiently rapidly as $\bar{x}^2 + \bar{t}^2 \rightarrow \infty$, so that all boundary terms resulting from integration by parts vanish at infinity. Note that the enthalpy is not uniquely defined since $\bar{h}$ may be multivalued at T_0. Viewed from the context of the weak formulation, the Stefan condition becomes a relationship between the changes in enthalpy and temperature gradient across the solid–liquid interface. It is completely analogous to the Rankine–Hugoniot equation in compressible flow problems with shock waves. Thus, the weak formulation brings out a strong likeness in the nature of these two types of problems [28]. More fundamentally, the interpretation of the Stefan condition as a Rankin–Hugoniot equation leads to the use of enthalpy in the weak formulation. When using a weak formulation to solve the Stefan problem, the integrodifferential equation [Eq. (11b)] is not actually solved. Rather, the associated differential equation

$$\frac{\partial \bar{h}}{\partial \bar{t}} = k \frac{\partial^2 \bar{T}}{\partial \bar{x}^2} \tag{11c}$$

is solved subject to the equation of state [Eq. (11a)] and appropriate initial and boundary conditions.

The principal advantages of weak solutions are

1. Derivatives of the temperature do not appear in the formulation. Jump discontinuities are allowed in the solution. Thus, it is not necessary to consider the liquid and solid regions separately [48]. In this respect, a weak formulation is a type of embedding method.
2. The solid–liquid interface does not appear explicitly in the formulation [28] and is determined only after the solution has been obtained.
3. The solid–liquid interface does not have to be a sharp boundary, and may instead be an extended "mushy" zone. The solid–liquid interface may also vary in a discontinuous manner. An example is the melting of a block of ice that breaks up into several smaller pieces [48]. This multidimensional problem would be extremely difficult to solve using a classical formulation.
4. Questions on existence and uniqueness of solutions can be answered more easily for the weak formulation because the analysis may proceed without (or at least with less) interference from the unknown solid–liquid interface position. Uniqueness and global existence of the classical solution have been proven only for one-dimensional, two-phase Stefan problems, whereas, for the weak solution, existence has also been proven for multidimensional problems [28, 49].
5. Weak solutions may be constructed without explicitly having to describe the behavior of the solution near any of its singularities [48].
6. If a classical solution exists, it will also be a weak solution [48].
7. The approach is easily modified to allow for an inhomogeneous PCM by using Kirchoff's transformation and an appropriately modified form of Eq. (11a) [28]. This modification is also known as the enthalpy flow temperature method [50].
8. The existence theory for weak solutions results in the important practical

result that explicit and implicit finite-difference solutions of Eq. (11c) with central space differencing converge in a mean square sense to the weak solution [28]. Thus, the theory of weak solutions actually yields a numerical method for practical computations, known as the enthalpy method.

9. The method is readily extended to multidimensions [51, 52].

One may be tempted to believe that weak formulations appear quite superior to classical formulations. In particular, advantages (2), (3) and (8) appear to be outstanding. For example, the ability of enthalpy methods to predict phase-change problems with simultaneous multiple solid–liquid interfaces is demonstrated by Schneider and Raw [53].

Unfortunately, in actual performance, the enthalpy method has not lived up to expectations. One of the fixed-grid methods employed by Furzeland in his comparative study of numerical methods [41] used an enthalpy method. Although easy to program, the method gave pathological solutions for a sharply defined melting temperature. Nonmonotone temperature histories were produced. It was also found that the enthalpy required smoothing across the solid–liquid interface. That is, the enthalpy was set to vary continuously in a temperature interval of magnitude ε centered on T_0. Furzeland believed that this temperature behavior and required enthalpy smoothing were due to the intrinsic unsuitability of finite difference approximations for handling the step change in enthalpy across the solid–liquid interface. He determined that an optimum value of ε existed but could not determine it in advance. The solutions appeared very sensitive to the value of ε.

The anomalous behavior found by Furzeland is not an isolated occurrence. Voller and Cross [54] clearly point out that while enthalpy methods work well when phase-change takes place over an extended temperature range, oscillatory movement of the solid–liquid interface and temperature plateaus occur if the temperature range becomes small enough. They also point out that use of a fictitious temperature interval ε, while alleviating nonphysical behavior, causes the solution to become critically dependent upon the value of ε. Voller and Cross developed a cure to the nonphysical behavior by reinterpreting the enthalpy of the nodal cell containing the solid–liquid interface so that the enthalpy of this cell became a weighted average of the solid and liquid parts. The method works quite well, but is of an ad hoc nature. Although Voller and Cross successfully generalized their method and determined a numerical solution to a two-dimensional Stefan problem, the reinterpretation of enthalpy at nodal cells near the solid–liquid interface became even more arbitrary. From a more fundamental viewpoint, note that the enthalpy reinterpretation requires that the solid–liquid interface must be explicitly tracked. The ability to ignore the position of the solid–liquid interface, one of the major advantages of the enthalpy method, is forfeited in exchange for physically realistic solutions.

Closely allied to the enthalpy method is the apparent specific-heat method. In this method, the latent heat is approximated as taking place over a temperature range

$$h_{sl} = \int_{T_0 - \varepsilon}^{T_0 + \varepsilon} c_0 \, d\bar{T} \tag{11d}$$

where $c_0 \to \infty$ as $\varepsilon \to 0$ [52]. Unlike the enthalpy method, the temperature is the dependent variable that is determined. The apparent specific-heat method is similar to the enthalpy method in that the solid–liquid interface does not have to be tracked. As in the enthalpy method, the nonlinear behavior of the problem is concentrated in the equation of state. If ε is not too small, the method works very well, but if $\varepsilon \to 0$, the solid–liquid interface advances in a nonphysical oscillatory fashion and the temperature distribution becomes distorted near the solid–liquid interface [55]. This is precisely the same problem that occurs with the enthalpy method—it works well only if there is an extended mushy zone.

Goodrich [55] introduced a modified apparent specific-heat method that does not give solutions with the pathological behaviors mentioned previously. As in the modified enthalpy method of Voller and Cross [54], the key modification used by Goodrich is to use a special finite-difference equation for the node undergoing phase change. Once again, accuracy requires forfeiture of the method's advantage in not having to explicitly determine the position of the solid–liquid interface. Although Goodrich's method gives excellent results, the special finite-difference equation is of an ad hoc nature and is computationally elaborate. Extension to multidimensional problems seems difficult.

Pham [56] pointed out another problem with apparent specific-heat methods, termed "jumping of the latent heat peak." Simply put, Eq. (11d) shows that a pronounced peak in specific heat exists near T_0, especially if the melting range is small. It is possible for a nodal temperature to jump past the melting range if the time step is too large, with the net result that the latent heat contribution is never felt. Although enthalpy methods do not have this problem, they generally are computationally less efficient in terms of computing time. Pham introduced a two-step procedure that attempts to combine the best elements of enthalpy and apparent specific-heat methods. It is essentially a predictor–corrector type of algorithm for the temperature. Pham's method does not require that the solid–liquid interface be tracked explicitly, but it does require something equivalent—that node temperatures constantly be compared to T_0 in order to avoid jumping past the melting temperature. Generalizations to multidimensions and finite-element methods are given in a second work [57]. The two-step method works well for either an extended mushy zone or a sharp melting point. Some oscillatory behavior in the solid–liquid interface position does appear.

In summary, methods based upon a weak formulation appear to work well only if phase-change occurs over an extended mushy zone. Quite generally, the nonlinear behavior of $\bar{h}(\bar{T})$ or $c(\bar{T})$ near T_0 presents a difficult problem for most numerical techniques [48]. There are additional disadvantages. Qualitative properties of weak solutions are relatively obscure. Asymptotic analyses, which have great value in enhancing physical understanding, are more readily carried out with classical formulations. Finally, the enthalpy equation of state [Eq. (11a)] must be a monotone function if a unique solution is to exist. In multidimensional situations, there is no obvious way to guarantee uniqueness of weak solutions [28].

G. Asymptotic Expansions

In 1931, Leibenson proposed a method of approximate solution for Stefan problems in which the temperature distribution in each phase was found by solving Laplace's equation with respect to the spatial coordinates, that is, the unsteady terms in the heat diffusion equations were dropped [4]. Unsteady behavior appeared only in the Stefan condition. This assumption, now known as the quasi-steady assumption along with a second one called the quasi-stationary assumption are frequently used in the phase-change literature for numerical and approximate analytical solutions. In a quasi-stationary method, the unsteady terms in the temperature equations are retained, whereas convective terms arising from motion of the solid–liquid interface are dropped. Thus, the temperature distributions are determined assuming that the solid–liquid interface is stationary. Because the unsteady temperature terms are retained, initial temperature distributions can be satisfied when using the quasi-stationary assumption. This is not the case for the quasi-steady assumption [58]. The widespread employment of these assumptions is seldom accompanied by any type of formal justification or understanding of how these approximations may be refined. Perturbation theory provides a vehicle by which one may observe the role of these assumptions from a more global viewpoint.

In particular, Duda and Vrentas [46] demonstrated that the quasi-stationary solution is actually the zero-order solution of a regular perturbation solution to the Stefan problem. In a study on the growth or dissolution of a pure fluid into a binary fluid mixture, they determined the zero-order, quasi-stationary solution and the two first-order corrections of the perturbation solution. Duda and Vrentas's work firmly established the quasi-stationary assumption as a rational approximation to the Stefan problem.

The solution developed by Duda and Vrentas is actually a short-time solution because the time scale that is used, l^2/α, is much less than the time scale needed for significant interface motion provided that $\mathrm{Ste} \ll 1$. The proper time scale needed for the latter is $l^2/(\alpha \cdot \mathrm{Ste})$. Weinbaum and Jiji [58] demonstrated that if a long-time perturbation solution based upon the dimensionless time $\tau = \mathrm{Ste}\, t$ is constructed, then the lowest order term of the solution is a quasi-steady approximation. Although the long-time solution is much simpler than the short-time solution, it cannot satisfy any initial conditions. Moreover, it becomes singular as $\tau \to 0$.

Weinbaum and Jiji [58] treated the initial singularity for a two-phase problem in an infinite slab using the method of matched asymptotic expansions. Isothermal cooling along one face of the slab was used to freeze an initially superheated liquid. Isothermal and adiabatic boundary conditions were used along the other face of the slab. Outer expansions based upon the long-time scale were constructed according to

$$T(x,\tau;\mathrm{Ste}) = \sum_{n=0}^{\infty} \mathrm{Ste}^{n/2}\, T_n(x,\tau)$$
$$B(\tau;\mathrm{Ste}) = \sum_{n=0}^{\infty} \mathrm{Ste}^{n/2}\, B_n(\tau) \tag{12a}$$

while inner expansions based upon the short-time scale were constructed according to

$$\hat{T}(x,t;\mathrm{Ste}) = \sum_{n=0}^{\infty} \mathrm{Ste}^{n/2}\, \hat{T}_n(x,t)$$
$$\hat{B}(t;\mathrm{Ste}) = \sum_{n=0}^{\infty} \mathrm{Ste}^{n/2}\, \hat{B}_n(t) \tag{12b}$$

Equation (12) is substituted into a classical formulation based upon the coordinate transformation [Eq. (8a)], and terms up to second order are determined. The asymptotic matching principle is then used to produce composite solutions that are uniformly valid in time. An interesting observation made by Weinbaum and Jiji is that the quasi-stationary approximation for interface motion (zero-order term of inner solution) is, to $0(\mathrm{Ste}^{1/2})$, uniformly valid for all time, and that the outer expansion provides no new information to this order. Also for very short times, the solutions are independent of the boundary conditions on the far side of the slab, and exhibit the similarity behavior of the Stefan–Neumann solution. With the adiabatic boundary condition, the slab freezes in a finite time. Insufficient terms are determined, however, to make a prediction of freezing time, which is different from the Stefan–Neumann solution. With the isothermal boundary condition (whereby the face is held at the initial superheated temperature), the freezing process decays into a steady state with both liquid and solid phases present. By letting $t \to \infty$ in the expansion for $\hat{B}$, Weinbaum and Jiji determined the final equilibrium position of the solid–liquid interface. They obtained a result that is in accord with the result from solving the steady heat diffusion equation directly.

In an effort to obtain higher order terms for the freezing of an infinite slab with an adiabatic boundary condition, Charach and Zoglin [59] developed a perturbation solution to a heat-balance integral formulation of the Stefan problem. The ordinary differential equations resulting from the integral method are less accurate than the exact formulation, but they allow higher order terms in a perturbation expansion (uniformly valid in time) to be determined more easily. The freezing process is subdivided into two parts, according to whether or not a thermal penetration depth has reached the adiabatic surface. Charach and Zoglin found that the temperature along the adiabatic face drops by an order of magnitude over an interval of time equal to the heat diffusion time in the liquid. The time for complete freezing is found to be

$$\tau_f = \tfrac{1}{2} + \tfrac{1}{8}\,\mathrm{Ste} + 0(\mathrm{Su}\cdot\mathrm{Ste}^{3/2}) \tag{12c}$$

This solution indicates that the one- and two-phase Stefan–Neumann solutions slightly underpredict and overpredict the freezing time, respectively. For a case with Ste $= 0.1$, Su $= 1$, $\Delta = 1$, and $\lambda = 1$, the under- and over-predictions are about -3.3% and $+14\%$, respectively. It is of interest to observe that the solution begins as the two-phase Stefan–Neumann solution and approaches the one-phase solution as $\tau \to \tau_f$.

Pedroso and Domoto [60] developed a perturbation solution for the freezing of a semi-infinite saturated liquid. A convective boundary condition was used along the cooled surface. The algebraic manipulations were rearranged in such a way that only the coefficients of higher order terms had to be determined. An algorithm was then programmed on a computer to do the calculations, and the first nine terms of the perturbation series for the solid–liquid interface position were determined. By examining the sequences of coefficients that resulted, Pedroso and Domoto believe that their series converge for Ste < 1.

Jiji and Weinbaum [61] have also used the method of matched asymptotic expansions to determine a perturbation solution for freezing in an annular region. A sudden isothermal cooling was applied along the outer cylinder. Initially, the PCM was uniformly superheated. Both adiabatic and isothermal boundary conditions were used for the inner surface of the annulus. As in an earlier study of Weinbaum and Jiji [58], inner and outer expansions were determined that reflect the boundary-layer-like structure of the governing equations in time as $t \to 0$. Unlike that study [58], the asymptotic matching failed to produce a composite expansion that is uniformly valid in time. The problem occurred not in the initial singularity, but in a final singularity, occurring as the last bit of liquid froze. Although it is natural to anticipate this singularity because the liquid phase degenerates at this point, singular behavior has only been observed in cylindrical and not in rectangular geometries, suggesting that the singularity is of a geometric nature [29]. Jiji and Weinbaum found that the effects of the final singularity became more significant as the inner cylinder decreased in radius. With the isothermal boundary condition along the inner cylinder, the final singularity was of no consequence because the freezing process decayed into a steady state with both PCM phases present. The final degenerate state was never reached. Their solution correctly indicated the equilibrium position of the solid–liquid interface, as determined from an elementary steady-state analysis. The effect of the initial superheat was minor with the insulated boundary condition, but significant for the isothermal boundary condition.

The method of strained coordinates was used by Pedroso and Domoto [62] in an attempt to overcome the final singularity in the inward freezing of spherical masses of liquid PCM initially at the freezing temperature. Although Pedroso and Domoto claim that their expansions are uniformly valid in time, they suspect that their series diverge as the final singularity is approached, and they used a Shanks transformation in order to obtain improved accuracy. Their values for the time of complete freezing match very well with a numerical solution by Tao [63]. The match in temperature distribution is not quite so good for larger values of Ste, however. It is interesting to note that the method of strained coordinates appears to have worked fairly well on the parabolic heat diffusion equation—despite the warning by Van Dyke [64] that this singular perturbation technique may have to be limited to hyperbolic equations.

Pedroso and Domoto's solution, although extending the range of validity of perturbation expansions much closer to the final singularity than any previous work, was not able to clearly reveal the essential physics at work in the freezing problem. For example, values of the final time were determined implicitly; no explicit representation was given. Riley *et al.* [65] used the method of matched asymptotic expansions to treat the final singularity. In the region near the solid–liquid interface, heat transfer is controlled by the latent heat condition, and far away from the solid–liquid interface, the transient variation of temperature plays a key role in determining the final development of the solid–liquid interface. For the sphere, the inner radial length scale is found to be $\sim l\,\mathrm{Ste}^{1/2}$ and for the cylinder $\sim l\gamma$, where $\gamma^2 \ln \gamma = -\mathrm{Ste}$ and l is the radius. Ockendon [29] pointed out that the solid–liquid interface approaches these length scales for times such that $(t_f - t) \sim \mathrm{Ste}$. Riley *et al.* determined the terms of the inner expansions appropriate for the central regions of the cylinder and sphere, and then matched these with the outer expansions based upon traditional scales. The composite expansions yielded final solidification times of

$$\tau_f = \tfrac{1}{4} + \tfrac{1}{4}\mathrm{Ste} + 0(\gamma^2) \qquad \text{(cylinder)} \tag{12d}$$

$$\tau_f = \tfrac{1}{6} + \tfrac{1}{6}\mathrm{Ste} + 0(\mathrm{Ste}^{3/2}) \qquad \text{(sphere)} \tag{12e}$$

The two terms of Eq. (12d) appear to give results accurate to within 5% for Ste < 0.25. Although the composite expansions extend the range of asymptotic solutions closer to the final singularity (closer than when using the method of Pedroso and Domoto [62]), a logarithmic behavior in the temperature distribution causes them to break down at the last moment. This irregularity appears exponentially small, and Riley *et al.* believe that the required corrections to their solutions are not significant. Nevertheless, this indicates that the nature of the final singularity is still not clearly understood [29].

Tokuda [21] developed an asymtotic, large time solution for the one-phase problem in a semi-infinite solid. The phase change is driven by convective cooling or heating. By recognizing that the solution to this problem must approach the Stefan–Neumann solution as $t \to \infty$, Tokuda was able to assume appropriate asymptotic expansions for the solid–liquid interface position and wall temperature. The transformation [Eq. (8a)] was used to immobilize the solid–liquid interface. The final asymptotic solution was obtained by requiring that the assumed asymptotic expansions be consistent with interfacial Lagrange–Burmann expansions. The leading terms correspond to the Stefan–Neumann solution. The solution breaks down as $t \to 0$ since logarithmic terms appear in the third terms of the expansions. Comparison of the asymptotic solution for the solid–liquid interface position with a numerical solution revealed excellent agreement as Ste $\to 0$ and $t \to \infty$. For Ste < 2 and $\tau > 5$, the maximum difference is about 5%. While no complete analytical solution exists for Stefan problems with convection boundary conditions, short-time series solutions (Section II,C) coupled with the long-time asymptotic solution do manage to point out most of the physics of these problems.

A fourth type of boundary condition that has yet to be discussed is the nonlinear radiation boundary condition. Applications in high-speed aerodynamics led Seeniraj and Bose [66] to consider one-phase problems for spheres and cylinders with convective and radiative heating. For the melting of initially solid spheres of PCM, the PCM was assumed to be opaque and gray. The radiative heating was incorporated as an additional term in the surface boundary condition. To leading order, their solutions for solid–liquid interface position gave a final melting time of

$$\tau_f = \frac{1}{4} + \frac{1}{2\,\mathrm{Br}} + 0(\mathrm{Ste}) \qquad \text{(cylinder)} \tag{12f}$$

$$\tau_f = \frac{1}{6} + \frac{1}{3\,\mathrm{Br}} + 0(\mathrm{Ste}) \qquad \text{(sphere)} \tag{12g}$$

where Ste is based upon the absolute temperature of the melting point (rather than on a temperature difference), and Br is a radiation Biot number, $\mathrm{Br} = \mathrm{Bi} + 4\,\overline{\mathrm{St}}$. Here, $\overline{\mathrm{St}}$ is the radiation Stefan number, $\sigma\varepsilon F l T_0^3/k$, and is physically quite different from Ste. Note that, as the Biot number increases, the time for complete melting decreases. A fourth dimensionless parameter of importance is the temperature ratio T_h/T_0. The radiation effect is appreciable only if $\mathrm{Bi} < 1$. Since the expansions break down as the final singularity is approached, Eqs. (12f) and (12g) cannot be improved much because the higher order terms are unbounded.

Earlier, Yan and Huang [67] considered the same type of phase-change problem for an infinite slab geometry. The slab was insulated on one side. Before applying the perturbation method, they used two types of approximations for the full nonlinear radiation term. These are a linear approximation for small temperature differences and a quadratic approximation for moderate temperature differences. With the linearized radiation condition, their perturbation solution led to a result of the form

$$\tau_f = \frac{1}{2} + \frac{1}{\mathrm{Br}} + 0(\mathrm{Ste}) \tag{12h}$$

for the time of complete melting of the slab. Here, Br is similar in definition to that found in Ref. 66. Yan and Huang found that the linearized result was indeed adequate for $T_0/T_h > 0.75$, but that the quadratic approximation was required for larger temperature differences, $T_0/T_h > 0.5$. For still larger temperature differences, even the quadratic approximation was insufficient. Seeniraj [68] later pointed out that the approximations to the radiation term in the Stefan condition that were used by Yan and Huang were not necessary for the success of the perturbation method. Although correct, Yan and Huang felt that the approximations simplified the resulting algebra to the point that higher order terms in the expansions could be determined more easily in closed form.

Asymptotic methods are ideally suited to the study of Stefan problems. In particular, singular perturbation techniques, already highly developed by applications in fluid mechanics, may be immediately applied to treat the singularities that occur as a phase degenerates. Applications to multidimensional problems are straightforward. Perturbation methods also provide one of the few analytical tools for handling additional nonlinear variations, such as radiation boundary conditions or convective motion in the melt. Perhaps most important of all, perturbation methods are invaluable tools in illuminating the physics of the problem. Balanced against these advantages is the fact that higher order terms become increasingly difficult to obtain. Even in relatively simple problems, the determination of first-order terms in the inner expansion of the temperature distribution has proven to be an intractable problem [58, 61].

H. APPROXIMATE INTEGRAL METHODS

In contrast to the exact integral methods discussed earlier in Section II,D,3 are the approximate integral methods of which the heat-balance integral is a well-known example. In fluid mechanics, the Karman–Pohlhausen method provides another example. As in the case of perturbation expansions, applications of approximate integral methods in fluid mechanics have caused these methods to become highly developed. Goodman has already given a comprehensive treatment on the application of approximate integral methods to heat transfer [3]. As a result, we will merely point out that the basic ideas of these methods are to make an *educated guess* about the functional form of the temperature distribution in one or more of the independent variables and then to integrate the governing equations in these variables to eliminate them from the problem. Thus, the outstanding advantage of approximate integral methods is that they reduce the number of independent variables that are involved. Quite generally this simplifies the governing equations so much that it becomes relatively easy to incorporate a number of nonlinear effects (such as property variation or radiation effects) into the model. Goodman demonstrates how the heat-balance integral method may be applied to one-phase Stefan problems in Ref. 3. He gives solutions for ablation problems as well as for melting with isothermal and time-dependent, heat-flux boundary conditions. In Ref. 69, Goodman also considers one-phase melting with a convective boundary condition.

Although Goodman's solutions provide reasonably accurate close-form expressions for the melting rate and temperature history, it may be possible to significantly increase their accuracy if the coordinate transformation [Eq. (8a)] is used to immobilize the solid–liquid interface. As an example, consider the one-phase problem of melting in a semi-infinite solid because of an isothermal heating at $x = 0$ (see Fig. 1). In order that the formulation be nonsingular at $t = 0^+$, we begin with the dimensionless equations [Eqs. (8b) and (10a)–(10e)]. A parabolic profile that satisfies boundary conditions [Eq. (10e)]

$$T_1 = 1 - (1 + \Gamma)x + \Gamma x^2 \tag{13a}$$

is substituted into the integrodifferential equation

$$\int_0^1 \left\{ 2tX^2 \frac{\partial T_1}{\partial t} - x(X^2 + 2tX\dot{X})\frac{\partial T_1}{\partial x} - \frac{\partial^2 T_1}{\partial x^2} \right\} dx = 0 \tag{13b}$$

to produce the coupled set of nonlinear ordinary differential equations

$$2tX^2\dot{\Gamma} + \{12 + \mathrm{Ste}(4 - \Gamma)\}\Gamma = 3\,\mathrm{Ste} \tag{13c}$$

and

$$X^2 + 2tX\dot{X} = \mathrm{Ste}(1 - \Gamma) \tag{13d}$$

Note that $-(1 + \Gamma)/(X\sqrt{2tX})$ is the dimensionless heat-transfer rate at $x = 0$, and that $B = X\sqrt{2t}$ gives the dimensionless solid–liquid interface position. Unique initial conditions may be determined for Eqs. (13c) and (13d) by examining their limiting forms as $t \to 0^+$, with the results

$$\Gamma = (2 + 6/\mathrm{Ste}) - \{(2 + 6/\mathrm{Ste})^2 - 3\}^{1/2}$$

and

$$X = \{\mathrm{Ste}(1 - \Gamma)^2\}^{1/2} \qquad \text{at} \qquad t = 0^+ \tag{13e}$$

Integration of Eq. (13c) and (13d) reveal that the solutions for Γ and X are actually constants. Therefore, Eq. (13e) gives the complete solution for all time. Goodman's solution [3], using a parabolic profile for T_1, gives the following result for X:

$$X = \sqrt{6}\left\{\frac{1 + 2\,\mathrm{Ste} - (1 + 2\,\mathrm{Ste})^{1/2}}{5 + 2\,\mathrm{Ste} + (1 + 2\,\mathrm{Ste})^{1/2}}\right\}^{1/2} \tag{14a}$$

Comparison between the predictions of Eqs. (13e) and (14a) demonstrates that for Ste < 1 the integral method with the coordinate transformation is roughly three times more accurate. The comparison is summarized in Table II.

Mennig and Ozisik [70] developed an approximate integral solution for one-phase melting in a semi-infinite solid resulting from a specified time-dependent temperature at $x = 0$. Their technique departs from the more

TABLE II

COMPARISON OF PREDICATED X VALUES FOR ONE-PHASE ISOTHERMAL MELTING OF THE SEMI-INFINITE SOLID

Ste	Exact X value, Eq. (4c) ($X = \mu\sqrt{2}$)	Goodman [3], Eq. (14a)	Error	Coordinate transformation, Eq. (13e)	Error	Menning and Ozisik [70]	Error
.01	.09984	.1000	+.16%	.09988	+.04%	.09988	+.04%
.03	.1723	.1732	+.48%	.1726	+.12%	.1726	+.12%
.10	.3112	.3157	+1.4%	.3124	+0.39%	.3124	+0.39%
.30	.5231	.5412	+3.5%	.5287	+1.1%	.5283	+1.0%
1.0	.8769	.9334	+6.4%	.9001	+2.6%	.8944	+2.0%
3.10	1.292	1.384	+7.1%	1.348	+4.3%	1.309	+1.3%
10.0	1.777	1.825	+2.7%	1.841	+3.6%	1.690	−5.1%

traditional heat-balance integral method in that they introduce the temperature gradient as an additional unknown function. As a result, they end up with a system of three governing equations rather than two [as in Eqs. (13c) and (13d)]. They assume a linear distribution for the gradient and obtain the following result for the case of isothermal heating:

$$X = \left(\frac{4\,\mathrm{Ste}}{4 + \mathrm{Ste}}\right)^{1/2} \tag{14b}$$

This result for the solid–liquid interface position is slightly better than the result from Eq. (13e) for smaller values of Ste and is somewhat worse for larger values. A summary of the comparison is also included in Table II.

Approximate integral methods have also been used to study phase change in spherical and cylindrical geometries. Stewart and Smith [71] determined a solution for inward solidification in cylinders driven by isothermal cooling. The PCM was uniformly superheated initially. The resulting ordinary differential equations did not have simple close-form solutions and were integrated numerically. Stewart and Smith used a least-squares fit to determine the following correlation for final freezing times for cases with Su = 0:

$$\tau_{\mathrm{f}} = 0.252 + 0.550\,\mathrm{Ste} \tag{14c}$$

The first coefficient compares well with the Riley *et al.* [65] perturbation result [Eq. (12d)], but the second coefficient is roughly two times larger. Some of the discrepancy may be due to higher order terms that were not used in the least squares fit. Additional errors may be caused by the fact that the integral formulation becomes singular as $\tau \to 0$ or $\tau \to \tau_{\mathrm{f}}$ (the initial singularity forced the use of a false starting method). Results were also presented for cases with a significant superheating effect, but the information was not organized in a physically meaningful way. Overall, Stewart and Smith's predictions for τ_{f} were within 7% of the experimental values that they determined. Bell [72] developed a heat-balance integral method using linear piecewise continuous profiles for solidification around an isothermally cooled cylinder. Lunardini [73] analyzed the two-phase problem around cylinders and used the Stefan–Neumann solution for a small initial interval of time—bypassing the initial singularity.

Milanez and Ismail [74] used first- and second-order rational functions for the temperature distribution in a study of one-phase solidification in spheres. Convective cooling was used to drive the phase-change process. They dealt with the initial singularity by deducing from the boundary conditions and the Stefan condition that $dB/d\tau \to \mathrm{Bi}$ as $\tau \to 0$, in accord with Eq. (6c). Using first-order profiles simplifies the algebra to the point where a closed-form solution may be obtained. Their result for complete solidification is

$$\tau_f = \frac{1}{6} + \frac{1}{3\,\text{Bi}} \tag{14d}$$

which compares very favorably with the result by Seeniraj and Bose [66], given by Eq. (12g). Also note that as $\text{Bi} \to \infty$, $\tau_f \to \frac{1}{6}$, in agreement with the leading term given by Riley *et al.* [65] for isothermally driven freezing [see Eq. (12e)]. Numerical integration of the governing equation is required when the second-order temperature distribution is used, however, so that no closed form expression for τ_f is presented. This equation becomes singular as $\tau \to \tau_f$, so substantial differences in the prediction of solid–liquid interface position occur, compared to the first-order result. Rao and Sarma [75] studied phase change for cool, metallic, spherical masses immersed in their own melt. With significant subcooling of the sphere and small values of Ste, it is actually possible for additional melt to freeze onto the sphere before melting commences.

Radiation effects on phase change have also been studied using approximate integral methods. Chung and Yeh [76] considered solidification of an infinite slab of saturated liquid. One face of the slab was subjected to convective and radiative cooling, while the opposite face was insulated. The cooled face was assumed to be perfectly opaque and black (this is the same problem studied by Yan and Huang [67]).

A more interesting radiation problem was studied by Habib [77, 78] for semitransparent materials. The processing of materials at high temperatures [79], laser annealing of semiconductors [80], the manufacture of glass and plastics [81], latent heat-of-fusion thermal storage systems [82], and melting of ice layers [83] are all involved with phase changes of semitransparent materials. Habib considered the one-phase solidification of a semi-infinite liquid due to a sudden isothermal cooling at $x = 0$. He assumed that the solid was a semitransparent, gray, nonscattering medium. Thus, radiation escaping outward from the solid at $x = 0$ could originate anywhere in the interior of the solid. Both the heat diffusion equation and the Stefan condition were modified to include a radiative heat flux term. In addition to the equations for solid–liquid interface position and temperature, this method required an additional equation to determine the radiative heat flux. Since the radiative flux is temperature dependent, an iterative method of solution was required. Habib found that, as the absorption coefficient $\kappa \to \infty$ (opaque medium), the solidification rate approaches a lower bound equal to a pure conduction rate. As $\kappa \to 0$ (nonparticipating medium), the solidification rate approaches an upper bound. The quadratic temperature profile that was used resulted in governing equations that had to be numerically integrated. From the data, it appears that the solid–liquid interface position has the form $B \sim t^p$, where $p = p(\text{Ste}, \overline{\text{St}}, T_c/T_0, n, \varepsilon_1, \varepsilon_2) > \frac{1}{2}$. Here, n is the refractive index, and ε_1 and ε_2 are the emissivities of the isothermally cooled surface and solid–liquid interface, respectively. Since internal radiative transfer is not modeled in the liquid, it is implicitly assumed that the solid–liquid interface is opaque.

Experiments that study the melting of an ice layer by radiant heat sources [84] indicate that short waves can penetrate the ice while long waves are mostly absorbed on the surface. The surface of melting, cloudy ice, produced by fast freezing of aerated water, becomes rougher with a shorter wave heat source. Chan *et al.* [85] proposed a theoretical model in which solid ice is partially melted by internal absorption of radiation near the solid–liquid interface. The sharp and smooth ice surface was therefore replaced by an isothermal two-phase region. They constructed a one-phase model for a semi-infinite liquid. An isothermal cooling was imposed at $x = 0$, the albedo of scattering of the PCM was assumed to be negligible, the PCM was assumed to be gray, and the PCM was free to lose heat to the environment by radiation. The existence of the two-phase zone requires that two important changes be made in the Stefan condition. First, a void fraction appeared in the coefficient of the $\dot{B}$ term. Quite generally, the void fraction will be discontinuous at the solid–liquid interface, and its value on the side of the solid–liquid interface facing the two-phase zone has to be determined. Note that by solid–liquid interface we mean the boundary between the solid (new phase) and two-phase zones. Second, the radiative flux term that appears should actually be replaced by the change in radiative flux across the solid–liquid interface. If one of the phases is opaque, then this difference is zero. In this case, the two-phase zone vanished and the model reduced to the classical one with a sharp solid–liquid interface. The void fraction of the two-phase zone was determined by a macroscopic balance between the radiant and latent heats. The detailed microscopic mechanisms of internal melting were not considered. The net result was a first-order, linear, unsteady, partial differential equation for the void fraction. Chan *et al.* used an integral method with a quadratic temperature distribution to study their model when both radiation and conduction were important. They used an additional assumption of an optically thin solid PCM layer. As done by Habib [77], an additional equation must be added to the analysis to determine the radiative flux terms. Chan *et al.* linearized the T^4 terms in this equation in order to keep the algebraic manipulations tractable. The results show a clear discontinuity in void fraction across the solid–liquid interface. The solidification in the two-phase zone is also much greater than that in the completely solid zone.

Oruma *et al.* [86, 87] extended the model [85] by allowing the PCM to have a nonzero albedo of scattering. They then investigated the effects of anisotropic scattering. Again, freezing in a saturated semi-infinite solid was considered. An isothermal cooling was used, and the PCM was free to lose heat to the environment through radiation. An approximate integral method with a quadratic temperature distribution was used to determine the solid–liquid interface position, while the radiative heat flux was determined using the F_n method. Oruma *et al.* found that backward scattering retards solidification whereas forward scattering enhances it. With forward scattering, the albedo of the PCM has little effect on the solidification rate until it approaches a value close to one. At that point, the freezing rate drops considerably and approaches the rate corresponding to pure conduction. With isotropic or backward scattering, the freezing rate drops consistently with increasing albedo.

As with other methods that we have discussed, the approximate integral methods have their disadvantages. Chief among these is that the success of the method hinges largely upon good choices for the assumed temperature distributions. Progress in obtaining more accurate solutions is not always guaranteed by using higher order approximations for the temperature distribution [3], and although these methods offer great simplifications in algebra when compared to more exact methods, the amount of work required for realistic problems may still be considerable [17].

I. OTHER METHODS

A number of additional methods exist that are useful in solving Stefan problems. In this section, we will highlight a few of them.

Certain two-dimensional, phase-change problems with stationary solid–liquid interfaces can be conveniently treated by methods of *complex variables.* Continuous casting of a slab [88] and the control of the solification-interface shape by applied sinusoidal surface heating [89] are representative cases. The use of methods such as conformal mapping and hodograph transformation has been reviewed by Crank [7].

In addition to using an approximate integral method, Chung and Yeh [76] use *Biot's variational method* to solve the melting problem of an infinite slab with a mixed convection–radiation boundary condition on one surface. Using an assumed temperature profile, the method results in ordinary differential equations for the solid–liquid interface position and the temperature of the cooled surface. These equations are coupled and nonlinear and are solved numerically. The results compare very closely with the integral solution. Agrawal [90] uses the variational method to solve convection driven melting in the semi-infinite solid using a linear temperature distribution. The result matches very well with Goodman's integral solution [69].

Biot's variational method [90] is based upon the concepts of

1. a heat flow vector $\langle \boldsymbol{H} \rangle$ defined by $\nabla \cdot \langle \boldsymbol{H} \rangle = -c\bar{T}$,
2. a thermal potential Φ defined by $\Phi = \frac{1}{2}\int_t c\bar{T}^2\, d\bar{t}$,
3. a dissipation function D defined by $D = \frac{1}{2k}\int_t \langle \dot{\boldsymbol{H}} \rangle \cdot \langle \dot{\boldsymbol{H}} \rangle\, d\bar{t}$,
4. generalized coordinates $q_i(\bar{t})$ defining the thermal field,
5. a generalized thermal force Q_i defined by $Q_i = \int_S \bar{T} \frac{\partial H_n}{\partial q_i}\, dS$, where S is

the boundary of the PCM.

In terms of these variables, Biot's variational principle becomes

$$\delta\Phi + \delta D = \int_S \bar{T}\, \delta(\langle \dot{\boldsymbol{H}} \rangle) \cdot (\langle \boldsymbol{n} \rangle\, dS) \tag{15a}$$

where $\langle n \rangle$ is the unit inward normal. Physically, this is a minimum dissipation principle [90]. The solution to the problem is found by assuming a temperature distribution in terms of the generalized coordinates and independent spatial variables, substituting this distribution into Eq. (15a), and then solving the Euler–Lagrange equations that are generated (the q_i are the unknowns to be determined). The solid–liquid interface position provides a natural choice for a generalized coordinate. The addition of more generalized coordinates will systematically improve the accuracy of the solution (unlike approximate integral methods); however, this will be balanced by a rapidly increasing complexity in the analysis. In theory, an infinite number of q_i would lead to the exact solution; in practice, one would obtain an intractable set of Euler–Lagrange equations [90]. The variational method may readily be generalized to include heat sources, temperature dependent properties, and even moving fluids with viscous dissipation [90]. Additional references on the use of variational methods are given by Bankoff [2]. Note that no true variational formulation exists for Stefan problems—the method outlined here is actually a quasi-variational principle [29]. More generally, solutions to Stefan problems may be obtained using variational inequalities [91]. A more precise theoretical framework for the application of these latter variational methods to Stefan problems has been provided by Elliot and Ockendon [28], including information on existence and uniqueness of solutions.

Another method that has been developed is the *isotherm migration method.* The objective is to calculate the trajectories of isotherms rather than determine the temperature histories as a function of position. The solid–liquid interface position is easily obtained by watching the isotherm with a temperature of T_0. Another advantage is that there is no need to determine temperature dependent variables at each time step [92]. The formulation is developed by considering spatial position to be a dependent variable of time and temperature, $\bar{x} = \bar{x}(\bar{t}, \bar{T})$. Thus, in an isothermal migration method, temperature is an independent variable. The governing equation for spatial position can be developed from the heat diffusion equation, and the Stefan condition is slightly altered to reflect the interchange of dependent and independent variables. The result for a one-phase melting problem is [92]

$$\frac{\partial \bar{x}}{\partial \bar{t}} = \alpha \left(\frac{\partial \bar{x}}{\partial \bar{T}_l} \right)^{-2} \frac{\partial^2 \bar{x}}{\partial \bar{T}_l^2} \tag{15b}$$

$$\rho_s h_{sl} \dot{R} = -k_l \left(\left. \frac{\partial \bar{x}}{\partial \bar{T}_l} \right|_{\bar{T}_l = T_0} \right)^{-1} \tag{15c}$$

subject to

$$\bar{x} = 0 \quad \text{at} \quad \bar{T}_l = T_h \quad \text{and} \quad \bar{x} = R \quad \text{at} \quad \bar{T}_l = T_0 \tag{15d}$$

The first boundary condition in Eq. (15d) corresponds to an isothermal heating. One disadvantage is now obvious—the governing equation is made more complex and nonlinear. Another problem that may occur with the isothermal migration method is that boundary conditions may degenerate, especially in multidimensional applications [93]. The method also cannot handle any type of uniform-temperature initial condition. This causes the formulation to become singular and requires that some other method be used to begin the solution.

Stefan, in one of his pioneering techniques, assumed that how the solid–liquid interface position varied with time was known *a priori* [12, 14]. This information was then used to determine an acceptable temperature distribution. Such a technique is known as an *inverse method*. Frederick and Greif [94] developed an inverse method for solving two-phase melting problems. They assumed a known temperature distribution for the solid phase that satisfied a specified initial condition and far-field boundary condition. By setting $T_s = T_0$, the location of a solid–liquid interface was then obtained. The melting problem then reduced to the task of determining a temperature distribution for the liquid that satisfied the solid–liquid interface boundary condition, $\bar{T}_l(\bar{x} = R) = T_0$. The boundary condition that $\bar{T}_l$ satisfied at the heated surface $\bar{x} = 0$ was unspecified. This technique has much in common with embedding methods [27]. A general series solution method was presented for the determination of $\bar{T}_l$. Once the liquid temperature distribution was obtained, the boundary condition that $\bar{T}_l$ observes at the heated surface was determined, completing the solution. The method is somewhat similar in spirit to the application of sinks and sources in hydrodynamics, whereby knowledge about elementary solutions to the governing equations is used to construct more complicated flows. Frederick and Greif developed a general solution for problems in which $R \sim \bar{t}^p$. With $p = \frac{1}{2}$, the classical Stefan–Neumann solution was recovered. While the method can provide solutions more easily than direct methods, it is limited by the necessity to choose the functional form of $R(\bar{t})$. It is entirely possible that in more complicated variations of the Stefan problem $R(\bar{t})$ is not expressible in terms of a small number of elementary functions.

Still other methods for the solution of Stefan problems use no special analytical or semi-analytical techniques and are essentially *brute-force numerical* solutions of the governing equations. Such methods are helpful in understanding the physics of more complicated problems that have yet to become amenable to analysis. In particular, for one-dimensional Stefan problems, substantial effort has been made in the numerical solution of problems with radiation effects.

Goodling and Khader [95] used a finite-difference method to solve one-dimensional freezing problems in plane, cylindrical, and spherical geometries. The PCM is initially saturated liquid and is opaque. The radiation effect occurs in a mixed radiation and convection boundary condition at the cooled surface of the PCM container. The model is similar to others that have been used [66, 67, 76]. Goodling and Khader found for $\mathrm{Bi} < 1$, the radiative effect can easily decrease the freezing time by a factor of two or more. They also determined from their data that if $\overline{\mathrm{St}} < \mathrm{Bi}^2/3$, the solidification time τ_f would be within 10% of the values obtained assuming pure convective cooling only.

The effect of internal radiative transfer on one-dimensional melting and freezing was considered by Abrams and Viskanta [96]. Phase-change was assumed to occur at a sharply defined solid–liquid interface. All interfaces and boundaries were diffuse, the solid PCM was considered to be isotropic, and the index of refraction was uniform in each phase. The model assumes that there are no scattering effects. The modified Stefan condition is similar to that proposed by Chan *et al.* [85], with the radiative flux term being set equal to the difference between its values on each side of the solid–liquid interface. A finite-difference method was used to solve several model problems. A one-phase melting example used gray optical properties for the PCM. Abrams and Viskanta found that as $t \to 0$, the solid–liquid interface position approaches the value resulting from pure conduction asymptotically (result of the initial singularity). Curiously, however, an anomalous bump was seen in the solid PCM temperature, next to the solid–liquid interface, for large values of Ste. This bump actually raised the solid temperature above T_0 for a small interval. Abrams and Viskanta attribute this result to a local buildup of energy due to excess absorption over emission of radiation. They believe that this anomalous behavior is physically realistic and may promote unstable interfacial growth. Another interpretation is that the bump indicates the existence of a two-phase zone (not allowed by their model) as proposed by Chan *et al.* [85]. The excess energy would then contribute to phase change rather than instability. Information for a second example, a freezing problem, using a band model for the absorption coefficients was also presented. The results indicate that spectral characteristics are important if quantitative information is desired.

Seki *et al.* [97] used a finite-difference method to study back melting due to a reradiating surface. Back melting is the melting of an ice layer immediately adjacent to a surface that is heated because of the absorption of short-wave radiation transmitted through the PCM. They considered a one-phase problem where the bottom wall (reradiating surface) of the PCM container was black and the PCM was radiated from above. Thus, their model contained two liquid layers separated by a solid layer. The PCM was cloudy ice, which is a strongly scattering medium because of trapped air bubbles. The scattering was assumed to be isotropic but wavelength dependent. Spectral effects were also considered for the absorption and extinction coefficients, as well as for the radiation source (assumed to be black). They found that melting in the upper layer was driven primarily by the direct absorption of radiant energy and was relatively unaffected by scattering, whereas the back melting was strongly affected by scattering. Also, the amount of back melting increased, relative to the top layer, as the temperature of the source increased. The model assumed the existence of one-phase zones and exhibited singular behavior as $t \to 0$. A false-starting method was used to begin the computations.

Melting with internal radiative effects was also studied by Diaz and Viskanta [98]. A finite-difference method was used to solve a model incorporating spectral effects for the absorption coefficient and emissivity. The radiative heat-flux terms were modeled using a forward scattering approximation. The solid was assumed to be initially subcooled. A planar, one-dimensional geometry was considered with a mixed convection and radiation boundary condition on one surface. Diaz and Viskanta found that the melting rate was, to a first approximation, constant in time: $B \sim \tau$, where $\tau = t\,\text{Ste}$; t is the Fourier number, and Ste is based upon the incident radiative flux. The subcooling effect was not as strong as the value of Ste on the melting rate, but it was found that it could still be significant. Out of a list of 12 dimensionless groups, Ste and Sb were the most important. A spectrally averaged absorption coefficient was found to lead to reasonably accurate values of solid–liquid interface position but not temperature distributions. An experimental study was also conducted and was in fair agreement with the numerical solutions. Experimentally, Diaz and Viskanta observed the back-melting phenomenon considered by Seki *et al.* [97] whenever the incident radiative flux was large and the slab thin. It is also of interest that in all of their experiments, the solid–liquid interface was a smooth and flat surface and never spread into a two-phase region. This was used as justification for incorporating only one-phase zones in their model.

The *boundary element method* (or *boundary integral method*) is a subset of the popular finite-element methods [99, 100]. An integral equation is formulated in terms of the fundamental solution (Green's function) of the heat diffusion equation (see Section II,D). By using a boundary discretization scheme, the effective dimensionality of the problem is reduced by one for steady problems. This advantage is lost for transient problems, which cannot be approximated as quasi-steady. The method has been successfully applied to study the freezing of soil around a buried pipe with quasi-steady approximation [101].

Although there are many papers in the literature (seemingly) testifying to the efficacy of numerical methods, the reader should be very cautious in their application to Stefan problems. Boley [27] indicates that numerical solutions are often found to be extremely sensitive to initial conditions (near an initial singularity), and that it is often essential that a brute-force numerical method not be used to start a solution. A further disadvantage of such methods is that they generally do not promote physical understanding of the problem at hand.

III. Convection-Dominated Melting–Freezing Problems

A. NATURAL CONVECTION

1. *Initial Studies*

General awareness of the importance of convection effects in phase-change problems has taken more than a century to develop. Only in recent times has it been appreciated that convection effects may be stronger than conduction effects by an order of magnitude or more. Tien and Yen [102] recognized that a melting problem with heating from below would be subject to Benard convection in the melt when the melt thickness exceeded a critical value. They assumed that the onset of convective motion occurred at a critical Rayleigh number of 1720 and used standard natural convection correlations to determine the heat flux at the solid–liquid interface. The convective heat flux was modeled by incorporating an additional term in the Stefan condition [Eq. (2c)]. An integral heat balance method was used to solve the problem, assuming that the heat transfer was one-dimensional. Tien and Yen's results showed that convection significantly enhanced the melting rate.

Boger and Westwater [103] experimentally investigated the problem treated by Tien and Yen [102]. Their results indicated that the value of the critical Rayleigh number was indeed about 1700 and that the magnitude of the convective effect agreed fairly well with values determined for natural convection without phase change. Large-scale oscillations in the solid–liquid interface position, because of strong flow-field oscillations, were observed for very large values of the Rayleigh number. Boger and Westwater also found that natural convection would significantly enhance melting or retard freezing, compared to conduction heat transfer only. This result was used to justify the use of an effective thermal conductivity in the Stefan condition to account for natural convection effects in a numerical solution to the problem. One-dimensional heat transfer was assumed. Agreement between the experimental and numerical results was quite good provided that the effective conductivity was appropriately chosen. Several difficulties encountered with the numerical method were improved by Heitz and Westwater [104]. They investigated the effects of the false starting procedure that was required (Boger and Westwater [103] used the method of Murray and Landis [43]). Heitz and Westwater found that the error generated by arbitrary initial conditions typically decreased rapidly with time. They also made some progress in determining the effective conductivity *a priori* by using flat plate correlations for natural convection heat transfer.

Hale and Viskanta [105] also investigated melting and freezing in horizontal plate geometries. Experimental and analytical methods were used to clearly reveal that natural convection effects are of great importance for melting with heating from below and for freezing with cooling from above. An approximate integral method was used to determine the analytical results. Heat transfer was assumed to be one-dimensional, and the natural convection effect was incorporated into the model by placing a convection term into the Stefan condition. The melt volume was found to increase almost linearly with time with isothermal heating from below—far in excess of the pure conduction result.

Each of these studies served to point out that natural convection could cause tremendous global changes in melting and freezing processes, but they failed to reveal that large, local changes could also occur. Such local effects were clearly observed in a series of experiments, conducted by several different teams of investigators, depicting phase change around cylinders.

Bathelt *et al.* [106] experimentally observed the effects of natural convection in melting around a horizontal cylinder. The cylinder was designed to generate a uniform surface heat flux. Beginning with saturated solid PCM, a thin, concentric, annular melt region initially formed around the heated cylinder. After a short time, the interface developed asymmetrically, with most of the melting occurring above the cylinder; below the cylinder, melting was observed to decrease. This behavior was interpreted by Bathelt *et al.* to be the result of natural convection. At first, the melt region is thin and viscous effects keep the convection contribution small while a negligible thermal resistance through the melt keeps the conduction effect large, forming a concentric annular melt region. Note that, unlike a horizontal plate geometry, there is no critical melt thickness required for the onset of convection. Fluid motion occurs the moment that a melt exists; initially, its effect on the heat transfer is simply overwhelmed by that of conduction. As the annular region thickens, viscous effects decrease whereas the thermal resistance increases. Ultimately, convection becomes the dominant mode of heat transfer. A thermal plume rises above the cylinder and impinges upon the uppermost portion of the solid–liquid interface, causing a maximum local head-transfer rate there. Below the cylinder, the melt separates away from the bottom of the interface, causing a minimum local heat-transfer rate. Two antisymmetric convection cells develop. Further evidence of the importance of natural convection was given by the observation that the average heat-transfer rate leveled off with time; that is, the process became quasi-steady. This implied that the thermal resistance of the melt region remained constant while the volume of the melt region increased. Such behavior is characteristic of natural-convection heat transfer. Interferograms clearly depicting the thermal plume in the melt region have been developed by Bathelt and Viskanta [107].

In a second closely related experiment, Bathelt *et al.* [108] observed melting induced by three horizontal heated cylinders placed at the vertices of an isosceles triangle. Initially, the melt regions developed identically and independently of each other. After a sufficient time had elapsed, the separate melt regions merged together, allowing a complicated natural-convection flow pattern to appear. While different spacings of the cylinders resulted in somewhat different shapes for the merged melt regions, only very minor changes in the average heat-transfer rates were observed to occur.

Sparrow *et al.* [109] also experimentally demonstrated local as well as global effects of natural convection in the melting problem around a heated horizontal cylinder. The study was essentially identical to that of Bathelt *et al.* [106]. All of the major qualitative results of the two investigations reinforce each other. Sparrow *et al.* found that the heat-transfer rate was only 12% less than the natural-convection value from a horizontal cylinder in an infinite fluid. They also observed the existence of an overshoot phenomenon in heat-transfer rates (a minimum with respect to time), occurring as the magnitude of the natural convection effects approached those of conduction. These results were expanded by Abdel-Wahed *et al.* [110] by including the effects of subcooling, which was found to decrease the melt volume for given heat fluxes and elapsed times. Compared to the melting of saturated solid PCM, it was also observed that subcooling caused narrower melt regions to form.

A numerical solution for melting around a vertically heated cylinder incorporating the effects of natural convection was determined by Sparrow *et al.* [111]. A primitive variable formulation was used to determine the fluid flow field in the melt. They used the coordinate transformation [Eq. (8a)] to immobilize the solid–liquid interface and, consequently, appear to have been the first investigators to apply it to a multidimensional convection-dominated Stefan problem. However, Sparrow *et al.* threw away the higher order curvature terms for the interface, limiting their method to cases where the radius of the interface varied only slightly in the vertical direction. This occurs only when buoyancy forces are small. Furthermore, a quasi-steady assumption was invoked (the convection terms resulting from interface motion were ignored), limiting the maximum heat transfer that could accurately be modeled. The initial singularity was not removed from the axisymmetric governing equations, so a false-starting procedure was required. Enhanced melting was found to occur near the top of the heated tube, producing a cone-like melt region, which was narrower at its base. An overshoot phenomenon for heat transfer rates, analogous to that observed by Sparrow *et al.* [109] for horizontal cylinders, was also observed. In a closely related work, Kemink and Sparrow [112] experimentally observed the effects of subcooling on melting around a heated vertical cylinder. They found that subcooling tends to delay the onset of the natural-convection dominance of the heat-transfer process. The heat-transfer coefficients with subcooling were found to be 10–15% lower than those occurring for saturated melting, *directly opposite* of the conduction dominated result. The weakened convective flow was attributed to the smaller melt region resulting from subcooling. The Nusselt numbers for convection-dominated, saturated melting were found to agree fairly well with standard natural-convection correlations for vertical, parallel-walled enclosures. The effect of an opened or closed container along the top of the PCM was negligible.

Goldstein and Ramsey [113] examined the melting of a saturated solid PCM about a heated horizontal cylinder. Their experimental results compare very favorably with those of Bathelt *et al.* [106] and Sparrow *et al.* [109]. Goldstein and Ramsey observed that the development of the solid–liquid interface is very sensitive to initial conditions and to any perturbations that occur during the experiment. Using a constant heat-flux, thermal boundary condition for the cylinder, they noticed that the melt volume grows linearly with time. They also observed that the temperatures on the heated cylinder exhibit maxima near the transition from conduction to convection-dominated heat transfer.

The effects of natural convection on freezing problems were also studied during this initial period of investigation. In one sense, natural convection is not so important in freezing as in melting because no buoyancy force exists if the melt is saturated. Convection can be important if the melt is superheated, however.

Freezing around cooled vertical tubes was experimentally studied by Sparrow *et al.* [114]. With an initially superheated melt, natural convection was found to significantly decrease the freezing rate compared to conduction-dominated phase change. In fact, the mass of frozen PCM asymptotically approached a constant value as $t \to \infty$; that is, the freezing process decayed into a steady-state, heat-transfer process. Sparrow *et al.* reasoned that the interface was being heated by natural convection as rapidly as it was being cooled by the cooling tube. From this they inferred that natural-convection heat transfer causes the freezing process to ultimately terminate. While the presence of natural convection has a great deal to do with the ultimate location of the solid–liquid interface, it has nothing to do with the fact that the freezing process decays into a steady state. This can be readily understood if a two-phase, conduction-dominated, freezing problem is considered (conceptually this presents no difficulty; in actual practice, a gelling agent may be added to the PCM to retard all convective motions in the melt). Let the geometry be the same as in the study by Sparrow *et al.* A cylindrical containment vessel has a vertical, cooled tube running along its axis, forming an annular PCM region. The inner tube is cooled isothermally to $T_c < T_0$, while the surface of the containment vessel is held at the superheated temperature $T_h > T_0$ by a water bath. It is obvious that, given these boundary conditions, the solid–liquid interface must come to rest between the cooled cylinder and the container wall. In fact, the exact conduction solution gives

$$B = L^{1/(1+\mathrm{Su})} - 1 \tag{16}$$

for the final interface position. Here $L = S/l$, where S and l are the outer and inner radii of the annular PCM region. With the interface at this location, the thermal resistances through the solid and liquid regions match. Natural convection will act to decrease the thermal resistance of the melt and, consequently, will force the interface inward until the resistances are once again in balance. The data from Sparrow *et al.* [114] provide a nice example of this phenomenon. From Fig. 7 of Sparrow *et al.* [114], an average terminal solid PCM thickness of 0.69 cm may be determined for the case $T_c = 8.2°C$, $T_0 = 36.0°C$, and $T_h = 53.8°C$, assuming that the PCM *n*-eicosane used has a specific gravity of 0.789. If $k_s/k_l \approx 2$ (as is typical for paraffins), then $\mathrm{Su} = 0.32$ and Eq. (16) predicts a much greater terminal thickness of 3.66 cms. In the experimental configuration, $L = 6.08$.

Since natural convection causes descending currents along the solid–liquid interface, Sparrow *et al.* found freezing was enhanced along the bottom of the cooled tube. Conical solid regions which are narrower at the top, formed. The behavior is analogous to that found for melting but is more subdued. Saturated freezing processes, in which heat transfer occurred by conduction only, were also studied. The frozen mass was observed to increase more rapidly than analytical predictions based upon pure conduction. This increase was because of the appearance of surface irregularities—dendrites—on the solid–liquid interface. These dendrites increased the heat-transfer rate, because the thermal conductivity of a single dendrite crystal is normally higher than that of the amorphous material and also because they increase the surface area along the solid–liquid interface.

The effects of natural convection in two-phase freezing processes around cooled horizontal tubes was experimentally investigated by Bathelt *et al.* [115]. Natural convection was found to significantly lower the freezing rate. Measurements at the end of the experimental run showed that the solid layer was about 10% thicker along the bottom of the cylinder than along the top. An approximate integral method was also developed to predict freezing rates. Heat transfer was assumed to be one-dimensional, and the natural convection effect was again modeled by incorporating a convection term into the Stefan condition. Circumferentially averaged values of the convection heat-transfer coefficient, measured experimentally, were used to evaluate this term. For large degrees of liquid superheat and for sufficiently large values of time, the solidification was found to stop completely, and the solid would actually begin melting. Bathelt *et al.* attributed this behavior to natural convection. Once again, however, we point out (as in the previous discussion of the study of Sparrow *et al.* [114]) that the cessation of freezing is actually due to the superheated boundary condition that exists. The role of natural convection in ultimately terminating a freezing process is also promulgated by Viskanta [9].

In summary, it seems that while the global effects of natural convection were becoming apparent over two decades ago, local effects were not appreciated until about 10 years ago. Global effects in both melting and freezing are significant. While local effects can be found in freezing processes, such effects generally are more exaggerated in melting processes. Finally, dendritic formation may be important in freezing processes. In light of the observations, it is clear that the fundamental symmetry exhibited by melting and freezing when they are conduction dominated is lost when natural convection becomes an important heat-transfer mechanism.

2. *Phase-Change around Cylinders*

Natural convection in multidimensional Stefan problems causes the solid–liquid interface to advance unevenly, depending upon whether or not the melt impinges upon or separates away from the interface. The resulting irregular growth or decay of the melt region probably presents the most outstanding complication that arises when the Stefan problem is generalized from one dimension to multidimensions. Nevertheless, present state-of-the-art modeling is advanced enough that reliable predictions of phase-change processes can be made for two-dimensional systems with an accuracy that exceeds what may be obtained experimentally. A number of boundary conditions may be accommodated, with the only restriction being that the phase-change process be of the Class 1 type (see Section II,D), as characterized by Boley [27]. Sparrow *et al.* [111] took the first step in the development of this model. The remainder of this section will highlight the continued recent development of the model into a mature form for melting around a heated horizontal cylinder.

Yao and Chen [116] treated the effects of natural convection as a perturbation on a conduction-dominated melting process. They determined a double perturbation series solution in terms of Ra and Ste. Since natural convection ultimately dominates the heat transfer, this analysis yielded a short-time solution. The solid was assumed to be saturated. Yao and Chen immobilized the solid–liquid interface by using the coordinate transformation [Eq. (8a)]. Unlike the earlier work [111], they retained all terms introduced into the governing equations by the transformation. Thus, this work was the first to correctly use the transformation in a multidimensional Stefan problem with convection effects. The first three terms of the perturbation series were determined, and they indicated that the solid–liquid interface would melt faster and slower, above and below the heated cylinder, respectively, when natural convection effects began to appear. The short-time solution was extended by Yao and Cherney [18] to incorporate the effects of subcooling. Using an approximate integral method to determine the successive terms of the perturbation solution, it was found that subcooling acted to suppress natural-convection heat transfer. For the case of zero subcooling, the results agreed very well with those of Yao and Chen [116]. In both of these studies, a stream function was introduced to account for the fluid flow in the melt. Although all terms introduced by the coordinate transformation were retained by Yao and Chen [116] and Yao and Cherney [18], it was not possible to verify the general use of the transformation for a highly irregular geometry since these solutions were limited to short initial time intervals. This was done by Prusa and Yao [117] by using a numerical method to determine the natural-convection heat transfer in eccentric annular regions.

A somewhat different model was independently developed by Rieger *et al.* [118] by using the method of body-fitted coordinates, as proposed by Thompson *et al.* [119]. The initial singularity was not removed from the formulation; hence, a false-starting procedure was required. Rieger *et al.* also encountered rezoning problems with their natural coordinate system. They found that the natural coordinates tended to cluster in certain areas. This undesirable behavior was corrected at each time level by using spline interpolation to produce more uniform coordinates. Despite these difficulties, their solution was the first to model strong natural-convection effects in a multidimensional melting problem and thus contributed to our understanding of the role of buoyancy forces in phase-change processes.

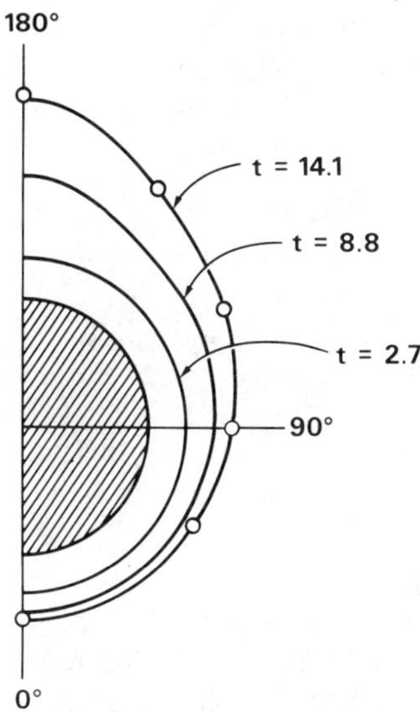

FIG. 3. Melt region growth resulting from an isothermally heated cylinder: position of the solid–liquid interface versus time. Numerical results from Prusa and Yao [121] (——) and Rieger *et al.* [118] (○). Ste = 0.02; Ra = 37,500; Pr = 50.

The extended coordinate transformation [Eqs. (8a), (8b), (10a), and (10f)], as generalized by Yao and Cherney [18], was used by Prusa and Yao [120, 121] to study the melting process around a heated horizontal cylinder. Natural-convection effects were incorporated into the model by using a stream function-vorticity formulation to model the fluid flow in the melt. Zero-velocity boundary conditions were used along the solid–liquid interface. These conditions *should not* be interpreted as no-slip conditions but rather are due to the fact that solid PCM at the interface is motionless as it melts [34]. The initial state used for the model was a saturated solid. A constant heat-flux boundary condition was used in one study by Prusa and Yao [120], whereas in another study [121] Prusa and Yao used an isothermal condition along the heated cylinder. Using both numerical and perturbation methods, Prusa and Yao determined that the melting process could be divided into three stages.

1. Conduction stage: The melt region is a concentric annulus. Convection effects exist but are insignificant. The melting process is conduction dominated and all characteristics of the melting process are functions of Ste only.

2. Transition stage: The melt region is still very nearly a perfect concentric annulus, but now conduction and convection effects are of the same magnitude. Local heat transfer rates begin to become nonuniform ($t = 2.7$ in Fig. 3).

3. Convection stage: The melt region is elongated, being thicker above the heated cylinder than elsewhere. Melting below the cylinder decreases to a very low level. Natural convection is the dominant mode of heat transfer ($t > 3$ in Fig. 3), which becomes quasi-steady.

The duration of the first two stages decreases with increasing values of the Ste and Ra, where Ra is the Rayleigh number defined by

$$\mathrm{Ra} = \frac{g\beta l^4 q}{\alpha_1 \nu k_1} \qquad \text{(constant heat flux)} \tag{17a}$$

or

$$\mathrm{Ra} = \frac{g\beta l^3 (T_\mathrm{h} - T_0)}{\alpha_1 \nu} \qquad \text{(isothermal)} \tag{17b}$$

and l is the heated cylinder radius. The Prandtl-number dependence is insignificant for $\mathrm{Pr} > 7$ [111, 120].

The rate of melting was found to depend most strongly upon the value of Ste. The value of Ra has a weaker, though still significant, effect. These effects can be observed in Fig. 4, where melt volume is plotted as a function of

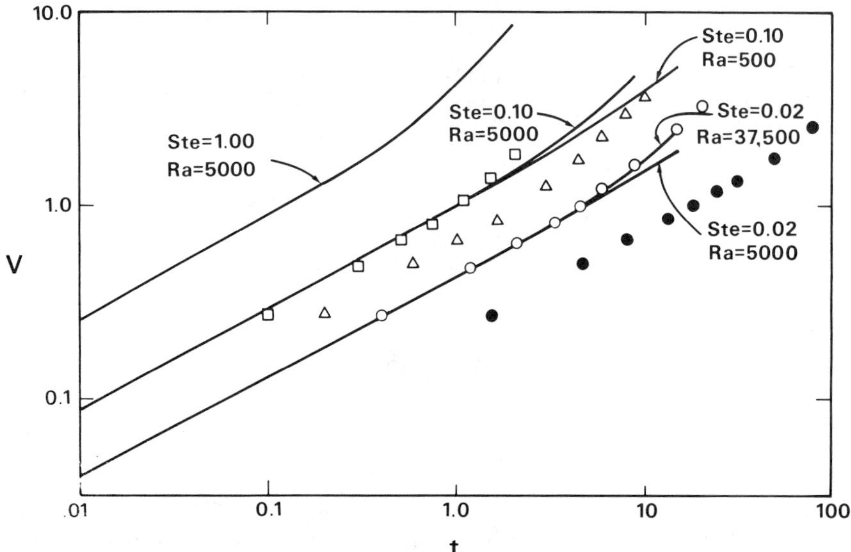

FIG. 4. Melt volume versus time for an isothermally heated cylinder: effects of Ste and Ra. The Ste and Ra values for the numerical results from Prusa and Yao [121] (——) are indicated on the figure. The Ste and Ra values for the numerical results of Rieger *et al.* [118] are as follows: ●, Ste = 0.005, Ra = 10,000; ○, Ste = 0.02, Ra = 37,500; △, Ste = 0.04, Ra = 75,000; □, Ste = 0.08, Ra = 150,000.

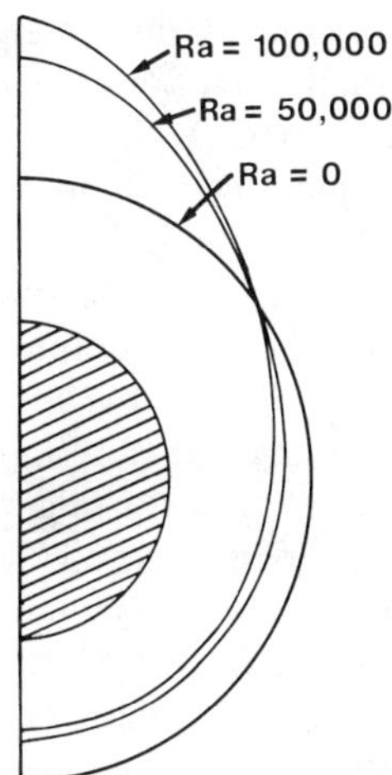

FIG. 5. Effect of Ra on the shape of the melt region for constant flux heating (numerical results from Prusa and Vandekamp [124]). Ste = 0.46, Pr = 54, t = 2.88.

dimensionless time. A straight line with a slope of $\frac{1}{2}$ corresponds to conduction-dominated melting. Away from the initial singularity, the numerical results of Rieger *et al.* [118] (isothermally heated cylinder) compare very well with those of Prusa and Yao [121], as can be observed in Figs. 3 and 4. A further comparison of the study of Prusa and Yao [121] with experimental results was made by Roberts and Rothgeb [122]. Their results verified the qualitative trends in the melting rate, which were given by Prusa and Yao [121]. Within the limits of experimental error, quantitative verification was also obtained.

The shape of the melt region is determined largely by the value of Ra, with larger values corresponding to more elongated melt regions. Figure 5 depicts this effect clearly. Although the value of Ra was found to be significant in determining the overall melt volume for the isothermally heated cylinder (see Fig. 4), virtually no effect at all occurred with the constant heat-flux boundary condition. Because the heat flux is specified, the effect of fluid motion on the melting rate is limited to its influence on reducing the degree of liquid superheating.

Since the initial singularity is extracted from the governing equations in the studies of Prusa and Yao [120] and [121], these models actually begin their solutions with the exact initial conditions and can thus be used to systematically study the errors resulting from false-starting methods, which are required when the singularity is not properly considered. Such a comparison is made for the isothermally heated cylinder in Prusa and Yao [121]. The false-starting model was conduction dominated and, consequently, could only provide reliable information on the error propagation in the conduction stage of melting. Although the initial errors, which are introduced by false-starting methods, were found to decay as melting progressed, they may last for a significant interval of time in the conduction stage. If the false-starting process is applied near the convection stage, these errors may be amplified by the convection. Thus, one cannot take it for granted that the initial error will always decay. A remark by Fox [17] that the effects of a singularity may not always remain localized in numerical solutions provides an additional warning on the possible danger of false-starting methods.

The results for melting due to isothermal heating were generalized to include the effects of subcooling and density change by Prusa and Yao [123]. The dimensionless formulation used in this study required an important modification in order to ensure nonsingular behavior at the beginning of melting. If the PCM changes density upon melting, then additional convection singularities appear in the governing equations at time zero. For example, if $\Delta > 1$, then excess liquid volume rushes out of the melt region in order to satisfy continuity. Since the volume of the melt region decreases to zero like $t^{1/2}$ as $t \to 0$, this liquid motion becomes unbounded. In particular, scale analysis reveals that the stream function and vorticity become unbounded like $t^{-1/2}$ and $t^{-3/2}$, respectively, as $t \to 0$. In order to remove these singularities, $t^{1/2}$ and $t^{3/2}$ were factored out of the dimensionless stream function and vorticity. Subcooling acts to decrease the melting rate, just as in conduction-dominated problems, but it does so in a more complicated way since smaller melt regions will retard fluid motion and decrease the effects of natural convection [18]. This interaction compounds the effects of subcooling, making it even more important in the convection stage than in the conduction stage of melting. Because of the increase in thermal potential, subcooling also caused the average heat-transfer rate along the heated cylinder to increase during the conduction and transition stages. During the convection stage, though, this rate was reduced compared to the saturated case because of the decreased convective flow. The effects of density change on the heat-transfer characteristics of the problem were found to be quite minor. The maximum effect, occurring during the initial conduction stage of melting, never amounted to more than a fraction of a percent. As melting progressed into the convection stage, the effects of density change smoothly decreased until they were imperceptible.

Prusa and Vandekamp [124] generalized the model of Prusa and Yao [120] to predict the melting process around multiple heated horizontal cylinders using a numerical method. In particular, a configuration with three cylinders arranged in an isosceles triangle (the geometry of the experimental study [108]) was considered. The coordinate transformation was again used, albeit in a more general form, to handle the complex, multiply connected melt region that ultimately forms. This melt region was first divided into three doubly connected regions by introducing two virtual boundaries. The extended coordinate transformation was then used in each doubly connected region. As in the studies of Bathelt *et al.* [108] and Prusa and Yao [120], constant heat-flux boundary conditions were used along the cylinders. The PCM was initially assumed to be a saturated solid, and the heating of each cylinder began simultaneously. The numerical solution indicates that when the separate melt regions merge together, no significant changes in global melting rate occur, in complete agreement with the experimental results [108]. The numerical solution also predicts that the growth rate of the melt volume is nearly linear in time and independent of the value of Ra. For example, with Ste = 0.46, at a dimensionless time of 2.88 (well into the convection stage), the melt volume is only 1.3% greater for Ra = 50,000 compared to Ra = 0 (conduction heat transfer only). Increasing the Rayleigh number to 100,000 further increases the melt volume by only another 0.1%. From these results follows the startling observation that Eq. (6c) (which is formally correct only for conduction-dominated phase change) gives an excellent prediction of melt volume well into the convection stage of melting. For a cylindrical geometry, it follows from Eq. (6c) that, to leading order, the dimensionless melt volume is given by

$$V = 2 \cdot \mathrm{Ste} \cdot t + 0(\mathrm{Ste}^2) \tag{17c}$$

where the volume is normalized with respect to the cylinder volume. For the case with Ste = 0.46, this compact formula predicts a melt volume that is high by only 4.8% at the end of the numerical solution, at $t = 4.64$—far along into the convection stage (with Ra = 50,000, the transition stage occurs roughly at $t = 1$).

Figure 6 clearly shows the linearity of the numerical solution for melt volume [124]. It also compares the numerical solution with data from three different experimental studies [106, 108, 113]. The experimental results are all for the same value of Ste = 0.46 (a personal communication between R. Viskanta and the authors pointed out that the value of Ste = 0.587 listed in the work of Bathelt *et al.* [106] should be corrected to 0.46). It is obvious that the experimental data, although qualitatively similar, show quantitative differences on the order of 100%. Most likely, these differences, as well as those between the experimental and numerical results, can be explained by considering end effects, subcooling, and uncertainty in the values of the thermophysical properties of the PCMs.

References that list PCM properties often deviate from each other by as much as 25%. In particular, if the value of the specific heat used by Bathelt *et al.* [106, 108] is decreased by 22% (a value in accord with modern references

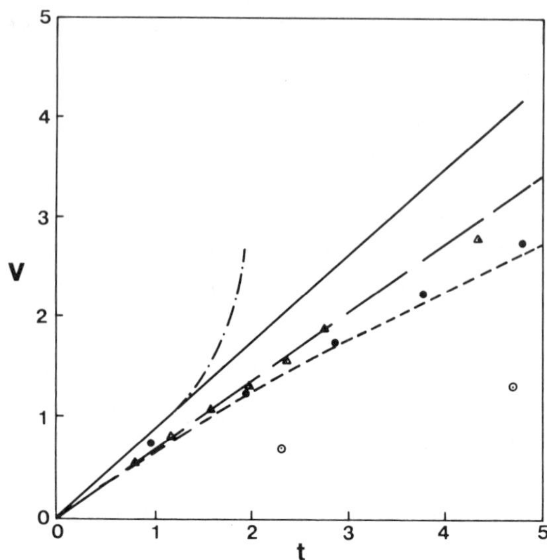

FIG. 6. Volume versus time for a constant heat-flux heating: comparison between theory and experiment. (——), numerical result of Prusa and Vandekamp [124], Ste = 0.46, Sb = 0; (— · — ·), perturbation result of Prusa and Yao [120], Ste = 0.46, Sb = 0; (⊙), experimental result of Bathelt *et al.* [106], Ste ~ 0.46, Sb ~ 0; (●), experimental result of Bathelt *et al.* [108], Ste ~ 0.46, Sb ~ 0; (△), experimental result of Goldstein and Ramsey [113], Ste ~ 0.46, Sb ~ 0; (— — —), numerical result of Prusa and Yao [120], Ste = 0.374; (— — —), integral result, Ste = 0.374, Sb = 0.05.

[120]), then the Stefan number for these studies [106, 108] drops to a value of 0.37. A numerical solution from Prusa and Yao [120] for this value of Ste is included in Fig. 6, and it is clear that this result compares much better with that of Bathelt *et al.* [108]. Nominally, each of the experimental studies began with a saturated solid, but, in practice, some amount of subcooling always exists. Such effects would increase as the melting progresses into the convection stage. This kind of increasing discrepancy is exactly what is observed in Fig. 6 between the numerical result [120] and the experimental result [108]. An approximate integral result, similar to Eq. (13e) but for two-phase melting, was used to determine a subcooling correction to the melting rate predicted by Prusa and Yao [120]. The result for a (seemingly) small value of Sb = 0.05 is drawn in Fig. 6. With a small amount of subcooling added, the comparison with the results of Bathelt *et al.* [108] now appears quite good—the two curves lie within about 4% of each other.

The comparisons with the results of Bathelt *et al.* [106] and Goldstein and Ramsey [113] are not so convincing, with the former being by far the worse. In the study conducted by Goldstein and Ramsey, [113], the subcooling was kept within 0.2°C of T_0, whereas in that of Bathelt *et al.* [106], it was estimated that no more than a couple of degrees of subcooling existed. From this statement and the thermophysical properties of the PCM, a subcooling number of Sb = 0.2 may be estimated for Bathelt *et al.* [106]. It thus appears that subcooling effects were an order of magnitude stronger in the experiment of Bathelt and co-workers [106] than in the Goldstein and Ramsey experiment [113]. End effects will also act to decrease the melting rates because of thermal losses and viscous damping of the fluid flow induced by buoyancy forces. The cylinder aspect ratios are 2.1 [106, 108] and 3.7 [113]. Because of the larger aspect ratio, end effects in the study of Goldstein and Ramsey [113] are probably less important. We believe that these two additional complications correctly explain the differences between the experiments and the numerical results, as well as between the two experiments.

Thompson's body-fitted coordinates [119] are certainly capable of immobilizing any irregularly shaped domain that the coordinate transformation [Eq. (8a)] can handle. In fact, it may be better suited for generating natural coordinates in some geometries, such as highly elongated domains, where Eq. (8a) might have trouble with skewness. Yet, the coordinate transformation as advanced in this section has some important advantages. It does not suffer from zoning problems, as mentioned by Rieger *et al.* [118]. Since Eq. (8a) is algebraic in nature, it is possible to tailor it to the specific geometry at hand, resulting in fewer transformation terms being added to the governing equations. Also, the method of body-fitted coordinates is restricted to numerical methods, whereas Eq. (8a) is useful even with analytical methods. Finally, since the natural coordinates have to be generated for each time step, Thompson's method results in a substantial increase of computing time compared to the coordinate-transformation method [Eq. (8a)].

3. *Other Studies in Cylindrical Geometries*

The effects of fins on phase-change processes around cylinders have been investigated in several studies. Fins have been experimentally observed to substantially enhance both freezing and melting around a horizontal cylinder [125, 126]. Bathelt and Viskanta [125] found that the average Nusselt numbers were within 20% of unfinned phase-change values, implying that the increased melting or freezing rates were primarily due to the increased heat-transfer surface area. Orientation of the fins, which affected the natural convection flow pattern, was found to be more important in melting than in freezing. Betzel and Beer [126] examined the effect of the type of fin material as well as fin orientation on melting rates with axial fins. Highly conducting copper fins were found to increase the melting rate roughly by a factor equal to the increase in heat-transfer surface area. With PVC fins of very low conductivity, the melting rate remained about the same as in the unfinned case. Orientation was observed to be important only for the PVC fins. In the convection stage of melting, the average Nusselt numbers were found to be functions of Ra and Ste, with the latter dependence being very weak. Betzel and Beer also indicated the flow patterns that occurred during some of the experimental runs. Gortler vortices were observed to occur along concave portions of the solid–liquid interface.

The effects of axial fins on phase change around a vertical cylinder have been studied [127, 128]. Sparrow *et al.* [127] experimentally investigated the effects of fins on the freezing process. For conduction-controlled freezing, they found that fins significantly enhance the freezing rate. Sparrow *et al.* also observed the formation of dendrites. For convection-dominated freezing, a decay of the phase-change process into a steady state was observed. As in an earlier study [114], Sparrow *et al.* fallaciously attributed this to natural convection rather than the superheated boundary condition that was applied. The increase in freezing was found to be roughly equal to the increase in the heat-transfer surface area. Dendrites were not observed to form for convection-dominated freezing. Kalhori and Ramadhyani [128] experimentally observed that subcooling lowered the total heat-transfer rate in the convection-dominated freezing. Kalhori and Ramadhyani [128] experimentally observed that subcooling lowered the total heat-transfer rate in the interesting result can be seen to be due to the finite size of their PCM container—with the fins, the heat-transfer area was increased so much (by a factor of about 5) that the solid was raised to the melting point in a relatively short time.

The effects of natural convection on the melting of a PCM inside a horizontal tube have been examined by a number of investigators [129–133]. Rieger *et al.* [129] used a numerical method very similar to that of a previous analysis [118] to determine the details of the heat transfer and fluid-flow fields. Improved grid-point distribution was obtained by using a local orthogonality condition along the PCM boundaries. They observed a short initial conduction period, characterized by a circular shape for the solid phase, which became shorter with increasing values of Ra. As time increased, natural convection became important and caused the solid to melt uniformly but more rapidly along the top surface of the solid phase. Experimental observations, which were also made, indicated that, for large values of Ra ($\geq 10^6$), three-dimensional, unsteady roll cells developed underneath the solid phase, causing locally nonuniform melting rates, and resulting in an inverted, flattened, pear-shaped, solid PCM region (cusp pointed down). For smaller values of Ra, no secondary vortices were observed and a more streamlined shape for the solid PCM resulted. The numerical solution by Rieger *et al.* captured the unsteady secondary flow for large values of Ra and, although in good general agreement, predicted more extreme local melting rates under the solid than was experimentally observed. This difference was thought to be due to the two-dimensional nature of the model. Natural convection was found to be very important during most of the melting process. Additional experimental data has been reported by Bareiss and Beer [130]. Ho and Viskanta [131] also studied melting inside horizontal cylinders using experimental and numerical methods. They again observed the secondary vortices induced by thermal instability in the bottom of the melt region for larger values of Ra, but unlike the results of Bareiss and Beer [130], Ho and Viskanta found that the convection-dominated stage occurs earlier for higher values of Ste (not Ra). Subcooling was observed to cause larger values of heat transfer at the cylinder surface, particularly at early times (in accord with the pure conduction result). The effect of natural convection on the local heat-transfer rates was determined using a shadowgraph technique. The numerical method was based upon the coordinate transformation [Eq. (8a)], but not all transformation terms were retained. For early times, a quasi-stationary assumption was used, while for late times, a quasi-steady assumption was employed. These assumptions, coupled with a coarse computational grid may be largely responsible for the poorer comparison (than the study of Rieger *et al.* [129]) between the numerical and experimental results that occur once convection heat transfer becomes important. The melting rate on top of the solid is predicted to be too high. Ho and Viskanta's numerical method also fails to predict the secondary vortices. Earlier numerical studies by Pannu *et al.* [132] and Saitoh and Hirose [133] resulted in solutions at variance with the numerical solution [129] and the experimental observations [129–131]. The Pannu *et al.* method used an expanding mesh to follow the growth of the melt region. This required interpolation of all variables from the past computational grid to the newly expanded grid. Although their results appear reasonable for smaller times, for later times, far into the convection stage, enhanced melting has occurred along the top of the solid phase, causing a local depression there. Saitoh and Hirose's method was based upon the coordinate transformation [Eq. (8a)]. It has a strong numerical instability, which they suppressed by employing a smoothing technique using a least-square, sixth-order polynomial. Saitoh and Hirose claimed that the smoothing allowed them to determine solutions for unusually large values of Ra. Their solutions show the secondary vortices in the bottom of the melt region, but the local melting rate is anomalously high there. A good comparative study of the numerical results [129, 132, and 133] is presented by Rieger *et al.* [129]. Although each of these studies indicates that natural convection is of critical importance for the internal melting of horizontal cylinders, it is interesting to observe that an experimental correlation for the melt volume in the study of

Ho and Viskanta [131] may be inverted and (slightly) extrapolated to give the following correlation for the time of complete melting:

$$t_f = 1.43\left(\frac{1 + \lambda\,\mathrm{Sb}\,\mathrm{Ste}}{\Delta\,\mathrm{Ste}}\right)^{0.58} \mathrm{Ra}^{-0.055} \quad . \tag{18}$$

Clearly the Ra dependence is slight.

Inward solidification of horizontal cylinders has also been studied [134–136]. Natural convection can have a substantial effect on freezing if the melt is superheated. As the solid–liquid interface moves inward, the melt is simultaneously cooled. The initial, natural-convection-dominated freezing gradually becomes conduction controlled when the melt is being cooled to its freezing temperature. The duration of natural-convection-dominated freezing depends on the Stefan number, which measures the freezing speed versus cooling rate of the melt. For small Ste's, the duration of the natural-convection dominated freezing can be very short. Viskanta and Gau [134] experimentally demonstrated that the liquid cools down to its freezing temperature very rapidly for Ste < 1. Consequently, a circular shape is maintained by the melt region as it shrinks in size. Without a natural-convection influence, the numerical results of their one-dimensional conduction model agree well with their experimental data. They also found that dendritic growth was suppressed in their experiments. Guenigault and Poots [135] used a regular perturbation method (valid for small times) to examine inward solidification in spheres and cylinders. They assumed the limacon of Pascal as a functional form for the solid–liquid interface shape and determined terms up to fifth order in the perturbation series. They found that the flow field in the melt consisted of an upward rising jet along the vertical axis of symmetry. This resulted in reduced freezing at the top and enhanced freezing at the bottom of the PCM container.

Natural-convection effects are found to be important in the beginning of a horizontal continuous casting process. Yao [136] investigated this problem with an isothermal boundary condition by using a singular perturbation method to solve asymptotic boundary-layer equations that are valid near the beginning of the cylinder. Three regions were identified: a solidified crust, a cooled melt layer, and an uncooled core. The solid crust was also shown to grow faster along the bottom of the cylinder because of the effects of natural convection. It is obvious that the importance of natural convection depends on the Rayleigh number, which is evaluated in terms of the thickness of the cooled melt layer. Initially, this layer is thin; consequently, freezing is conduction dominated. Natural convection can become the dominant heat-transfer mode if the melt region cannot be cooled to the phase-change temperature quickly enough. In other words, natural convection becomes important if the degree of superheating and the Stefan number are both not small. The perturbation results indicated that the natural-convention controlled solidification rate grew downstream like $z^{5/2}$, where z is the axial position, rather than like $z^{1/2}$ (pure conduction case). Since, if one moves with the cylinder, the casting process becomes an unsteady, two-dimensional freezing problem, a simple transformation, which measures the required time as the distance divided by the constant casting speed, can be used to demonstrate that the solution presented by Yao [136] is identical to the short-time solution of inward freezing in horizontal cylinders under strong cooling.

Inward melting of vertical cylinders has been investigated [132, 137–139]. Sparrow and Broadbent [137] found that such melting processes exhibited the same general characteristics as melting in other cylindrical geometries. An initial conduction-dominated process, characterized by a concentric annular melt region, was ultimately replaced with a melting process dominated by natural convection. The top of the remaining solid PCM melted faster than the bottom, producing a bullet-shaped solid region. Initial rounding off of the top of the solid PCM was also attributed to melt overflow because of density change. The usual effects of subcooling were observed. A one-dimensional numerical solution was also determined for comparison purposes. The rate of melting was enhanced by natural convection by about 50%. Sparrow and Broadbent found that natural convection increases the sensible heat of the melt by about 50% when compared to the pure conduction value. A correlation for dimensionless melt volume of the form $V = f(\tau \, \mathrm{Gr}^{1/4})$, where Gr is the Grashof number, was found to represent the data well. In particular, an error function was used for f. Unfortunately, this had the drawback of predicting that melting would never be completed in a finite amount of time. The numerical study by Pannu *et al.* [132] also considered vertical cylinders. The calculations clearly show that natural convection causes enhanced melting at the top of the solid PCM. The interface was not rounded near the top as in the experiment of Sparrow and Broadbent [137], presumably because the melt overflow due to density change was not modeled. Pannu *et al.* found that the average heat flux exhibits an overshoot behavior (see [109]) and has a local minimum in the transition stage of melting. They also found that as the cylinder aspect ratio $\to \infty$ (very long cylinder), the heat flux decreases strongly and asymptotically approaches the pure conduction value. The results of Sparrow and Broadbent [137] have been extended to the melting of impure substances in an experimental study by Sparrow *et al.* [138]. Such substances have two temperatures that define a melting range: T_{01}, which is the lowest temperature at which any melting begins to occur, and T_{02}, which is the lowest temperature at which complete melting can occur. For a pure substance, $T_{01} = T_{02} = T_0$. In the presence of a temperature gradient, an extended "mushy" zone will appear. A correlation for the dimensionless melt volume of the form $V = f(\tau \, \mathrm{Gr}^{1/8})$ was found to work very well, both for pure and impure PCMs. Both τ and Gr for this correlation are based upon T_{01}. Note that the exponent of Gr conflicts with the value of $\frac{1}{4}$ reported by Sparrow and Broadbent [137], and suggests a reduced role for the effect of natural convection on the overall melting rate. Another study by Menon *et al.* [139] experimentally investigated melting inside circular and square vertical ducts. An elementary, quasi-steady, melting model employing an effective thermal conductivity was also developed to compare with the experimental data. Information on the flow pattern induced by buoyancy force was also given. Partitioned ducts were found to reduce melting rates since they suppress natural convection. Menon *et al.* found that their melting data are functions of $Ra^{1/4}$, in support of the results of Sparrow and Broadbent [137].

As in the case of horizontal cylinders, for Ste < 1, natural convection has been found to have an almost negligible role in freezing processes inside vertical cylinders. Sparrow and Broadbent [140] experimentally observed that natural convection in the central melt region decayed very rapidly as the initial superheat dropped to zero. The melt region retained its initial concentric annular shape as it shrank in size, indicating that conduction was the dominant mode throughout. Although convection was present, a decrease in freezing rate was observed. The volumetric decrease of PCM during freezing caused the level of the remaining liquid to drop, generating a curved top surface that was orientated concave up. As expected, a one-dimensional, pure conduction, numerical solution compared fairly well with the experimental results, although it underpredicted the thickness of the solid layer. Several possible sources for the discrepancy were discussed, including the effects of dendrites, which were observed to form when natural convection decayed to an insignificant level.

It is to be expected that inward freezing will progress, more or less, independently of any angle of inclination of the containment tube, if natural convection is unimportant. This has been confirmed by Larson and Sparrow [141]. The only discernable effect of inclination (and hence of natural convection) was in the circumferential thickness of the solid layer. Only a slight thinning was observed to occur along the top of the tube for more horizontal configurations.

4. *Melting and Freezing in Rectangular Geometries*

A number of experimental [130, 142–146] and analytical/numerical [147–152] studies have clearly demonstrated the importance of natural convection effects on melting processes in rectangular geometries. As in cylindrical geometries, typically the beginning of melting is conduction controlled, with the melt region forming a thin rectangular slab adjacent to the heated surface. As the melt region widens, conduction gives way to convection-dominated heat transfer.

Hale and Viskanta [142] examined melting along a vertical heated wall. When the convection stage of melting was reached, melting progressed more rapidly near the top of the heated wall and was retarded near the base. Consequently, the overall melt shape was similar to that observed in melting around a vertical heated cylinder. Overall, the local enhancement in melting near the top of the wall dominates the decrease occurring near the bottom, and the melt volume increases almost linearly with time [130]. Interferograms clearly showing the S-shaped isotherms which are characteristic of natural convection may be found in the results of Van Buren and Viskanta [143]. Thermal boundary layers are observed along the heated wall and solid–liquid interface, with a central stratified zone in between, once the melt region is sufficiently large [144]. Subcooling is found to significantly impede the melting process because of the required sensible heating of the solid and the delay in the occurrence of the convection stage which the reduced melt region causes [145]. Gau and Viskanta [146] considered the melting of an anisotropic metallic PCM. Despite the low Prandtl number, natural convection was observed to significantly increase the melting rate over the pure conduction value. Once past the initial conduction stage, the melt volume increased almost linearly with dimensionless time τ. Enhanced melting occurred at the top of the enclosure, just as for higher Pr substances, but much more melting, because of the metal's high thermal diffusivity, occurred at the bottom. Consequently, the solid–liquid interface curved more gently than for higher Pr substances. Correlations developed to predict melt volume indicate that it is only a weak function of Ra, Ste, and the cavity aspect ratio.

Huang [147] has developed a close-form solution for natural-convection melting along a vertical wall early in the melting process when the melt region is thin and conduction controlled. He recognized that the buoyancy force induces a flow the instant that melting begins. As in melting around horizontal cylinders, there is no critical melt thickness that is required for the onset of convective motions. However, since the process is conduction controlled, the energy equation is decoupled from the fluid motion. Only the vertical component of velocity is significant, and it is a function of both time and the similarity variable η [see Eq. (1)]. The shear stress at the wall is found to be slightly greater than that at the solid–liquid interface.

In the transition and convection stages, the melt thickness becomes nonuniform, and the fluid flow two dimensional, making the problem far less amenable to analytical attack. Okada [148] developed a numerical method to determine solutions for saturated melting that remained valid when the heat and fluid flow became two dimensional. He used the coordinate transformation [Eq. (8a)] to immobilize the solid–liquid interface and made a quasi-steady assumption. Adiabatic boundary conditions along the top and bottom of the PCM enclosure were used. An initial conduction-dominated stage followed later on by a convection-dominated stage was clearly observed.

During the convection stage, the melt region was seen to grow much more rapidly near the top of the heated wall, and the overall melt volume grew nearly linearly with time, in close agreement with experimental observations. The development of hydrodynamic boundary layers was clearly shown. The average Nusselt number along the heated wall decreased rapidly during the conduction stage, reached a minimum value during transition, and then approached a slowly increasing, almost linear variation in time. For given values of Ste, Pr, and the cavity aspect ratio, Okado determined that by increasing Ra, the average Nusselt number rose above the pure conduction value earlier and more steeply. The average Nusselt number along the solid–liquid interface decreased with increasing Ra, since the interface length increased in time. For $Pr > 7$, the Prandtl-number variation was found to be insignificant.

Other numerical solutions for saturated melting have been determined. Benard *et al.* [144] also used the coordinate transformation [Eq. (8a)] but dropped higher order curvature terms introduced into the governing equations by the transformation, despite clear experimental evidence that the interface curvature becomes significant at the top of the melt region for large values of Ra. Nevertheless, Benard *et al.* claim that the extra terms are negligible and believe that the method gives good results for $Ste < 0.2$. A quasi-steady assumption was used in determining the flow field. Locally, the predicted interface position matches experimental results fairly well, until the interface curvature increases sharply near the top of the PCM enclosure. In this region, the numerical solution underpredicts the melting rate. It is interesting to observe that Okada's numerical solution [148], which retains all transformation terms, matches experimental interface positions much better in this region. As soon as distinct boundary layers developed along the heated wall and solid–liquid interface, Benard *et al.* observed that standard heat-transfer correlations for rectangular enclosures could be used to predict the local heat-transfer rates. They also conducted a study on the effect of the aspect ratio and found that for large Ra, as the aspect ratio was decreased (toward a short, wide enclosure), the melting rate increased. A final interesting feature of this work is that melting was followed in the numerical solution well beyond the point where the solid–liquid interface contacts the vertical wall opposite the heated one (held at T_0 for saturated melting). Further details and results from this model are given by Gadgil and Gobin [149]. Convective motions due to density change are found to be small, using a simple order-of-magnitude analysis. For use in the fluid-flow calculation, a cubic spline interpolation method was developed for predicting the solid–liquid interface position. The effect of a slip versus a no-slip boundary condition along the top of the PCM container was found to be negligible. Once the far wall was encountered, the total heat-transfer rate into the remaining solid PCM suddenly dropped. The decrease became more sudden and irregular for larger values of Ra. It is not clear if the lack of smoothness was a real phenomenon or an artifact of the numerical method. Ho and Viskanta [145] presented another numerical solution for the saturated melting problem. They used both the quasi-stationary (for small times) and quasi-steady (for large times) assumptions. They also used the coordinate transformation [Eq. (8a)], and although all curvature terms were retained, the numerical solution for the interface position did not appear to match experimental results significantly better than those of Benard *et al.* [144] and Gadgil and Gobin [149]. Again, the numerical solution underpredicts the melting rate at the top of the enclosure. The underprediction is attributed to excess melt volume, due to density change, overflowing into the air space and over the top of the remaining solid PCM. The numerical solution for the average Nusselt number also exhibits a damped oscillation in time once convection becomes the dominant mode of heat transfer. Such an oscillation was also observed by Gadgil and Gobin [149]. It is known that during the transition stage, when convection first competes with conduction, that an overshoot phenomenon with a local minimum in the average Nusselt number exists (see Ref. [109]). Once convection is the dominant mode, however, it is not clear how further local maxima and minima in the average Nusselt number may occur. The experimental results by Ho and Viskanta [145] do not indicate this type of oscillatory behavior; instead, the average Nusselt number decreases sharply during the conduction stage, and then attains an approximately constant quasi-steady value in the convection stage.

The saturated melting problem was also solved numerically using the enthalpy method by Schneider [150]. The method is very efficient and appears to result in an order-of-magnitude speedup in computing time compared to classical methods. A smoothing of the enthalpy across the solid–liquid interface was used. Unfortunately, the effects of the enthalpy smoothing, known to be significant, were not addressed (see Section II,F). The solution predicts that sometimes natural convection induces double vortices in the melt, a behavior that has apparently not been predicted by any of the other numerical methods or seen experimentally. Also, the predicted interface is not smooth, but irregular and jagged in shape. Schneider used three different computational grids of increasing resolution. He found that a grid with 20×20 nodes agreed well with the numerical result [145]. However, both numerical results predicted a melt volume noticeably less than experimentally observed by Ho and Viskanta [145]. A grid with 40×40 nodes matched the data much better. It is not obvious how good the 40×40 node result is, however, because the melt volumes predicted using the three grids do not clearly show convergence with increasing grid resolution.

The effect of density change in the two-phase melting problem along a vertical heated wall was examined by Kassinos and Prusa [40], using the model and numerical method of Prusa and Yao [123]. As in the study of Prusa and Yao [123], the effects of density change on the heat transfer were found to be very small, and these effects decreased rapidly with time. Subcooling was again found to be quite important, especially in the convection stage. With the subcooling effect, convective cooling was used as the boundary condition for the enclosure wall opposite the heated side. It was found that the convective loss from the solid PCM through this wall could equal the heat transfer into the solid through the solid–liquid interface. In such a case, a steady-state, heat-transfer process results. Using the Biot number to characterize the magnitude of the convective cooling, they pointed out the existence of a critical Biot number Bi_c such that if $Bi/Bi_c > 1$, then the melting process decays into a steady state. The decay of the melting process into a steady state is analogous to the decay of a freezing process with a superheated boundary condition. The solution for the average Nusselt number along the heated wall shows the only extremum to be the local minimum of the transition stage. Thereafter, the value is a very slowly increasing function of time and appears very nearly constant far into the convection stage. This matches the experimental results of Ho and Viskanta [145] very well. The oscillatory prediction of the numerical solutions [145, 149] is not confirmed. One possible source of the difference in behavior is numerical error. Both Ho and Viskanta [145] and Gadgil and Gobin [149] used computational grids of similar resolution, 13×21 and 12×20 nodes, respectively. Both studies found the improvement resulting from a finer grid to be negligible. However, the results by Kassinos and Prusa, based upon a grid with 41×51 nodes, indicate that results using coarser grids are in error by more than 5% once the convection stage is reached, for $Ra > 10^6$ and $Ste \geq 0.1$. Another source of the difference may be the false-starting methods used in both studies [145, 149]. The method used by Kassinos and Prusa [40] was based upon the extended coordinate transformation [Eqs. (8a), (8b), and (10a)], and thus began with the exact initial conditions for the melting problem. The initial error in the average Nusselt number introduced by the false-starting method of Ho and Viskanta [145] is plainly revealed in Fig. 3 of Kassinos and Prusa [40]. Benard *et al.* [151] have also developed a numerical method for two-phase melting problems. They used a subcooled isothermal boundary condition for the wall opposite the heated surface, quite different from the convective condition of Kassinos and Prusa [40]. With the isothermal condition, decay of the melting process into a steady state will automatically occur (see the discussion of Sparrow *et al.* [114]). The numerical method is similar to that of Benard *et al.* [144] and Gadgil and Gobin [149]. It is also pointed out that natural convection does not take place in the melt layer during the conduction stage, when it is very thin, and that a critical melt thickness must be attained before convection commences; but, in fact, convection does begin along the vertical heated wall when melting begins. Convection effects are simply overpowered by conduction because the latter effects become infinite like $t^{-1/2}$ as $t \to 0$.

Freezing along vertical cooled walls was studied by Ramachandran *et al.* [152, 153]. In the first study, the coordinate transformation [Eq. (8a)] with a quasi-stationary approximation was used in a numerical work to study two-phase solidification in molds. Higher order curvature terms were dropped from the formulation. Large values of Ra inhibited freezing moderately but only early in the process. This was because of transient cooling of the mold. No PCM could be frozen until the inside surface of the mold was cooled to the freezing temperature. As the remaining liquid approached a saturated state, the effects of natural convection decayed. Residual effects, particularly with regard to the solid–liquid interface shape could exist throughout the phase change. They also found that increasing the Ste enhances freezing. This agreed with the results of Yao [136]. Note that the definition of Ste is inverted [152]. Experimental results of Ramachandran *et al.* [153] compared well with the numerical predictions. Note also that the captions for Figs. 2 and 3 of Ramachandran *et al.* [153] are interchanged. An anisotropic metal was considered in the investigation of Gau and Viskanta [146]. Considerable irregularity in the shape, because of random crystal orientation, occurred locally along the solid–liquid interface. A slight increase in the freezing rate along the bottom of the enclosure was discernable.

Melting and freezing inside a rectangular cavity with conducting vertical walls has also been studied [154, 155]. A shadowgraph technique was used to determine heat-transfer information for melting and freezing from below [154]. For the melting process, a conduction stage clearly occurred. This was characterized by a uniform melt layer along the heated bottom and another melt layer of decreasing thickness running up along the highly conducting side walls. Melting along the vertical walls proceeded more rapidly toward the bottom of the enclosure except in the region of the free surface. Here, excess melt volume due to density change overflowed the top of the remaining solid PCM. Locally, the resulting convective motion enhanced the melting rate, causing the melt layer to increase with height. Eventually a global natural-convection flow developed, and the interface behaved more like that in a vertical melting problem. Underneath the solid PCM, along the heated bottom surface, Benard convection cells arose (clearly depicted in an interferogram). The unsteady vortices that were generated caused the solid–liquid interface to take on a wavy appearance. The local Nusselt number became spatially quasi-periodic. Subcooling was found to have a large initial effect, but this subsided as the remaining solid became saturated. Large aspect ratios were found to decrease melting rates sharply, in agreement with the results for melting along vertical walls [144]. For the freezing process, the effects of natural convection dissipated quickly. Because of the volume decrease of the PCM upon freezing, the level of the remaining liquid PCM dropped, causing a curved top boundary, concave side up. In the study of Gau and Viskanta [155], a metal eutectic PCM melted and froze uniformly, resulting in a flat solid–liquid interface at all times. Curiously, however, the overall melting and freezing rates were significantly influenced by natural convection, as is typical for higher Pr PCMs.

Webb and Viskanta [156] experimentally investigated melting in inclined rectangular enclosures. They found that the effect of inclination from the vertical is to create three-dimensional, natural-convective motion, which considerably enhances the overall melting rate compared to a vertical melting case. Once again, the melt volume was seen to increase in an approximately linear fashion with time once the convection stage was reached. For a negligibly subcooled solid, heat transfer rates are reasonably predicted by a correlation for natural convection in inclined enclosures in the absence of phase change.

5. *Additional Complications*

All of the natural convection effects so far considered in Section III have been for PCMs whose melts are *normal* fluids; that is, the density of the melt decreases monotonically with increasing temperature. Despite the substantial complications that arise even with these substances, nature taunts us further in that the most common liquid, water, exhibits the phenomenon of density anomaly (or inversion) near its freezing temperature. Maximum density occurs approximately at 4°C. The availability of water speaks loudly for its consideration as the PCM in many latent energy systems. Also, freezing and melting of water are necessarily of great importance in polar marine environments.

For phase-change processes with moderate amounts of superheating, a density anomaly leads to considerably more complicated flow patterns than appear in normal melts. Such effects have been studied by a few investigators [133, 157–159]. Wilson and Lee [157] studied the melting of a vertical ice wall facing into a body of fresh water. They began their work with a historical survey of early works on similar problems. The ice in their problem was considered to be saturated and to be melting uniformly at a constant rate. As a result, the entire melting process was modeled as a steady-state process. A numerical method was used to determine the fluid and energy flow for the water near the ice wall. Far from the wall, the fluid became quiescent and isothermally superheated. Wilson and Lee found that three basic types of flow pattern occurred, depending upon the superheated temperature, T_h. If $T_h < 4.5°C$, then the water was less dense near the ice wall and rose up along it. If $T_h > 6.0°C$, then water near the ice wall was observed to descend. For the temperature range $5.7°C < T_h < 6.0°C$, a bidirectional flow resulted. In this last case, water attained a temperature of 4.0°C some distance from the wall, and since this is the temperature of maximum density, at this location the flow was downward. Closer to the ice wall the density decreased with temperature, so the flow was upward. For T_h slightly less than 6.0°C, the downward flow dominated, whereas for T_h slightly more than 4.5°C, the upward flow dominated. In the superheated range $4.5°C < T_h < 5.7°C$, the numerical method failed to produce converged solutions. This was attributed to unsteady flow behavior, which had been noted experimentally.

This sensitivity of the basic flow pattern to the magnitude of the superheat temperature has also been observed in other geometries. White *et al.* [158] experimentally investigated the melting of saturated ice around a heated, horizontal cylinder. Photographs depicting the flow motion in the melt as well as interferograms were presented. The temperature difference, $\Delta T = T_h - 8°C$, was found to be a major parameter characterizing natural convection flow in the melting process. Here, T_h was the superheated temperature of the cylinder. If $\Delta T < 0$, denser water in contact with the cylinder would flow downward, increasing the melting rate *below* the cylinder. The resulting melt shapes looked like upside down versions of Figs. 3 and 5 of this chapter. For $\Delta T > 0$, the opposite effect occurred and enhanced melting was observed above the heated cylinder. As $\Delta T \to 0$, convection was increasingly suppressed, and conduction tended to dominate the melting process longer. White *et al.* pointed out that this behavior most likely occurs because the density of water at 8°C and 0°C is practically the same. Thus, as $\Delta T \to 0$, the density difference driving the flow decreases. Correlations developed for predicting melt volume indicate that the Ra dependence is stronger for $\Delta T > 0$ than for $\Delta T < 0$. Secondary vortices appeared in the melt, adjacent to the heated cylinder for $-4°C < \Delta T < 0°C$, and adjacent to the solid–liquid interface if $0°C < \Delta T < 2°C$. Complex, three-dimensional, unstable flow motions were also noted. In one case, these motions were observed to actually cause the local Nusselt number along one point of the cylinder to almost vanish for an instant. Quasi-steady, heat-transfer conditions were not reached in the experiment.

A numerical study on the effects of density anomaly on melting inside a horizontal cylinder was made by Rieger and Beer [159]. They found that natural-convection flow in the melt is characterized by multiple vortices, depending upon the value of the cylinder temperature T_h. Basically, for $6°C < T_h < 10°C$, the flow consisted of an inner and outer vortex. For $T_h = 8°C$, the density of water was less along the cylinder and the solid–liquid interface, leading to upward flow there. In between, the density maximum was attained, and the water flowed downward. As T_h decreased from 8°C, the inner vortex along the interface grew stronger. As T_h increased from 8°C, the outer vortex was found to dominate the flow field. Superimposed upon this basic flow structure may be a number of secondary vortices along the top (noted for $T_h = 6°C$) and bottom (noted for $T_h = 15°C$) of the melt region. Rieger and Beer characterized these secondary vortices as roll cells due to thermal instability. The smallest melting rate was obtained when $T_h = 8°C$; in this case, the inner and outer primary vortices appeared to be of the same strength. The maximum melting rate was obtained for $T_h = 15°C$; in this case, the effect of the density anomaly was weakest. Experimental observations were also made and generally agreed with the numerical predictions. Significant discrepancies were noticed only when the roll cells predicted numerically were experimentally observed to be three-dimensional. Saitoh and Hirose [133] also presented some information on the melting of ice inside a horizontal cylinder. Their isotherm and stream-function plots for the case $T_h = 8°C$ appeared quite similar to those presented by Rieger and Beer [159]. A strong outer vortex enhanced melting and flattened the ice top.

In the horizontal cylinder melting problems that have been discussed in this article, it has been implicitly assumed that the PCM and heated surfaces are held fixed in position. If the PCM is initially held in place by the surface that becomes the heat source, then the solid PCM mass will fall (provided $\Delta > 1$) as melting proceeds so that its weight will continue to be borne by some part of the heated surface. This is known as close-contact melting. In the region where the solid falls upon the heated surface, a thin layer of melt separates the two. Throughout the melting process, conduction remains the dominant mode of heat transfer in this layer, even when natural convection becomes important elsewhere.

Moore and Bayazitoglu [160] studied close-contact melting inside a sphere. The solid PCM was saturated initially and was denser than the liquid phase. Experimentally, they observed that the bottom of the solid region melted so as to remain almost spherical, matching the shape of the heated surface closely. The thickness (or gap width) of the thin melt layer was found to be nearly constant (spatially) and equal to about 2% of the sphere's radius. In time, the gap width changed only slowly. Consequently, the process appeared quasi-steady. The bulk of the melt accumulated in the upper part of the spherical enclosure between the heated surface and the descending solid mass. The top surface of the solid also remained roughly spherical in shape. A numerical method using the coordinate transformation [Eq. (8a)] was developed to investigate the effects of fluid motion caused by the falling PCM mass in the bulk melt region. Since the gap was very thin, Moore and Bayazitoglu assumed a one-dimensional parabolic velocity profile for the flow in the gap. The profile was determined by zero-velocity conditions along the heated surface and solid–liquid interface and by requiring that the mass flow rate be determined by the melting rate. The position of the solid was found using a simple force balance—the weight of the solid was matched to the vertical components of the shear and pressure forces existing in the gap. Since the thermal resistance in the narrow gap remained small throughout the melting process, the rate of conduction heat transfer in this region remained very high and eventually became the dominant mode. In the bulk melt region above the solid, convective motions arising from the dropping of the solid mass were found to be significant for Ste > 0.1. For Ste = 0.5, an enhancement in the overall melting rate of about 25% was determined, as compared to a pure conduction result.

Close-contact melting inside horizontal tubes was investigated by Bareiss and Beer [161] and by Sparrow and Geiger [162]. Despite the difference in geometry, almost every characteristic feature described by Moore and Bayazitoglu [160] was confirmed. The most significant difference was that natural-convection heat transfer was seen to dominate the melting process above the descending solid mass despite the fact that the solid maintained a cylindrical shape for its top surface. Analysis of experimental data indicated that about 10% of the overall melting occurred in this region. An approximate analytical model reveals that the Archimedes number, which is the ratio of buoyant weight of the solid PCM to viscous force, may be an important parameter in close-contact melting [161]. A comparative experimental study with inward melting of a fixed solid [162] clearly indicates that close-contact melting increases the overall melting rate 50–100%. Also, the dimensionless time τ becomes an almost perfect similarity variable for melt volume for all values of Ste. The small residual effect of Ste that does exist in a plot of $V(\tau)$ is that the melting rate $dV/d\tau$ decreases with increasing Ste (note that this does not imply that dV/dt decreases with increasing Ste, which seems to have been suggested [162]). An extension of close-contact melting to melting problems in concentric annular regions was made by Betzel and Beer [163]. Two conduction-dominated, narrow melt gaps were observed to appear, one along the bottom of the outer heated cylinder and the other along the top of the inner heated cylinder. In the bulk melt region above the descending PCM mass and below the top of the outer cylinder, natural-convection heat transfer was found to be important. A second bulk melt region, below the inner cylinder and interior to the solid mass, was stably stratified and thus conduction dominated. Melting along the interface in this region contributed about 5% of the total melt. As in the study of Sparrow and Geiger [162], close-contact melting was observed to greatly enhance the melting rate compared to either natural-convection or conduction-dominated melting for fixed PCMs. An approximate analytical solution was also developed, and it compared reasonably well with the experimental data.

If the solid phase is less dense than the liquid ($\Delta < 1$), then an unfixed solid mass will rise because of the buoyancy force. Now the narrow melt gap appears above the solid mass, along the top of the heated cylinder. Webb *et al.* [164] experimentally investigated such behavior in the melting of ice. For cylinder temperatures under 4°C, the overall behavior appeared to be a mirror image of that found for normal close-contact melting where the solid descends. However, for cylinder temperatures above 4°C, typical density anomaly effects, as discussed earlier, appeared. The maximum anomaly effects were observed for cylinder surface temperatures near 8°C.

Close-contact melting may occur in a second class of problems in which the heat source itself sinks through the solid PCM. Initially, the surface that will be heated is held in place by the surrounding solid. Once the surface is heated, melting begins and a thin melt layer forms along its bottom. Moallemi and Viskanta [165] investigated the descent of a heated horizontal cylinder through solid PCM in an experimental study. As in close-contact melting due to the motion of the solid PCM, the thickness of the gap was found to be very small. Conduction appeared to be the sole mode of heat transfer in the center of the gap and controlled the terminal speed of descent, which appeared to be independent of the weight of the source (provided it was significantly different from being neutrally buoyant) and proportional to Ste. In all cases, the terminal speed exhibited an overshoot behavior, reaching a maximum value as the heat source was first completely covered with the melt. This overshoot behavior was attributed to the interaction between circumferential conduction in the source (which was adjusting from an initially isothermal state) and external cooling by the surrounding melt. The source always reached its terminal speed quickly, at which point a quasi-steady state was reached. The thin melt gap was observed to widen with increasing distance from its center point, and convection effects were believed to then become important. The resulting two streams of melt, one on each side of the cylinder, entered the bulk melt region above the heated cylinder and interacted with the global flow pattern, causing an abrupt local change in the shape of the solid–liquid interface. It was felt that the global flow pattern was mainly induced by source motion, and that natural-convection effects were minor, in direct contrast to the close-contact results for a moving solid PCM [161–163]. A detailed numerical model for this problem has also been developed by Moallemi and Viskanta [166]. The formulation is based upon quasi-steady and boundary-layer assumptions for the liquid PCM in the narrow gap. Shear and pressure forces existing in the bulk melt region are also accounted for (in addition to those in the narrow gap) in the force balance governing the position of the source. The thickness of the gap is found to increase as one moves from the center of the gap outward to its edges. This increases the thermal resistance through the layer causing a gradual decrease in melting outward from the center of the gap. This behavior is exactly what is required in order to maintain the constant cylindrical shape of the solid surface adjacent to the source.

Detailed temperature and flow-field solutions indicate that the linear temperature and one-dimensional parabolic velocity distributions used in more approximate analyses may be reasonable assumptions if high accuracy is not required. The effects of density change due to phase change were also modeled and found to be extremely small (as observed by Prusa and Yao [123]). For $0.95 \leq \Delta \leq 1.05$, the magnitude of such effects on the speed of descent are about 0.01%. The speed of descent for the heated cylinder was found to be almost independent of its mass and linearly proportional to Ste, in agreement with Moallemi and Viskanta [165]. However, the numerical predictions are up to 40% lower than the experimental values. In contrast to these experiments of Moallemi and Viskanta [165], additional model details and numerical results given in a later study by these investigators [167] indicate that natural convection in the bulk melt region is very important in controlling the global flow pattern.

Close-contact melting for a hot sphere melting its way through a solid has also been analyzed. Emerman and Turcotte [168] used a Stokes-flow approximation for the melt motion in the narrow gap. This simplified the equation of motion sufficiently, so that a close-form solution for the velocity distribution could be obtained. The energy equation was solved by assuming a quadratic polynomial for temperature in an approximate integral method. The model ignored all effects in the bulk melt region above the sphere. The result for the speed of descent was that it was linearly proportional to Ste and independent of the weight of the sphere, in the limit Ste $\rightarrow 0$. For larger values of Ste, the descent speed did depend on the weight of the sphere and was somewhat less than predicted by the small Ste approximation. Emerman and Turcotte used their model to predict the time that a nuclear reactor core could melt its way through the solid earth to the Earth's core. However, the solution is not realistic, one of the reasons being that the tremendous changes in physical properties that would occur during the descent (in the Earth's core, the specific gravity and pressure are estimated to be about 13 and 3.5 million atmospheres, respectively [169]) are not considered.

B. FORCED CONVECTION

A considerable amount of information on the effects of forced convection on freezing and melting can be found in the review articles by Epstein and Cheung [10] and Cheung and Epstein [11]. As a result, we will make no attempt to give a comprehensive treatment of forced convection effects but will instead focus upon the phenomenon of solid–liquid interface *instability*, particularly in regard to freezing processes. This is a very interesting phenomenon that has been observed experimentally in several investigations but has almost been completely ignored in all modeling attempts. It is our belief that the techniques and methodology that have been developed to build natural-convection, phase-change models are also equal to the task of building models to study this instability.

Cheung and Epstein [11] pointed out that two simplifying assumptions are generally made in forced-convection freezing problems. First, the solid layer is usually considered to be either very thin or of uniform thickness. Consequently, conduction heat transfer through the solid can be considered to be one-dimensional. Second, a quasi-steady assumption is usually made for the fluid flow. Often the quasi-steady assumption is also used to simplify the energy equations in the solid and flowing liquid PCM. These assumptions are very reasonable if the Stefan number is small. Thus, one is typically led to a model for freezing inside cooled tubes that has a solid layer on the inside surface whose thickness is a monotone increasing function of axial position. Moreover, the solid thickness is a very weak function of axial position [170–174].

A study by Sadeghipour, *et al.* [170] determined a Graetz-type solution for the problem using Laplace transforms. The tube is convectively cooled on its outside surface. Uniformly superheated liquid PCM enters with a constant specified flow rate. An initial cooling zone is required in which the superheated liquid cools enough to begin freezing. Since the problem is steady, the axial coordinate is timelike, and this initial zone is completely analogous to the prefreezing period that is required in an unsteady freezing problem (see study of Yao [136]). The length of the cooling zone decreases with increasing convective cooling (characterized by the Biot number) and decreasing liquid superheat. Beyond the initial cooling zone, a thin solid layer forms, and the diameter of the liquid region decreases. The coordinate transformation [Eq. (8a)], with the higher order curvature terms ignored, is used to normalize the liquid region in the radial coordinate. An elementary logarithmic temperature distribution is used for the solid phase. The solution predicts that an asymptotic solid thickness is approached as the flow progresses further and further downstream. Since the flow rate is specified, complete blockage cannot occur. An interesting result is that the asymptotic thickness has a maximum value for $\mathrm{Bi} \approx 1.5$, rather than for $\mathrm{Bi} \to \infty$. Sadeghipour *et al.* attributed this to the interaction between an enhanced cooling effect at the surface and a warmer melt in the core due to a shorter cooling zone, as Bi increases from zero. Cervantes *et al.* [171] used the two basic simplifying assumptions to construct a regular perturbation solution to the freezing problem inside a pipe. Their formulation requires that $\mathrm{Ste}/\mathrm{Pe} \ll 1$, where Pe is the Peclet number. The perturbation parameter is $1/(1 + \mathrm{Su})$; the zero-order solution is the classical Graetz solution. Thus, to first order, the thickness of the solid layer increases like the cube root of the axial distance from the leading edge of the layer. A numerical solution for the problem was given by Toda *et al.* [172]. As in the preceding two studies, laminar flow for the melt was assumed. However, the quasi-steady assumption was relaxed, and axial conduction in the solid layer was allowed. The flow was still assumed to be fully developed, and natural convection was assumed to be negligible. The radial extent of both solid and liquid regions was normalized using Eq. (8a). Toda *et al.* found that axial-conduction effects can be very important, especially near the thermal entrance where the solid increases in thickness rapidly. The effects of turbulent melt flow were considered by Epstein and Cheung [173] and Sampson and Gibson [174]. Each study used all of the typical simplifying assumptions for laminar flows, with the addition of empirical correlations for turbulent friction factors or Nusselt numbers. Sampson and Gibson suggested that, in determining blockage by freezing, it is more appropriate to use a specified axial pressure gradient rather than a specified mass flow rate (as used by Sadeghipour *et al.* [170]). With the latter type of condition, unbounded axial velocities will occur downstream as the tube freezes solid. In reality, the most appropriate condition is one that would take into consideration the pressure-discharge characteristic of the melt source [175]. Thus, the freezing problem in the tube would have to be viewed in the context of the whole flow system.

One of the main objectives of these studies is to predict under what conditions a tube flow might freeze shut, and yet each of them is incapable of making this type of prediction accurately except for a limited range of flow parameters. The difficulty is fundamental, and it lies in the two basic assumptions that are usually made in analysis. When a pipe actually freezes shut, the process is not necessarily quasi-steady, and the solid thickness is usually not a slowly varying, monotone function of axial position.

Gilpin [175] experimentally observed the freezing of pipes that were convectively cooled on the outside. When a laminar flow began to freeze, it did so in the uniform, quasi-steady manner typically assumed in analytical studies—*initially*. However, after a certain interval of time, a sudden expansion in the melt passageway was observed to form near the tube exit. As time advanced further, this sudden expansion slowly migrated upstream. Downstream of the expansion, the solid was completely melted away. Eventually the expansion point stopped migrating upstream, and the solid reappeared downstream of it. Another sudden expansion appeared, and the whole process occurred over again. Thus, a slowly increasing, monotone function for solid thickness results in a solid–liquid interface that is unstable with respect to large disturbances. Ultimately, the tube settled into a steady state in which the solid layer showed a periodic thickening and thinning with axial distance, known as the *ice-band* structure. Gilpen considered the phenomenon to be due to an interaction between the fluid flow and heat transfer. If the flow passage undergoes a rapid expansion at some point, flow separation occurs. This generates a region of enhanced turbulent mixing and, hence, heat transfer downstream, causing melting. The increased heat transfer is sufficient to undercut the sudden expansion, until it moves far enough upstream that the convective heat transfer from the melt into the solid matches the loss through the tube wall. In the experiment, the initial disturbance was provided by the sudden expansion of the flow at the pipe exit. Gilpin was also able to introduce disturbances by partially blocking the pipe in other locations, however. The expansion angles were measured to range from almost 90° for migrating disturbances to about 20° for the steady-state, ice-band structure. If the melt flow was initially turbulent, then many sudden irregularly spaced expansions were observed to appear and grow simultaneously. These multiple expansions were observed to migrate somewhat, though ultimately a steady-state pattern very similar to that for the initially laminar flow occurred. This similarity demonstrated that the period of the ice bands was insensitive to the value of the Reynolds number Re. Gilpin showed that the period did depend strongly on the value of Su and that only for very large Su (Su $\rightarrow \infty$ appears to be the trend, in fact) was the thickness monotone. Once the minimum openings for the melt flow were about 20% of the tube diameter, sporadic fluctuations in the flow and solid thickness would occur. When the tube froze shut, it was observed to do so rapidly and unexpectedly at conditions for which the system had run for an extended period of time. Complete freezing would first occur in the regions of maximum solid thickness, trapping liquid regions in between. The pressure drop through the ice-band structure has two contributions: viscous drag along the interface and losses due to the sudden expansions. The latter causes the pressure loss to be one to two orders of magnitude larger than would ordinarily be expected.

Hirata and Ishihara [176] also investigated the freezing problem experimentally. They confirmed every major result of Gilpin [175]. In fact, they found that the ice-band period is well correlated by the relation (using $k_l/k_s = 3.5$ for water)

$$\lambda/l = 11\,\mathrm{Su}^{3/2} + 4.5 \qquad \text{for} \qquad 0.11 < \mathrm{Su} < 2.2 \tag{19}$$

where λ is the disturbance wavelength and l is the tube diameter. Hirata and Ishihara also developed a modified Re for the ice-band structure which led to a criterion for determining whether or not steady-state ice bands will occur or the tube will freeze shut. Increasing values of the Su and modified Re will tend to enhance the possibility of the steady-state occurring. An ice-band-like structure was also observed experimentally [172], leading Toda *et al.* to conclude that their numerical method, which was able to predict only a monotone increasing solid thickness, was valid for Su $> \frac{1}{2}$.

From the experimental results, it is very clear that the ice-band structure appears often and that the assumption of a slowly monotone increasing thickness for the solid layer is an asymptotic approximation, strictly valid only as Su $\rightarrow \infty$. Furthermore, when a pipe does freeze shut, the quasi-steady assumption may be unreasonable. These two commonly used assumptions must both be relaxed if accurate predictions of complete freezing are to be made. A controversy on whether or not natural convection effects are important has been noted [10, 11]. Toda *et al.* [172] stated that such effects will be suppressed by flow acceleration through converging passages in the ice-band structure. However, it is well recognized that the natural-convection effects are accumulative [136]. Also, Morton [177] demonstrated that the buoyancy force is actually proportional to Re · Ra and thus will be enhanced by an increase in axial velocity. It is likely that natural-convection effects are more significant than is generally appreciated. With natural convection added, the three-dimensional, steady model that would result would be similar to those discussed in Section III,A,2.

The unstable nature of the solid–liquid interface has been observed in other geometries and thus seems to be a general feature of forced convection freezing. Seki *et al.* [178] considered freezing in a flat, rectangular channel. Experimental observations revealed an ice-band-like structure. Sudden expansions in the melt passageway formed, and they migrated upstream as in tube flow. However, only one expansion was observed to form in the steady state. A criterion for the development of a smooth ice (thickness is a monotone increasing function) versus a transition ice (the ice-band-like structure) was developed in terms of Re and Su. As Su and Re increase, smooth ice is more likely to occur. A similar type of phenomenon is found in freezing on flat plates [10, 11]. Where transition to turbulent flow occurs, there is a substantial increase in heat transfer. Locally, the solid melts, causing a concave depression in the interface. An unfavorable pressure gradient develops, promoting turbulence or even boundary-layer separation. The point of transition migrates steadily upstream until the convective heat gain of the solid is again in balance with the conduction loss to the plate. Thus, a critical Re up to an order of magnitude less than what is typical may be observed. Cheung [179] presented evidence that the "sand dune" appearance of the solid near the leading edge of a cooled flat plate may also result from nonisothermal boundary conditions along the plate. He used a perturbation method to demonstrate that, with such boundary conditions, the solid thickness increases and then decreases with distance from the leading edge, when an isothermal condition would lead to a monotone increase in thickness. Other interesting examples of the interaction between solid–liquid interface shape and forced convection flow can be found [180, 181]. Typically, such interactions result in critical Re, which are well below the usual values, and in heat-transfer rates, which are considerably enhanced over conventional values.

The interaction between interface shape and the transition to turbulence presents a very tough challenge at the present time (and perhaps in the foreseeable future) to modeling efforts. Nevertheless, there appear to be a host of forced-convection problems, especially laminar flow, which are ripe for the methods of analysis advanced in Section III,A,2. In particular, the coordinate transformation [Eq. (8a)], with all terms retained, would be ideal in simplifying the ice-band geometry in an analysis. There is no need to make quasi-steady or one-dimensional assumptions. Similarly, solid–liquid interface shapes in a variety of other geometries do not have to be measured experimentally and input into multidimensional heat-conduction codes [182] in order to generate heat-transfer information. The methods of analysis that are now available allow us to determine the behavior of many of these problems from first principles alone.

IV. Additional Effects in Phase-Change Processes

A. Porous Media

The freezing and melting of liquid-saturated, or partially liquid-saturated, porous media [183–185] are phenomena commonly observed in nature or in human-designed processes. Successive freezing and melting of soils induces underground water flow from warm to cold regions [186–190]. The knowledge of phase change in porous media is also important for energy storage, pipeline transport in permafrost regions, and cryosurgery and in the transportation of coal in cold weather [101, 191–195].

On the one hand, analytical predictions of phase change in porous media generally agree well with experiments. On the other hand, discrepancies do exist and the agreement between predictions and experiments worsens when the ratio of the thermal conductivities of the solid matrix and the liquid deviates from one. The importance of these thermal conductivities can be elucidated by two limiting cases. If the conductivity of the solid matrix is much larger than that of the liquid, thermal waves travel much faster in the solid matrix than in the liquid. Thus, freezing can occur almost homogeneously in the porous medium and can be treated as a local phenomenon. However, for a solid matrix with a low thermal conductivity, heat transfer in the matrix can be ignored and the phase-change process is determined by the energy transfer in the liquid phase alone.

Nonuniform porosities have rarely been considered in studies of energy transfer in porous media. For a partially saturated porous medium, the migration and condensation of moisture become important mechanisms in changing phases. A meaningful approximation would consider the medium to have at least two groups of pore sizes. Small pores are filled with liquid and large pores are occupied by air and vapor. The freezing of liquid trapped in the small pores is determined by heat conduction in the solid matrix, and the growth of the solid phase is controlled by the migration of moisture.

It has been recognized that fluid motion has some influence on energy transfer in porous media. However, this effect has not been included in studies of the phase-change process in porous media.

B. INSTABILITY-INDUCED CONVECTION

The solidification of a binary, eutectic-forming solution has received considerable attention because of its relevance to metal processing [196], the cryopreservation of biological cells [197, 198], the desalination of seawater [199, 200], the "freezing-out" process [201], and nuclear reactor safety [202, 203]. Unlike the solidification of a pure substance in which phase change can occur isothermally at a sharp solid–liquid interface, freezing of a binary aqueous solution takes place over a temperature range within a two-phase region that corresponds to the area between the liquidus and solidus curves in the substance phase diagram. Typically, a single component is frozen until the concentration of the solution reaches the eutectic composition. Then, the component proceeds as a solid eutectic. A concentration gradient, established by the solute rejection near the solid–liquid interface, causes double-diffusive instability [204]. This instability can have profound effects on the solid as it crystallizes from the fluid state and is of significant technological importance in the processing of electronic materials and in the growth of single crystals. Many review articles have been devoted solely to the issue of how solute transport is modified by fluid motion. A useful start would be a reading of the review by Glicksman *et al.* [205].

A reasonable estimation of the macroscopic structure of the two-phase zone can be achieved by enforcing conservation of energy and mass and by satisfying phase equilibrium [206–211]. Analyses based on these ideas can provide sufficient information for most engineering applications. It is, however, recognized that fluid motion can have substantial effects on the phase-change process. Natural convection, as discussed in Section III, and flow instabilities induced by temperature gradients and density stratifications [212–222] can modify phase-change processes substantially. In particular, flow instability in the form of rolls and hexagonal cells causes local extremes in the heat-transfer rate along the solid–liquid interface. This results in multidimensional heat transfer. Several unstable convective modes may exist. The relative importance of these fluid motions and instabilities has not been clearly identified. Furthermore, most theoretical studies in this area have been simplified to avoid the difficulty of treating *unknown* moving boundaries. The importance of bulk fluid motions, *nonlinear* effects that have been demonstrated for other phase-change problems, should also be considered here.

The possibility of producing pure substances under reduced gravity in an orbiting spacecraft provides a completely different manufacturing environment. Surface-tension-induced convection and double diffusion still exist in a reduced gravity. Many review articles have been published within the past few years which summarize the extensive activity in this research area. Articles which would be good starting points for further study are Refs. 205, 216, 223, 224.

Similar to the thermal-energy storage system, the concept of latent heat of fusion has been used in the design of "core catchers" as a safety device for nuclear power reactors. Indeed, Russian nuclear engineers wish they had such a device at the Chernobyl reactor! Hot molten fuel from the reactor would be retained by melting "sacrificial" material in the catcher. The sacrificial materials would have a large latent heat in order to absorb the decay heat of the fission products in the molten fuel. Several experiments [225–228] have been conducted to simulate the phase-change process that could occur in the core catcher. A popular setup, without considering the effects of the decay heat of the molten fuel, pours a hot liquid on a less dense solid. Convection can be induced by the mixing of a lighter molten solid with the hot liquid and by the temperature gradient. Downward heat flux and the penetration rate of the molten pool are important for the design of the catchers. No definite conclusion has been reached as to what the proper dimensionless parameters are for this problem. Some experimental data suggest that the density difference between the fuel and the sacrificial material is important; others show that the temperature difference has a larger influence on the penetration rate. This is a typical example in which a proper analysis can advance our understanding of the physics in order to show the appropriate experimental parameters. Relevant studies with a different geometry involve the melting of a vertical solid surface by a surrounding hot liquid [229] and by condensation of a hot vapor [230–232].

C. ELECTROMAGNETIC EFFECTS

When a piece of metal is placed above a coil carrying a high-frequency current, the induced surface current in the metal provides a Lorentz force that can support it against gravity; at the same time, the heat produced by joule dissipation can melt the metal. This is the process of levitation melting. The most obvious advantage of this process over the usual method of crucible melting is that the liquid metal does not contact the wall of the crucible so that there is no danger of contamination [233]. A similar arrangement can use the electromagnetic force to stir the solution in controlled solidification or continuous casting processes [234]. In these processes, density stratification can introduce additional mixing during solidification [136, 235].

Welding is another process that requires phase changes. Studies of commonly used arc welding [236–241] and laser welding and cutting [242–246] began only within the past decade. Most theoretical models avoid the complex phenomena that are caused by convection induced by temperature gradients, surface tension, and moving reference frames. Irrespective of the simplifications of these models, they have provided useful relations between surface heat fluxes, heat losses, fusion boundaries, and thermal stresses. This knowledge has helped to improve the quality and productivity of welding.

D. FROST, ABLATION, AND SUBLIMATION

Frost, an unwelcomed but unavoidable phenomenon, can be formed on a cool surface [247–253]. It is frequently observed on the coils of refrigerators or air conditioning units and on cryogenic heat exchangers with atmospheric heat sources. The crust of frost can grow thick and form an insulating layer. Consequently, the efficiency of heat-transfer devices declines dramatically. Frost is responsible for the sudden decline in the output of an automobile air conditioner in hot and humid weather. It can also impede the capability of cryogenic heat exchangers and degrade their performance to the point where their use in continuous applications is impractical.

Frost formation is a complex transient heat- and mass-transfer process with moving boundaries. The crystals of frost appear first at locations where the surface contains preferred nucleation sites and where moisture condenses. The onset of frost is similar to the nucleate boiling on a hot surface. The number of nucleation sites increases quickly and the formation becomes almost uniform over the entire surface. Freezing of condensed water forms needle-like crystals that form a rough surface. These needle-like crystals branch out and form tree-like structures as the frost grows thicker. It has been observed that the density of the frost layer increases with respect to time and so does its thermal conductivity. This suggests that, in addition to deposition at the frost–air interface, appreciable moisture is diffused into frost and frozen inside it. If the ambient humidity is high, the process is not mass-transfer limited, but rather is controlled by heat transfer. Therefore, the thermal conductivity of frost plays a deterministic role in frost formation. For a dense frost, the temperature at the frost–air interface can drop to the triple-point temperature, and moisture can be condensed on the interface to form a smooth ice layer. The cycle can then be repeated and fresh crystals of ice can grow on the ice surface. For less humid air, the process would be limited by the mass transfer of moisture. Since the density of frost is less than that of ice and the frost thermal conductivity is low, the interface temperature is above the triple-point temperature. Moisture is condensed only on the solid matrix of the frost; consequently, no ice layer is formed.

Heat ablation is commonly used to provide the thermal shielding for a reentry space vehicle [254–258]. It also occurs in solid-fuel combustion. Thermal radiation is usually a dominant heat-transfer mode at the typically high temperatures of these applications. A similar process, sublimation, occurs at much lower temperatures and has applications in separation processes, in food preservation, and in underground storage of liquefied gases [259, 260].

E. GEOPHYSICAL PHASE-CHANGE PROBLEMS

The formation of Earth's mantle involves a freezing process. On geological time scales, the Earth's mantle exhibits a viscous fluid behavior. Thermal convection in the mantle drives plate tectonics and continental drift [261, 262]. If viscous heating decreases the fluid viscocity, thermal runaway, which is believed to be the mechanism for a substantial fraction of the observed surface volcanism, can occur when a constant stress is applied [263–266]. Viscous heating due to the motion driven by buoyancy can cause a solid or an immiscible fluid to melt its way through a host medium. This mechanism can also explain the Earth's core formation [267–269] and magma migration [270–273]. The importance of viscous heating on melting has been assessed [274].

The interaction of the ice cap near the Earth's poles and the atmosphere plays an important role in climate dynamics, especially in connection with the onset of glaciation cycles [275–279]. The interaction can be studied by energy balances, which include solar flux and radiative cooling, atmospheric convection, oceanic currents, falling snow, and conduction and ablation of the ice cap. An accurate estimation of these energy fluxes can predict the growth of the ice cap. Simplification is necessary for such a complex phenomenon. Models usually ignore the energy variations along the radial and longitudinal directions. The interaction of the ice cap with the ice-free part of the Earth's atmosphere is then treated as a moving-boundary problem. It has been found that the phase-change process of the ice cap can influence the Earth's climate.

NOMENCLATURE

B	R/l, dimensionless solid–liquid interface position
Bi	hl/k, Biot number
Br	radiation Biot number [see Eqs. (12f) and (12g)]
c	specific heat
F	view factor
g	acceleration of gravity
h	convection heat transfer coefficient
$\bar{h}$	enthalpy [see Eq. (11a)]
h_{sl}	latent heat of fusion
k	thermal conductivity
L	geometry parameter [see Eq. (16)]
l	characteristic length
n	refractive index
Pr	Prandtl number
q	heat flux
R	solid–liquid interface position
Ra	Rayleigh number [see Eqs. (17a) and (17b)]
S	position of PCM boundary [see Eq. (8a)]
Sb	subcooling number [see Eqs. (3) and (9b)]

$\overline{\mathrm{St}}$	radiation Stefan number (see Br)	T_c	cooling temperature
Ste	Stefan number [see Eqs. (3), (6c), (12f), and (12g)]	T_h	heating temperature
Su	superheating number [see Eqs. (7b) and (7c)]	$\bar{t}, t$	time, dimensionless time [see Eq. (8b)]
$\bar{T}, T$	temperature, dimensionless temperature [see Eqs. (3) and (10f)]	V	coordinate translation distance [see Eq. (8a)], dimensionless melt volume [see Eq. (17c)]
T_∞	bulk temperature of convective heat transfer fluid	X	gap function [see Eqs. (10a), (10d), (10f), (13e), (14a), and (14b)]
T_0	phase-change temperature	$\bar{x}, x$	position, dimensionless position [see Eq. (8a)]

Greek Symbols

α	thermal diffusivity	μ	growth constant for isothermal phase change [see Eqs. (3) and (5a)]
β	coefficient of expansion	ν	kinematic viscosity
Δ	ratio of solid/liquid PCM densities [see Eq. (3)]	ρ	density
ε	$\Delta - 1$ [see Eq. (3)], small temperature change [see Eq. (11d)], distance, emissivity	σ	Stefan Boltzmann constant
η	similarity variable [see Eqs. (1) and (3)]	τ	dimensionless time [see Eqs. (9c) and (9e)]
λ	ratio of solid/liquid PCM thermal diffusivities [see Eq. (3)]		

Subscripts

f	final instant of phase change	s	solid PCM
l	liquid PCM		

REFERENCES

1. V. J. Lunardini, "Heat Transfer in Cold Climate." Van Nostrand-Reinhold, Princeton, New Jersey, 1981.
2. S. G. Bankoff, Heat conduction or diffusion with change of phase. *Adv. Chem. Eng.* **5**, 75 (1964).
3. T. R. Goodman, Application of integral methods to transient nonlinear heat transfer. *Adv. Heat Transfer* **1**, 71 (1964).
4. L. I. Rubinstein, The Stefan problem. *Am. Math. Soc. Transl. Math. Monogr.* **27** (1971).
5. J. R. Ockendon and W. R. Hodgkins, "Moving Boundary Problems in Heat Flow and Diffusion." Oxford Univ. Press (Clarendon), London and New York, 1975.
6. D. G. Wilson, A. D. Solomon, and P. T. Boggs, "Moving Boundary Problems." Academic Press, New York, 1978.
7. J. Crank, "Free and Moving Boundary Problems." Oxford Univ. Press (Clarendon), London and New York, 1984.
8. R. Viskanta, Phase change heat transfer. In "Solar Heat and Storage: Latent Heat Materials" (G. A. Lane, ed.), p. 153. CRC Press, Boca Raton, Florida, 1983.
9. R. Viskanta, Natural convection in melting and solidification. "Natural Convection: Fundamentals and Applications" (S. Kakac, W. Aung, and R. Viskanta, eds.), p. 845. Hemisphere, Washington, D.C., 1985.

10. M. Epstein and F. B. Cheung, Complex freezing-melting interfaces in fluid flow. *Annu. Rev. Fluid Mech.* **15,** 293 (1983).
11. F. B. Cheung and M. Epstein, Solidification and melting in fluid flow. *Adv. Transp. Processes* **3,** 35 (1984).
12. H. S. Carslaw and J. C. Jaeger, "Conduction of Heat in Solids," 2nd Ed., pp. 283, 353. Oxford Univ. Press, London and New York, 1959.
13. J. C. Muehlbauer and J. E. Sunderland, Heat conduction with a freezing or melting. *Appl. Mech. Rev.* **18,** 951 (1965).
14. J. Stefan, Ueber die Theorie der Eisbildung, insbesondere ueber die Eisbildung im Polarmeere. *Ann. Phys. Chem., N. S.* **42,** 269 (1891).
15. L. Boltzmann, Zur Integration der Diffusiongleichung bei variabeln Diffusionscoefficienten. *Ann. Phys. Chem., S.* **3,** 53, 959 (1894).
16. A. B. Taylor, The mathematical formulation of Stefan problems. "Moving Boundary Problems in Heat Flow and Diffusion." *Proc. Univ. Oxford, March 25–27* 120 (1974).
17. L. Fox, What are the best numerical methods? "Moving Boundary Problems in Heat Flow and Diffusion." *Proc. Univ. Oxford, March 25–27* 210 (1974).
18. L. S. Yao and W. Cherney, Transient phase-change around a horizontal cylinder. *Int. J. Heat Mass Transfer* **24,** 1971 (1981).
19. H. Budhia and F. Kreith, Heat transfer with melting or freezing in a wedge. *Int. J. Heat Mass Transfer* **16,** 195 (1973).
20. S. H. Cho and J. E. Sunderland, Approximate temperature distribution for phase change of a semi-infinite body. *J. Heat Transfer* **103,** 401 (1981).
21. N. Tokuda, An asymptotic, large time solution of the convection Stefan problem with surface radiation. *Int. J. Heat Mass Transfer* **29,** 135 (1986).
22. J. Prusa, A spatial Stefan problem modified by natural convection: Melting around a horizontal cylinder. Ph. D. Dissertation, Univ. of Illinois at Urbana-Champaign, 1983.
23. L. N. Tao, On free boundary problems with radiation boundary conditions," *Q. J. Mech. Appl. Math.* **37,** 1 (1979).
24. L. N. Tao, The exact solutions of some Stefan problems with prescribed heat flux. *J. Heat Transfer* **48,** 732 (1981).
25. L. N. Tao, On free boundary problems with arbitrary initial and flux conditions. *J. Appl. Math. Phys.* (*ZAMP*) **30,** 416 (1979).
26. G. W. Evans, E. Isaacson, and J. K. L. MacDonald, Stefan-like problems. *Q. J. Appl. Math.* **8,** 312 (1950).
27. B. A. Boley, The embedding technique in melting and solidification problems. "Moving Boundary Problems in Heat Flow and Diffusion." *Proc. Conf. Univ. Oxford, March 25–27* 150 (1974).
28. C. M. Elliot and J. R. Ockendon, "Weak and Variational Methods for Moving Boundary Problems." Research Notes in Mathematics 59. Pitman Advanced Publishing Program, Boston, 1982.
29. J. R. Ockendon, Techniques of analysis. "Moving Boundary Problems in Heat Flow and Diffusion." *Proc. Conf. Univ. Oxford, March 25–27* 138 (1974).
30. D. L. Sikarskie and B. A. Boley, The solution of a class of two-dimensional melting and solidification problems. *Int. J. Solids Struct.* **1,** 207 (1965).
31. L. Prandtl, AurBerechung der Grenzschichten. *Z. Angew. Math. Mech.* **18,** 77 (1938). Also, "Laminar Boundary Layers," by L. Rosenhead. Oxford Univ. Press (Clarendon), London and New York, 1963.
32. C. Zener, Theory of growth of spherical precipitates from solid solution. *J. Appl. Phys.* **20,** 950 (1949).
33. H. G. Landau, Heat conduction in a melting solid. *Q. Appl. Math.* **8,** 81 (1949).

34. K. T. Yang, Formation of ice in plane stagnation flow. *Appl. Sci. Res.* **17,** 377 (1967).
35. J. L. Duda, M. F. Malone, R. H. Notter, and J. S. Vrentas, Analysis of two-dimensional diffusion-controlled moving boundary problems. *Int. J. Heat Mass Transfer* **18,** 901 (1975).
36. T. Saitoh, Numerical method for multidimensional freezing problems in arbitrary domains. *J. Heat Transfer* **100,** 294 (1978).
37. E. M. Sparrow, S. Ramadhyani, and S. V. Patankar, Effect of subcooling on cylindrical melting. *J. Heat Transfer* **100,** 395 (1978).
38. L. C. Tien and S. W. Churchill, Freezing front motion and heat transfer outside an infinite, isothermal cylinder. *AIChE J.* **11,** 790 (1965).
39. J. M. Hill and A. Kucera, Freezing a saturated liquid inside a sphere. *Int. J. Heat Mass Transfer* **26,** 1631 (1983).
40. A. Kassinos and J. Prusa, Effects of density change and subcooling on the melting of a solid in a rectangular enclosure. *Proc. Int. Heat Transfer Conf., 8th, San Francisco* **4,** 1787 (1986).
41. R. M. Furzeland, A comparative study of numerical methods for moving boundary problems. *J. Inst. Math. Appl.* **26,** 411 (1980).
42. L. S. Yao, Analysis of heat transfer in slightly eccentric annuli. *J. Heat Transfer* **102,** 279 (1980).
43. W. D. Murray and F. Landis, Numerical and machine solutions of transient heat-conduction problems involving melting or freezing; Part I—Method of analysis and sample solutions. *J. Heat Transfer* **81,** 106 (1959).
44. P. R. Pujado, F. J. Stermole, and J. O. Golden, Melting of a finite paraffin slab as applied to phase-change thermal control. *J. Space Rockets* **6,** 280 (1969).
45. D. R. Lynch and K. O' Neill, Continuously deforming finite elements for the solution of parabolic problems, with and without phase change. *Int. J. Numer. Methods Eng.* **17,** 81 (1981).
46. J. L. Duda and J. S. Vrentas, Perturbation solutions of diffusion-controlled moving boundary problems. *Chem. Eng. Sci.* **24,** 461 (1969).
47. J. Crank, Two methods for the numerical solution of moving-boundary problems in diffusion and heat flow. *Q. J. Mech. Appl. Math.* **10,** 220 (1957).
48. D. R. Atthey, A finite difference scheme for melting problems based on the method of weak solutions. "Moving Boundary Problems in Heat Flow and Diffusion." *Proc. Conf. Univ. Oxford, March 25–27* 182 (1974).
49. R. M. Furzeland, Symposium on free and moving boundary problems in heat flow and diffusion, University of Durham, July, 1978. *Inst. Math. Appl.* **15,** 172 (1979).
50. D. C. Baxter, The fusion times of slabs and cylinders. *J. Heat Transfer* **84,** 317 (1962).
51. N. Shamsundar and E. M. Sparrow, Analysis of multidimensional conduction phase change via the enthalpy model. *J. Heat Transfer* **97,** 333 (1975).
52. E. A. Boucheron and R. N. Smith, An enthalpy formulation of the simple algorithm for phase change heat transfer problems. *ASME/AIChE Natl. Heat Transfer Conf., 23rd, Denver, August 4–7* 85-HT-7 (1985).
53. G. E. Schneider and M. J. Raw, An implicit solution procedure for finite difference modeling of the Stefan problem. *AIAA J.* **22,** 1685 (1984).
54. V. Voller and M. Cross, Accurate solutions of moving boundary problems using the enthalpy method. *Int. J. Heat Mass Transfer* **24,** 545 (1981).
55. L. E. Goodrich, Efficient numerical technique for one-dimensional thermal problems with phase change. *Int. J. Heat Mass Transfer* **21,** 615 (1978).
56. Q. T. Pham, A fast, unconditionally stable finite-difference scheme for heat conduction with phase change. *Int. J. Heat Mass Transfer* **28,** 2079 (1985).
57. Q. T. Pham, The use of lumped capacitance in the finite-element solution of heat conduction problems with phase change. *Int. J. Heat Mass Transfer* **29,** 285 (1986).

58. S. Weinbaum and L. M. Jiji, Singular perturbation theory for melting or freezing in finite domains not at the fusion temperature. *J. Appl. Mech.* **44,** 25 (1977).
59. Ch. Charach and P. Zoglin, Solidification in a finite, initially overheated slab. *Int. J. Heat Mass Transfer* **28,** 2261 (1985).
60. R. I. Pedroso and G. A. Domoto, Exact solution by perturbation method for planar solidification of a saturated liquid with convection at the wall. *Int. J. Heat Mass Transfer* **16,** 1816 (1973).
61. L. M. Jiji and S. Weinbaum, Perturbation solutions for melting or freezing in annular regions initially not at the fusion temperature. *Int. J. Heat Mass Transfer* **21,** 581 (1978).
62. R. I. Pedroso and G. A. Domoto, Inward spherical solidification-solution by the method of strained coordinates. *Int. J. Heat Mass Transfer* **16,** 1037 (1973).
63. L. C. Tao, Generalized numerical solution of freezing in a saturated liquid in cylinders and spheres. *A. I. Ch. E. J.* **13,** 165 (1967).
64. M. Van Dyke, "Perturbation Methods in Fluid Mechanics." The Parabolic Press, Stanford, California, 1975.
65. D. S. Riley, F. T. Smith, and G. Poots, The inward solidification of spheres and circular cylinders. *Int. J. Heat Mass Transfer* **17,** 1507 (1974).
66. R. V. Seeniraj and T. K. Bose, One-dimensional phase-change problems with radiation convection. *J. Heat Transfer* **104,** 811 (1982).
67. M. M. Yan and P. N. S. Huang, Perturbation solutions to phase-change problem subject to convection and radiation. *J. Heat Transfer* **101,** 96 (1979).
68. R. V. Seeniraj, Perturbation solutions to phase-change problem subject to convection and radiation. Letter to the Editor, *J. Heat Transfer* **102,** 395 (1980).
69. T. R. Goodman, The heat-balance integral and its application to problems involving a change of phase. *ASME Trans.* **80,** 335 (1958).
70. J. Mennig and M. N. Ozisik, Coupled integral equation approach for solving melting or solidification. *Int. J. Heat Mass Transfer* **28,** 1481 (1985).
71. W. E. Steward and K. L. Smith, Inward solidification of water in a cylinder—An analytical and experimental study. *ASME/AIChE Natl. Heat Transfer Conf., 23rd, Denver, August 4–7* 85-HT-2 (1985).
72. G. E. Bell, Solidification of a liquid about a cylindrical pipe. *Int. J. Heat Mass Transfer* **22,** 1681 (1979).
73. V. J. Lunardini, Phase change around a circular cylinder. *J. Heat Transfer* **103,** 598 (1981).
74. L. F. Milanez and K. A. R. Ismail, Solidification in spheres—theoretical and experimental investigation. *In* "Multi-Phase Flow and Heat Tranfer III. Part B: Applications" (T. N. Veziroglu and A. E. Bergles, eds.), p. 565, Elsevier, Amsterdam, 1984.
75. V. D. Rao and P. K. Sarma, Direct contact heat transfer in spherical geometry associated with phase transformation—A closed-form solution. *Int. J. Heat Mass Transfer* **28,** 1956 (1985).
76. B. T. F. Chung and L. T. Yeh, Solidification and melting of materials subject to convection and radiation. *J. Spacecraft* **12,** 329 (1975).
77. I. S. Habib, Solidification of semi-transparent materials by conduction and radiation. *Int. J. Heat Mass Transfer* **14,** 2161 (1971).
78. I. S. Habib, Solidification of a semi-transparent cylindrical medium by conduction and radiation. *J. Heat Transfer* **95,** 37 (1973).
79. E. M. Breinan, B. H. Kear, and C. M. Banas, Processing materials with lasers. *Phys. Today* **29,** 44 (1976).
80. G. E. Giles, J. R. Kirkpatrick, and R. F. Wood, Laser annealing of solar cell wafers. *ASME Natl. Heat Transfer Conf., 19th, Orlando, July 27–30* 80-HT-13 (1980).
81. D. Gelder and A. C. Guy, Current problems in the glass industry. "Moving Boundary

Problems in Heat Flow and Diffusion" (J. R. Ockendon and W. R. Hodgkins, eds.). Oxford Univ. Press (Clarendon), London and New York, 1975.

82. R. N. Smith, T. E. Ebersole, and E. P. Griffi, Exchanger performances in latent heat thermal storage. *ASME J. Sol. Energy* **102** (1980).
83. Maeno, N., Integral and surface melting of ice. *Low-Temp. Sci.* **28,** 22 (1978).
84. N. Seki, M. Sugawara, and S. Fukusako, Radiative melting of ice layer adhering to a vertical surface. *Wärme-Stoffübertrag.* **12,** 137 (1979).
85. S. H. Chan, D. H. Cho, and G. Kocamustafaogullari, Melting and solidification with internal radiative transfer—A generalized phase change model. *Int. J. Heat Mass Transfer* **26,** 621 (1983).
86. F. O. Oruma, M. N. Ozisik, and M. A. Boles, Effects of non-grayness on melting and solidification of a semi-infinite, semi-transparent medium. *J. Franklin Inst.* **319,** 459 (1985).
87. F. O. Oruma, M. N. Ozisik, and M. A. Boles, Effects of anisotropic scattering on melting and solidification of a semi-infinite, semi-transparent medium. *Int. J. Heat Mass Transfer* **28,** 441 (1985).
88. R. Siegel, Boundary perturbation methods for free boundary problem in convectively cooled continuous casting. *J. Heat Transfer* **108,** 230 (1986).
89. R. Siegel, Free boundary shape of a convectively cooled solidified region. *Int. J. Heat Mass Transfer* **29,** 309 (1986).
90. H. C. Agrawal, Biot's variational principle for moving boundary problems. "Moving Boundary Problems in Heat Flow and Diffusion." *Proc. Conf. Univ. Oxford, March 25–27* 242 (1974).
91. G. Duvaut, The solution of a two-phase Stefan problem by a variational inequality. "Moving Boundary Problems in Heat Flow and Diffusion." *Proc. Conf. Univ. Oxford, March 25–27* 173 (1974).
92. J. Crank, Finite-difference methods. "Moving Boundary Problems in Heat Flow and Diffusion." *Proc. Conf. Univ. Oxford, March 25–27* 192 (1975).
93. J. Crank and R. S. Gupta, Isotherm migration method in two dimensions. *Int. J. Heat Mass Transfer* **18,** 1101 (1975).
94. D. Frederick and R. Greif, A method for the solution of heat transfer problems with a change of phase. *J. Heat Transfer* **107,** 520 (1985).
95. J. S. Goodling and M. S. Khader, Inward solidification with radiation-convection boundary condition. *J. Heat Transfer* **96,** 114 (1974).
96. M. Abrams and R. Viskanta, The effects of radiative heat transfer upon the melting and solidification of semi-transparent crystals. *J. Heat Transfer* **96,** 184 (1974).
97. N. Seki, M. Sugawara, and S. Fukusako, Back-melting of a horizontal cloudy ice layer with radiative heating. *J. Heat Transfer* **101,** 90 (1979).
98. L. A. Diaz and R. Viskanta, Experiments and analysis on the melting of a semi-transparent material by radiation. *Wärme Stoffübertrag.* **20,** 311 (1986).
99. Y. K. Chuang and J. Szekely, On the use of greens function for solving melting and solidification problems. *Int. J. Heat Mass Transfer* **14,** 1285 (1971).
100. P. K. Banerjee and R. P. Shaw, Boundary element formulation for melting and solidification problems. "Developments in Boundary Element Methods-2" (P. K. Banerjee and R. P. Shaw, eds.), p. 1. Applied Science Publ, London, 1982.
101. A. M. Sadegh, L. M. Jiji, and S. Weinbaum, Boundary integral equation technique with application to freezing around a buried pipe. *ASME Winter Annu. Meet., Miami Beach, November 17–22* 85-WA/HT-73 (1985).
102. C. Tien and Y.-C. Yen, Approximate solution of a melting problem with natural convection. *AIChE J.* **62,** 166 (1966).
103. D. V. Boger and J. W. Westwater, Effect of buoyancy on the melting and freezing process. *J. Heat Transfer* **89,** 81 (1967).

104. W. L. Heitz and J. W. Westwater, Extension of the numerical method for melting and freezing problems. *Int. J. Heat Mass Transfer* **13,** 1371 (1970).
105. N. W. Hale, Jr. and R. Viskanta, Solid-liquid phase-change heat transfer and interface motion in materials cooled or heated from above or below. *Int. J. Heat Mass Transfer* **23,** 283 (1980).
106. A. G. Bathelt, R. Viskanta, and W. Leidenfrost, An experimental investigation of natural convection in the melted region around a heated horizontal cylinder. *J. Fluid Mech.* **90,** 227 (1979).
107. A. G. Bathelt and R. Viskanta, Heat transfer at the solid-liquid interface during melting from a horizontal cylinder. *Int. J. Heat Mass Transfer* **23,** 1493 (1980).
108. A. G. Bathelt, R. Viskanta, and W. Leindenfrost, Latent heat-of-fusion energy storage: Experiments on heat transfer from cylinders during melting. *J. Heat Transfer* **101,** 453 (1979).
109. E. M. Sparrow, R. R. Schmidt, and J. W. Ramsey, Experiments on the role of natural convection in the melting of solids. *J. Heat Transfer* **100,** 11 (1978).
110. R. M. Abdel-Wahed, J. W. Ramsey, and E. M. Sparrow, Photographic study of melting about an embedded horizontal heating cylinder. *Int. J. Heat Mass Transfer* **22,** 171 (1979).
111. E. M. Sparrow, S. V. Patankar, and S. Ramadhyani, Analysis of melting in the presence of natural convection in the melt region. *J. Heat Transfer* **99,** 520 (1977).
112. R. G. Kemink and E. M. Sparrow, Heat transfer coefficients for melting about a vertical cylinder with or without subcooling and for open or closed containment. *Int. J. Heat Mass Transfer* **24,** 1699 (1981).
113. R. J. Goldstein and J. W. Ramsey, Heat transfer to a melting solid with application to thermal energy storage systems. "Heat Transfer Studies: Festschrift for E. R. G. Eckert." Hemisphere, New York, 1979.
114. E. M. Sparrow, J. W. Ramsey, and R. G. Kemink, Freezing controlled by natural convection. *J. Heat Transfer* **101,** 578 (1979).
115. A. G. Bathelt, P. D. Van Buren, and R. Viskanta, Heat transfer during solidification around a cooled horizontal cylinder. *AIChE J.* **75,** 103 (1979).
116. L. S. Yao and F. F. Chen, Effects of natural convection in the melted region around a heated horizontal cylinder. *J. Heat Transfer* **102,** 667 (1980).
117. J. Prusa and L. S. Yao, Natural convection heat transfer between eccentric horizontal cylinders. *J. Heat Transfer* **105,** 108 (1983).
118. H. Rieger, U. Projahn, and H. Beer, Analysis of the heat transport mechanisms during melting around a horizontal circular cylinder. *Int. J. Heat Mass Transfer* **25,** 137 (1982).
119. J. F. Thompson, F. C. Thames, and C. W. Mastin, Automatic numerical generation of body-fitted curvilinear coordinate system for field containing any number of arbitrary two-dimensional bodies. *J. Comp. Phys.* **15,** 299 (1974).
120. J. Prusa and L. S. Yao, Melting around a horizontal heated cylinder: Part I—Perturbation and numerical solution for constant heat flux boundary condition. *J. Heat Transfer* **106,** 376 (1984).
121. J. Prusa and L. S. Yao, Melting around a horizontal heated cylinder: Part II—Numerical solution for isothermal boundary condition. *J. Heat Transfer* **106,** 469 (1984).
122. A. S. Roberts and T. M. Rothgeb, An experimental investigation of melting about an isothermal horizontal cylinder. *ASME/AIChE Natl. Heat Transfer Conf., 22nd, Niagara Falls, August* 84-HT-9 (1984).
123. J. Prusa and L. S. Yao, Effects of density change and sub-cooling on the melting of a solid around a horizontal heated cylinder. *J. Fluid Mech.* **155,** 193 (1985).
124. J. Prusa and L. Vandekamp, A numerical solution for melting around multiple horizontal heated cylinders. *ASME/AIChE Natl. Heat Transfer Conf., 23rd, Denver August 4–7* 85-HT-12 (1985).

125. A. G. Bathelt and R. Viskanta, Heat transfer and interface motion during melting and solidification around a finned heat source/sink. *J. Heat Transfer* **103,** 720 (1981).
126. T. Betzel and H. Beer, Experimental investigation of heat transfer during melting around a horizontal tube with and without axial fins. *Tech. Note Inst. Tech. Therodyn.* (1987).
127. E. M. Sparrow, E. D. Larson, and J. W. Ramsey, Freezing on a finned tube for either conduction-controlled or natural-convection-controlled heat transfer. *Int. J. Heat Mass Transfer* **24,** 273 (1981).
128. B. Kalhori and S. Ramadhyani, Studies on heat transfer from a vertical cylinder, with or without fins, embedded in a solid phase change medium. *J. Heat Transfer* **107,** 44 (1985).
129. H. Rieger, U. Projahn, M. Bareiss, and H. Beer, Heat transfer during melting inside a horizontal tube. *J. Heat Transfer* **105,** 226 (1983).
130. M. Bareiss and H. Beer, Experimental investigation of melting heat transfer with regard to different geometric arrangements. *Int. Commun. Heat Mass Transfer* **11,** 323 (1984).
131. C. J. Ho and R. Viskanta, Heat transfer during inward melting in a horizontal tube. *Int. J. Heat Mass Transfer* **27,** 705 (1984).
132. J. Pannu, G. Joglekar, and P. A. Rice, Natural convection heat transfer to cylinders of phase change material used for thermal storage. *AIChE J.* **76,** 47 (1980).
133. T. Saitoh and K. Hirose, High Rayleigh number solutions to problems of latent heat thermal energy storage in a horizontal cylinder capsule. *J. Heat Transfer* **104,** 545 (1982).
134. R. Viskanta and C. Gau, Inward solidification of a superheated liquid in a cooled horizontal tube. *Wärme Stoffübertrag.* **17,** 39 (1982).
135. R. Guenigault and G. Poots, Effects of natural convection on the inward solidification of spheres and cylinders. *Int. J. Heat Mass Transfer* **28,** 1229 (1985).
136. L. S. Yao, Natural convection effects in the continuous casting of a horizontal cylinder. *Int. J. Heat Mass Transfer* **27,** 697 (1984).
137. E. M. Sparrow and J. A. Broadbent, Inward melting in a vertical tube which allows free expansion of the phase-change medium. *J. Heat Transfer* **104,** 309 (1982).
138. E. M. Sparrow, G. A. Gurtcheff, and T. A. Myrum, Correlation of melting results for both pure substances and impure substances. *J. Heat Transfer* **108,** 649 (1986).
139. A. S. Menon, M. E. Weber, and A. S. Mujumdar, The dynamics of energy storage for paraffin wax in cylindrical containers. *Can. J. Chem. Eng.* **61,** 647 (1983).
140. E. M. Sparrow and J. A. Broadbent, Freezing in a vertical tube. *J. Heat Transfer* **105,** 217 (1983).
141. E. D. Larson and E. M. Sparrow, Effect of inclination on freezing in a sealed cylindrical capsule. *J. Heat Transfer* **106,** 394 (1984).
142. N. W. Hale, Jr. and R. Viskanta, Photographic observation of the solid-liquid interface motion during melting of a solid heated from an isothermal vertical wall. *Lett. Heat Mass Transfer* **5,** 329 (1978).
143. P. D. Van Buren and R. Viskanta, Interferometric measurement of heat transfer during melting from a vertical surface. *Int. J. Heat Mass Transfer* **23,** 568 (1980).
144. C. Benard, D. Gobin, and F. Martinez, Melting in rectangular enclosures: Experiments and numerical simulations. *J. Heat Transfer* **107,** 794 (1985).
145. C. -J. Ho and R. Viskanta, Heat transfer during melting from an isothermal vertical wall. *J. Heat Transfer* **106,** 12 (1984).
146. C. Gau and R. Viskanta, Melting and solidification of a pure metal on a vertical wall. *J. Heat Transfer* **108,** 174 (1986).
147. S. C. Huang, Analytical solution for the buoyancy flow during the melting of vertical semi-infinite region. *Int. J. Heat Mass Transfer* **28,** 1231 (1985).
148. M. Okada, Analysis of heat transfer during melting from a vertical wall. *Int. J. Heat Mass Transfer* **27,** 2059 (1984).

149. A. Gadgil and D. Gobin, Analysis of two-dimensional melting in rectangular enclosures in presence of convection. *J. Heat Transfer* **106,** 20 (1984).
150. G. E. Schneider, Computation of heat transfer with solid/liquid phase change including free convection. *AIAA Aerospace Sci. Meet., 23rd, Reno, Nevada, January 14–17* AIAA-85-0404 (1985).
151. C. Benard, D. Gobin, and A. Zanoli, Numerical simulation of transient melting in two dimensions by coupling of natural convection in the melt and heat conduction in the solid phase. *ASME/AIChE Natl. Heat Transfer Conf., 22nd, Niagara Falls, August 5–8* 84-HT-5 (1984).
152. N. Ramachandran, J. P. Gupta, and Y. Jaluria, Thermal and fluid flow effects during solidification in a rectangular enclosure. *Int. J. Heat Mass Transfer* **25,** 187 (1982).
153. N. Ramachandran, J. P. Gupta, and Y. Jaluria, Experiments on solidification with natural convection in a rectangular enclosed region. *Int. J. Heat Mass Transfer* **25,** 595 (1982).
154. C. -J. Ho and R. Viskanta, Inward solid-liquid phase-change heat transfer in a rectangular cavity with conducting vertical walls. *Int. J. Heat Mass Transfer* **27,** 1055 (1984).
155. C. Gau and R. Viskanta, Melting and solidification of a metal system in a rectangular cavity. *Int. J. Heat Mass Transfer* **27,** 113 (1984).
156. B. W. Webb and R. Viskanta, Natural-convection-dominated melting heat transfer in an inclined rectangular enclosure. *Int. J. Heat Mass Transfer* **29,** 183 (1986).
157. N. W. Wilson and J. J. Lee, Melting of a vertical ice wall by free convection into fresh water. *J. Heat Transfer* **103,** 13 (1981).
158. D. White, R. Viskanta, and W. Leidenfrost, Heat transfer during the melting of ice around a horizontal, isothermal cylinder. *Exp. Fluids* **4,** 171 (1986).
159. H. Rieger and H. Beer, The melting process of ice inside a horizontal cylinder: Effects of density anomaly. *J. Heat Transfer* **108,** 166 (1986).
160. F. E. Moore and Y. Bayazitoglu, Melting within a spherical enclosure. *J. Heat Transfer* **104,** 19 (1982).
161. M. Bareiss and H. Beer, An analytical solution of the heat transfer process during melting of an unfixed solid phase change material inside a horizontal tube. *Int. J. Heat Mass Transfer* **27,** 739 (1984).
162. E. M. Sparrow and G. T. Geiger, Melting in a horizontal tube with the solid either constrained or free to fall under gravity. *Int. J. Heat Mass Transfer* **29,** 1007 (1986).
163. Th. Betzel and H. Beer, Prediction and measurement of melting heat transfer to an unfixed phase change material heated in a horizontal concentric annulus. *Proc. Int. Heat Transfer Conf., 8th, San Francisco, August 17–22* **4,** 1793 (1986).
164. B. W. Webb, M. K. Moallemi, and R. Viskanta, Phenomenology of melting of unfixed phase change material in a horizontal cylindrical capsule. *AIAA/ASME Natl. Thermophys. Heat Transfer Conf., Boston, June 2–4* 86-HT-10 (1986).
165. M. K. Moallemi and R. Viskanta, Experiments on fluid flow induced by melting around a migrating heat source. *J. Fluid Mech.* **157,** 35 (1985).
166. M. K. Moallemi and R. Viskanta, Analysis of close-contact melting heat transfer. *Int. J. Heat Mass Transfer* **29,** 855 (1986).
167. M. K. Moallemi and R. Viskanta, Analysis of melting around a moving heat source. *Int. J. Heat Mass Transfer* **29,** 1271 (1986).
168. S. H. Emerman and D. L. Turcotte, Stokes' problem with melting. *Int. J. Heat Mass Transfer* **26,** 1625 (1983).
169. M. H. P. Bott, "The Interior of The Earth: Its Structure, Constitution and Evolution, 2nd. Ed., p. 164. Elsevier, New York, 1982.
170. M. S. Sadeghipour, M. N. Ozisik, and J. C. Mulligan, Transient freezing of a liquid in a convectively cooled tube. *J. Heat Transfer* **104,** 316 (1982).

171. J. Cevantes, C. Trevino, and M. Rodriguez, Transient freezing and laminar flow in a circular pipe. *Int. Symp. Heat Mass Transfer Cryoeng. Refrig., 18th, Dubrovnik, September 1–5* (1986).
172. S. Toda, H. Sugiyama, H. Owada, M. Kurokawa, and Y. Hori, Laminar flow heat transfer in a tube with internal solidification. *Proc. Int. Heat Transfer Conf., 8th, San Francisco, August 17–22* **4,** 1745 (1986).
173. M. Epstein and F. B. Cheung, On the prediction of pipe freeze-shut in turbulent flow. *J. Heat Transfer* **104,** 381 (1982).
174. P. Sampson and R. D. Gibson, A mathematical model of nozzle blockage by freezing—II. Turbulent flow. *Int. J. Heat Mass Transfer* **25,** 119 (1982).
175. R. R. Gilpin, Ice formation in a pipe containing flows in the transition and turbulent regimes. *J. Heat Transfer* **103,** 363 (1981).
176. T. Hirata and M. Ishihara, Freeze-off conditions of a pipe containing a flow of water. *Int. J. Heat Mass Transfer* **28,** 331 (1985).
177. B. R. Morton, Laminar convection in uniformly heated horizontal pipes at low Rayleigh numbers. *J. Mech. Appl. Math.* **12,** 410 (1959).
178. N. Seki, S. Fukusako, and G. W. Younan, Ice-formation phenomena for water flow between two cooled parallel plates. *J. Heat Transfer* **106,** 498 (1984).
179. F. B. Cheung, Analysis of freeze coating on a nonisothermal moving plate by a perturbation method. *J. Heat Transfer* **107,** 549 (1985).
180. K. C. Cheng, H. Inaba, and R. R. Gilpin, An experimental investigation of ice formation around an isothermally cooled cylinder in crossflow. *J. Heat Transfer* **103,** 733 (1981).
181. G. S. H. Lock and T. M. V. Kaiser, Icing on submerged tubes: A study of occlusion. *Int. J. Heat Mass Transfer* **28,** 1689 (1985).
182. K. C. Cheng and P. Sabhapathy, Determination of local heat transfer coefficients at the solid-liquid interface by heat conduction analysis of the solidification region. *J. Heat Transfer* **107,** 703 (1985).
183. Y. Shiina, and P. G. Kroeger, Transient moisture migration and phase change front propagation in porous media. *ASME/AIChE Natl. Heat Transfer Conf., 22nd, Niagara Falls, August 5–8* 84-HT-7 (1984).
184. J. A. Weaver and R. Viskanta, Freezing of liquid-saturated porous media. *J. Heat Transfer* **108,** 654 (1986).
185. K. E. Torrance, Phase-change heat transfer in porous media. *Proc. Int. Heat Transfer Conf., 8th, San Francisco* **1,** 181 (1986).
186. P. E. Frivik and G. Comini, Seepage and heat flow in soil freezing. *J. Heat Transfer* **104,** 323 (1982).
187. M. A. Boles and M. N. Ozisik, Exact solution for freezing in cylindrically symmetric porous moist media. *J. Heat Transfer* **105,** 401 (1983).
188. H. Inaba, Heat transfer behavior of frozen soils. *J. Heat Transfer* **105,** 680 (1983).
189. R. J. Couvillion and J. G. Hartly, Drying front movement near low-density, impermeable underground heat sources. *J. Heat Transfer* **108,** 182 (1986).
190. M. Vauclin, S. Giakoumakis, J. P. Gaudet, and A. Albergel, Experimental and numerical analysis of coupled heat and mass transfer in a partially saturated frozen soil. *Symp. Heat Mass Transfer Cryoeng. Refrig., 18th, Dubrovnik, September 1–5* (1986).
191. G. -P. Zhang, L. M. Jiji, and S. Weinbaum, Quasi-three-dimensional steady-state analytic solution for melting or freezing around a buried pipe in a semi-infinite medium. *J. Heat Transfer* **107,** 245 (1985).
192. G. -P. Zhang, S. Weinbaum, and L. M. Jiji, An approximate three-dimensional solution for melting or freezing around a buried pipe beneath a free surface. *ASME Winter Annu. Meet, Miami Beach, November 17–22* 85-WA/HT-74 (1985).

193. B. Rubinsky, Solidification of a conglomerate of particles. *J. Heat Transfer* **104,** 193 (1982).
194. J. F. Raymond and B. Rubinsky, A numerical study of the thawing process of a frozen coal particle. *J. Heat Transfer* **105,** 197 (1983).
195. D. B. Moog and B. Rubinsky, An analytical model of thermal and vapor diffusion in freezing of wet coal. *J. Heat Transfer* **107,** 5 (1985).
196. R. Elliot, Eutectic solidification. *Int. Met. Rev.* **219,** 161 (1977).
197. K. Wollhover, Ch. Korber, M. W. Scheiwe, and U. Hartmann, Unidirectional freezing of binary aqueous solutions: An analysis of transient diffusion of heat and mass. *Int. J. Heat Mass Transfer* **28,** 761 (1985).
198. K. Wollhover, M. W. Scheiwe, U. Hartmann, and Ch. Korber, On morphological stability of planar phase boundaries during unidirectional transient solidification of binary aqueous solutions. *Int. J. Heat Mass Transfer* **28,** 897 (1985).
199. P. A. Weiss, Desalination by freezing. *In* "Practice of Desalination" (R. Bakish, R. Noyes, ed.) p. 260. Noyes Data Corp., Park Ridge, New Jersey, 1973.
200. V. P. Carey and B. Gebhart, Transport near a vertical ice surface melting in saline water: Some numerical calculations. *J. Fluid Mech.* **117,** 379 (1982).
201. L. Gradon and D. Orlicki, Separation of liquid mixtures in the freezing-out process-mathematical description and experimental verification. *Int. J. Heat. Mass Transfer* **28,** 1983 (1985).
202. L. J. Fang, F. B. Cheung, J. H. Linehan, and D. R. Pedersen, Selective freezing of a dilute salt solution on a cold ice surface. *J. Heat Transfer* **106,** 385 (1984).
203. T. C. Chawla, D. R. Pedersen, G. Leaf, W. J. Minkowycz, and A. R. Shouman, Adaptive collocation method for simultaneous heat and mass diffusion with phase change. *J. Heat Transfer* **106,** 491 (1984).
204. W. W. Mullins, and R. K. Sekerka, Stability of a planar interface during solidification of a dilute binary alloy. *J. Appl. Phys.* **35,** 4444 (1964).
205. M. E. Glicksman, S. R. Coriell, and G. B. McFadden, Interaction of flows with the crystal-melt interface. *Annu. Rev. Fluid Mech.* **18,** 307 (1986).
206. R. H. Tien and G. E. Geiger, A heat transfer analysis of the solidification of a binary eutectic system. *J. Heat Transfer* **89,** 230 (1967).
207. S. H. Cho and J. E. Sunderland, Heat conduction problems with melting and freezing. *J. Heat Transfer* **91,** (1969).
208. Y. Hayashi and T. Komori, Investigation of freezing of salt solutions in cells. *J. Heat Transfer* **101,** 459 (1979).
209. M. G. O'Callaghan, E. G. Cravalho, and C. E. Huggins, An analysis of the heat and solute transport during solidification of an aqueous binary solution. *Int. J. Heat Mass Transfer* **25,** 553 (1982).
210. B. W. Webb and R. Viskanta, An experimental and analytical study of solidification of a binary mixture. *Proc. Int. Heat Transfer Conf., 8th, San Francisco* **4,** 1739 (1986).
211. H. Kehtarnavaz and Y. Bayazitoglu, Solidification of binary mixture in a finite planar medium: Saline water. *J. Heat Transfer* **107,** 964 (1985).
212. E. M. Sparrow, L. Lee, and N. Shamsundar, Convective instability in the melt layer heated from below. *J. Heat Transfer* **98,** 88 (1976).
213. N. Seki, S. Fukusako, and M. Sugawara, A criterion of onset of free convection in a horizontal melted water layer with free surface. *J. Heat Transfer* **99,** 92 (1977).
214. L. S. Yao and I. Catton, The effects of phase change on the onset of thermal instability in an internally heated pool. *Natl. Heat Transfer Conf., 17th, Salt Lake City* Reprints of AIChe Paps. 176–181 (1977).
215. L. S. Yao and I. Catton, Thermal instability of a volumetrically heated pool with phase change and a free upper surface. *J. Heat Transfer* **100,** 376 (1978).

216. W. E. Langlois, Buoyancy-driven flows in crystal-growth melts. *Annu. Rev. Fluid Mech.* **17,** 191 (1985).
217. F. B. Cheung, Periodic growth and decay of a frozen crust over a heat generating liquid layer. *J. Heat Transfer* **103,** 369 (1981).
218. S. H. Davis, U. Muller, and C. Dietsche, Pattern selection in single-component systems coupling Bernard convection and solidification. *J. Fluid Mech.* **144,** 133 (1984).
219. C. Dietsche and U. Muller, Influence of Bernard convection on solid-liquid interfaces. *J. Fluid Mech.* **161,** 249 (1985).
220. Q. T. Fang, M. E. Glicksman, S. R. Coriell, G. B. McFadden, and R. F. Boisdert, Convective influence on the stability of a cylindrical solid-liquid interface. *J. Fluid Mech.* **151,** 121 (1985).
221. C. Gau and R. Viskanta, Effect of natural convection on solidification from above and melting from below. *Int. J. Heat Mass Transfer* **28,** 573 (1985).
222. W. Englberger and E. R. F. Winter, Onset of natural convection and heat transfer in a layer of water below melting ice. *Proc. Int. Heat Transfer Conf., 8th, San Francisco* **4,** 1799 (1986).
223. S. Ostrach, Low-gravity fluid flows. *Annu. Rev. Fluid Mech.* **14,** 313 (1982).
224. S. H. Davis, Thermocapillary instabilities. *Annu. Rev. Fluid Mech.* **19,** 403 (1987).
225. R. Farhadieh and L. Baker, Jr., Heat transfer phenomenology of a hydrodynamically unstable melting system. *J. Heat Transfer* **100,** 305 (1978).
226. I. Catton, W. A. Brinsfield, and S. M. Ghiaasiaan, Heat transfer from a heated pool to a melting miscible substrate. *J. Heat Transfer* **105,** 447 (1983).
227. G. Eck and H. Werle, Experimental studies of penetration of a hot liquid pool into a melting miscible substrate. *Nucl. Technol.* **64,** 275 (1984).
228. M. Epstein and M. A. Grolmes, Natural convection characteristics of pool penetration into a melting miscible substrate. *J. Heat Transfer* **108,** 190 (1986).
229. M. M. Chen, R. Farhadieh, and L. Baker, Jr., On free convection melting of a solid immersed in a hot dissimilar fluid. *Int. J. Heat Mass Transfer* **29,** 1087 (1986).
230. K. Taghavi-Tafreshi and V. K. Dhir, Analytical and experimental investigation of simultaneous melting-condensation on a vertical wall. *J. Heat Transfer* **104,** 24 (1982).
231. K. Taghavi-Tafreshi and V. K. Dhir, Shape change of an initially vertical wall undergoing condensation-driven melting. *J. Heat Transfer* **105,** 235 (1982).
232. D. Galamba and V. K. Dhir, Transient simultaneous condensation and melting of a vertical surface. *J. Heat Transfer* **107,** 812 (1985).
233. W. A. Peifer, Levitation melting, a survey of the state-of-art. *J. Met.* **17,** 487 (1965).
234. C. Vives and C. Perry, Effects of electromagnetic stirring during the controlled solidification of tin. *Int. J. Heat Mass Transfer* **29,** 21 (1986).
235. P. G. Kroeger and S. Ostrach, The solution of a two-dimensional freezing problem including convection effects in the liquid region. *Int. J. Heat Mass Transfer* **17,** 1191 (1974).
236. B. Rubinsky, Thermal stresses during solidification processes. *J. Heat Transfer* **104,** 196 (1982).
237. C. S. Landram, Measurement of fusion boundary energy transport during arc welding. *J. Heat Transfer* **105,** 550 (1983).
238. G. M. Oreper and J. Szekely, Heat and fluid-flow phenomena in weld pools. *J. Fluid Mech.* **147,** 53 (1984).
239. P. Wang, S. Lin, and R. Kahawita, The cubic spline integration technique for solving fusion welding problems. *J. Heat Transfer* **107,** 485 (1985).
240. M. Salcudean, M. Choi, and R. Greif, A study of heat transfer during arc welding. *Int. J. Heat Mass Transfer* **29,** 215 (1986).
241. A. Francis and R. E. Craine, On a model for frictioning stage in friction welding of thin tubes. *Int. J. Heat Mass Transfer* **28,** 1747 (1985).

242. J. Dowden, M. Davis, and P. Kapadia, Some aspects of the fluid dynamics of laser welding. *J. Fluid Mech.* **126,** 123 (1983).
243. P. M. Beckett, Laser heating of a solid with change of phase. *ASME/AIChE Natl. Heat Transfer Conf., 22nd, Niagara Falls, August 5–8* 84-HT-6 (1984).
244. M. F. Modest and H. Abakians, Evaporative cutting of a semi-infinite body with a moving cw laser. *ASME/AIChE Natl. Heat Transfer Conf., 23rd, Denver, August 4–7* 85-HT-25 (1985).
245. M. F. Modest and H. Abakians, Heat conduction in a moving semi-infinite solid subjected to pulsed laser irradiation. *J. Heat Transfer* **108,** 597 (1986).
246. J. Srinivasan and B. Basu, A numerical study of thermocapillary flow in a rectangular cavity during laser melting. *Int. J. Heat Mass Transfer* **29,** 563 (1986).
247. J. E. White and C. J. Cremers, Prediction of growth parameters of frost deposits in forced convection. *J. Heat Transfer* **103,** 3 (1981).
248. C. J. Cremers and V. K. Mehra, Frost formation on vertical cylinders in free convection. *J. Heat Transfer* **104,** 3 (1982).
249. I. Tokura, H. Saito, and K. Kishinami, Study on properties and growth rate of frost layers on cold surfaces. *J. Heat Transfer* **105,** 895 (1983).
250. H. P. Steinhagen and G. E. Myers, Numerical solution to axisymmetric thawing and heating of frozen logs. *J. Heat Transfer* **105,** 195 (1983).
251. D. R. Thompson, Frost formation on cryogenic heat exchangers with atmospheric heat sources. *ASME/AIChE Natl. Heat Transfer Conf., 23rd, Denver, August 4–7* 85-HT-3 (1985).
252. H. Auracher, Effective thermal conductivity of frost. *Int. Symp. Heat Mass Transfer Cryoeng. Refrig., 18th, Dubrovick, Sept. 1–5* (1986).
253. P. M. Chung and R. Bywater, Role of the liquid layer in ice accumulation on flat surfaces. *J. Heat Transfer* **106,** 5 (1984).
254. B. T. F. Chung, T. Y. Chang, J. S. Hsiao, and C. T. Chang, Heat transfer with ablation in a half-space subjected to time-variant heat fluxes. *J. Heat Transfer* **105,** 200 (1983).
255. B. T. F. Chung and J. S. Hsiao, Heat transfer with ablation in a finite slab subjected to time-variant heat fluxes. *AIAA J.* **23,** 145 (1985).
256. C. Park, J. H. Lundell, M. J. Green, W. Winovich, and M. A. Covington, Ablation of carbonaceous materials in a hydrogen-helium arcjet flow. *AIAA J.* **22,** 1491 (1984).
257. M. Hogge and P. Gerrekens, Two-dimensional deforming finite element methods for surface ablation. *AIAAJ.* **23,** 465 (1985).
258. C. Park and A. Balakrishnan, Ablation of Galileo probe heat-shield models in a ballistic range. *AIAA J.* **23,** 301 (1985).
259. S. Lin, An exact solution of the sublimation problem in a porous medium. *J. Heat Transfer* **104,** 808 (1982).
260. M. Sakly, J. Aguirre-Puente, and G. Lambrinos, Experimental study of the sublimation of pure ice under very low temperature and relative humidity conditions. *Int. Symp. Heat Mass Transfer Cryoeng. Refrig., 18th, Dubrovnik, Sept 1–5* (1986).
261. D. L. Turcotte, A. T. Hsui, K. E. Torrance, and G. Schubert, Influence of viscous dissipation on Benard convection. *J. Fluid Mech.* **64,** 369 (1974).
262. J. M. Hewitt, D. P. McKenzie, and N. O. Weise, Dissipative heating in convective flows. *J. Fluid Mech.* **68,** 721 (1975).
263. U. Nitsan, Viscous heat production in a slab. *J. Geophys. Res.* **78,** 1395 (1973).
264. D. A. Yuen and G. Schubert, On the stability of frictionally heated shaw flows in the asthenosphere. *Geophys. J. R. Astron. Soc.* **57,** 189 (1979).
265. M. J. Melosh and J. Ebel, A simple model for thermal instability in the asthenosphere. *Geophys. J. R. Astron. Soc.* **59,** 419 (1979).

266. J. R. Ockendon, A. B. Tayler, S. H. Emerman, and D. L. Turcotte, Geodynamic thermal runaway with melting. *J. Fluid Mech.* **152,** 301 (1985).
267. W. M. Oversby and A. E. Ringwood, Time formation of the earth's core. *Nature (London)* **234,** 463 (1971).
268. L. Grossman and J. W. Larimer, Early chemical history of the solar system. *Rev. Geophys. Space Phys.* **12,** 71 (1974).
269. A. E. Ringwood, "Composition and petrology of the earth's mantle." McGraw-Hill, New York, 1975.
270. F. J. Spera, Aspects of magma transport. "Physics of Magma Processes" (R. B. Hargraves, ed.), p. 263. Princeton University Press, New Jersey, 1980.
271. D. L. Turcotte, Magma migration. *Annu. Rev. Earth Planet. Sci.* **10,** 397 (1982).
272. B. D. Marsh and L. L. Kantha, On the heat and mass transfer from an ascending magma. *Earth Planet. Sci. Lett.* **39,** 435 (1978).
273. B. D. Marsh, On the mechanics of igneous diapirism, stoping and zone melting. *Am. J. Sci.* **282,** 808 (1982).
274. S. C. Huang, Melting of semi-infinite region with viscous heating. *Int. J. Heat Mass Transfer* **37,** 1337 (1984).
275. S. H. Schneider and R. E. Dickinson, Climate modeling. *Rev. Geophys. Space Phys.* **12,** 447 (1974).
276. P. G. Drazin and D. H. Griffel, On the branching structure of diffusive climatological models. *J. Atmos. Sci.* **34,** 1696 (1977).
277. D. Pollard, An investigation of the astronomical theory of the ice ages using a simple climate-ice-sheet model. *Nature (London)* **272,** 233 (1978).
278. C. Kallen, C. Craft, and M. Ghil, Free oscillations in a climate model with ice-sheet dynamics. *J. Atmos. Sci.* **36,** 2292 (1979).
279. C. Nicolis, A free boundary value problem arising in climate dynamics. *Int. J. Heat Mass Transfer* **25,** 371 (1982).

Heat Transfer between Immersed Surfaces and Gas-Fluidized Beds

S. C. SAXENA*

Department of Chemical Engineering and Technology, Institute of Technology, Banaras Hindu University, Varanasi-221 005, India

I. Introduction

The limited availability of petroleum in many countries and the uncertainties associated with the foreign markets have prompted, a concerted effort to explore and assess the various alternative energy sources, including the fossil energy reserves. The United States Geological Survey has estimated the U.S. coal resources to be about 80% of the total fossil energy reserves in the country. This has provided a rational and logical basis for developing technologies for the economic utilization of coal for power generation in an environmentally acceptable fashion. The fluidized-bed combustion of coal has emerged as one of the most successful technologies for the utilization of high-sulfur and low-ranking coals. The phenomenon of gas fluidization is basic to a number of diverse chemical, petroleum, metallurgical, and energy technologies. The high heat transfer rates achieved for boiler tubes immersed in fluidized-bed coal combustors, along with their *in situ* capability to remove SO_x and suppress the emission of NO_x in high-pressure systems, have led to their exploitation as a promising technology for the generation of electrical power from sulfur-rich coal. The atmospheric fluidized-bed combustion (AFBC) technology has been claimed to be ready for commercialization for some time. On the other hand, thermodynamic analysis has shown that pressurized fluidized-bed combustion (PFBC) of coal in the combined-cycle mode is very efficient and economically preferable. This technology, which deals with large particles at high temperatures and pressures, has presented some new design and scale-up problems that will be discussed in this article in the perspective of the work accomplished during the last few years.

* Present address: Department of Chemical Engineering, University of Illinois at Chicago, Box 4348, Chicago, Illinois 60680.

A large proportion of work performed toward the understanding of the heat transfer mechanism in gas-fluidized beds has been limited to small particles, and these investigations have been reviewed by many authors, such as Zabrodsky [1], Botterill [2], Saxena *et al.* [3], and Saxena and Gabor [4]. For such small particle systems, bubble dynamics and bubble-induced motion of solid particles play a very significant role inasmuch as the predominant mechanism of heat transfer is due to particle convection. Suitable modifications of the packet model [4, 5] have been found adequate to simulate the heat transfer process. The particle residence time at the heat transfer surface and the particle concentration in its vicinity are the controlling parameters. In such systems, the interstitial gas velocity is smaller than the bubble velocity, and the fluidization behavior is referred to as the fast-bubble regime. On the other hand, in beds of large particles, the interstitial gas velocity is much larger than the bubble velocity and, as the fluidizing velocity is increased, this slow-bubble regime changes to rapidly growing and finally to a turbulent regime [6]. It has been shown by Canada and McLaughlin [7] and Borodulya *et al.* [8] that, in the turbulent regime, the solids mixing is poor and the particle temperature remains essentially constant [1]. For such a case of large particles, when the temperature decrement of a particle during its contact with the heat transfer surface is negligible, the unsteady-state heat conduction from the hot surface degenerates to a steady-state one, and the gas convective heat transfer becomes important. For such large particle systems, the gas convection contribution is large compared to the particle convection contribution. So far, we have used the words small particles and large particles in a qualitative sense, but, in the next section, we present a criterion that distinguishes between these two types of particles in a quantitative manner. In this perspective of the particle classification scheme, we then examine the bed hydrodynamic behavior and the various theories and correlations that have been recommended for estimating the heat transfer rates from immersed surfaces in a gas-fluidized bed.

The presence of radiation as a mode of heat transfer in high-temperature, gas-fluidized beds, along with particle and gas convection, leads to an enhancement of the overall heat transfer rate between the bed and an immersed surface as compared to that in beds at low and moderate temperatures. This necessitates the development of reliable mechanistic models and simple correlations capable of adequately describing the heat transfer characteristics of the bed at elevated temperatures. High-temperature heat transfer is significantly more difficult to describe since the convective (particle and gas) and radiative modes occur simultaneously and in parallel, and the inclusion of the two processes in a single step becomes difficult because of the nonlinear dependence of radiation on temperature. Because of the industrial interest in fluidized-bed technology, the available experimental and theoretical information concerning the bed to surface heat-transfer coefficient is reviewed, discussed, and assessed in this chapter in some detail.

II. Bed Fluidization and Powder Classification

The fluidization of particles of a bed by a fluid brings about a state of matter that is very attractive for a variety of physical and chemical processes. In general, a good fluid–solid contacting and solids circulation are attractive features of fluidized beds, and these are qualitatively represented by the term "the quality of fluidization." Many attempts exist in the literature to quantify the concept by developing different criteria involving dimensionless groups that would distinguish between bubbling (aggregative or heterogeneous) and nonbubbling (particulate or homogeneous) fluidization. The dimensionless groups employed for characterization are the Froude number, Fr; the Reynolds number, Re; a density ratio, $(\rho_s - \rho_g)/\rho_g$; a length ratio, H_{mf}/D_b; a modified Froude number; a characteristic wavelength, λ_c; the kinematic gas viscosity, ν_g; the minimum fluidization velocity, u_{mf}; etc. These criteria have been reviewed by Geldart [9], Kunii and Levenspiel [10], Zenz and Othmer [11], Grace [12], Seki *et al.* [13], Saxena and Ganzha [14], and others.

Of all these investigations, the Geldart's scheme [9] considers the hydrodynamic behavior of solid particles forming a bed when fluidized by a gas in detail. Geldart [9] has classified the particles in four groups, namely, A, B, C and D. Groups B and D are most typical of many applications of fluidization technology. Group B comprises particles of mean diameter, d_p, and density, ρ_s, lying in the ranges of 40 μm $< d_p <$ 500 μm, and 1400 kg/m^3 $< \rho_s <$ 4000 kg/m^3. Bubbles form in beds of such powders at or only slightly above u_{mf}. The majority of the bubbles rise faster than the interstitial gas velocity, and bubble size increases linearly with bed height and excess gas velocity, $u - u_{mf}$. Bubble coalescence is the predominant phenomenon. Group D refers to powders whose particles are large in size and very dense. In this case, the majority of the bubbles rise slower than the interstitial gas. The gas velocity in the dense phase is high, and solids mixing is relatively poor. The flow regime around the particles may be turbulent, and these particles satisfy the criterion that

$$(\rho_s - \rho_g) d_p^2 \geq 10^{-3} \tag{1}$$

The majority of the information that exists on such large particle systems is qualitative [6, 15–19]. Slow bubbles grow by cross-coalescence in contrast to fast bubbles, which grow by vertical in-line coalescence [15]. The latter are more efficient in mixing particles of the bed than are the former.

Recent interest in fluidized-bed combustion and gasification of coal has renewed interest in hydrodynamic and heat transfer characteristics of large particle systems. Many investigations [12, 15, 16, 18, 19] have revealed that with an increase in particle size the flow in a packed bed just prior to the onset of fluidization may be either transitional or turbulent, and the pressure drop correlations are of the general type proposed by Ergun [20]. Saxena and Vogel [18] and Botterill *et al.* [17] have shown on the basis of their experimental data over a range of temperatures and pressures with particles belonging to Geldart's groups B and D that the Ergun correlation can represent the data quite well if appropriate values of ε_{mf} and particle sphericity are used. In view of the practical interest in the fluidized-bed behavior of large particles, many theoretical and experimental investigations have been undertaken from time to time [1, 15, 17, 21–32], and these shed some light on the hydrodynamic properties of the bed and mechanistic details of the heat transfer process for the immersed surfaces. Saxena and Ganzha [14] found that Geldart's powder classification scheme [9] fails to simultaneously characterize the hydrodynamic and thermal behavior of the bed. Their [14] argument is briefly elaborated in the following.

Let us recall that the heat transfer coefficient for an immersed surface in a fluidized bed of large particles is primarily controlled by gas convection, the particle convection contribution is relatively smaller but not necessarily negligible [1, 2, 5]. As a result, the heat transfer coefficient increases with particle diameter for large particle systems. In contrast, due to the preponderance of heat transfer by particle convection for smaller particle systems, the heat transfer coefficient decreases with an increase in particle diameter. Thus, as an example a bed of sand particles fluidized by air at ambient conditions of temperature and pressure, the relation of Eq. (1) would predict that for $d_p > 0.63$ mm, the bed would behave like a large particle bed, Geldart group D particles. The heat transfer for such sand particles is still controlled by particle convection, so that while from hydrodynamic considerations a sand bed of 0.63-mm-diameter particles will be regarded as a large particle system, from the heat transfer point of view, it will be a small particle system. The limit of 1-mm particle diameter, mostly quoted in numerous publications as being approximately the size demarcating the small and large particle systems, suffers from the same drawback as mentioned previously for sand particles of d_p equal to 0.63 mm. Saxena and Ganzha [14] proposed a particle classification scheme that is free from such a shortcoming and is developed by considering the fluid flow and heat transfer behaviors simultaneously. Their scheme is described in the following.

Saxena and Ganzha [14] characterized the powders by considering the Archimedes number and Reynolds number at minimum fluidization as appropriate parameters to simultaneously represent the thermal and hydrodynamic properties of fluidized beds. The rationale for this selection was developed from considerations that are briefly reproduced here. The appropriateness of the Ergun correlation [20] in representing the flow resistance of a fluid through a bed of particles prior to and up to the state of incipient fluidization has been demonstrated for a variety of particles of a wide size range and for diverse operating conditions.

The Ergun correlation [20] is

$$\left[\frac{1.75(1-\varepsilon_{mf})\rho_g u_{mf}^2}{\varepsilon_{mf}^3 \phi_s d_p}\right] + \left[\frac{150(1-\varepsilon_{mf})^2 \mu_g u_{mf}}{\varepsilon_{mf}^3 (\phi_s d_p)^2}\right] = \frac{\Delta P_{mf}}{H_{mf}} \tag{2}$$

The bed pressure drop at minimum fluidization, ΔP_{mf}, is given by

$$\Delta P_{mf} = H_{mf} g(1-\varepsilon_{mf})(\rho_s - \rho_g) \tag{3}$$

Equations (2) and (3) may be combined and rearranged to yield the following relation:

$$\left[\frac{150(1-\varepsilon_{mf})}{\phi_s^2 \varepsilon_{mf}^3}\right]\mathrm{Re}_{mf} + \left[\frac{1.75}{\phi_s \varepsilon_{mf}^3}\right]\mathrm{Re}_{mf}^2 = \mathrm{Ar} \tag{4}$$

where

$$\mathrm{Re}_{mf} = \frac{\mu_{mf} d_p \rho_g}{\mu_g} \tag{5}$$

and

$$\mathrm{Ar} = \frac{d_p^3 g \rho_g(\rho_s - \rho_g)}{\mu_g^2} = \frac{3}{4} C_D \,\mathrm{Re}^2 \tag{6}$$

Goroshko *et al.* [33] have shown that the Ergun correlation can be simplified to the following expression for computing the minimum fluidization velocity of a stationary bed of spherical particles:

$$\mathrm{Re}_{mf} = \frac{\mathrm{Ar}}{150\left[\dfrac{(1-\varepsilon_{mf})}{\varepsilon_{mf}^3}\right] + \left(\dfrac{1.75\,\mathrm{Ar}}{\varepsilon_{mf}^3}\right)^{1/2}} \tag{7}$$

It is recognized that in deriving Eq. (7), Goroshko *et al.* [33] neglected the product term in solving the quadratic equation, and that this approximation has a pronounced effect in the transitional flow regime as shown by Botterill *et al.* [17]. Putting an approximate value of 0.4 for ε_{mf}, Goroshko *et al.* [33] further simplified Eq. (7) to

$$\mathrm{Re}_{mf} = \frac{\mathrm{Ar}}{1400 + 5.22\sqrt{\mathrm{Ar}}} \tag{8}$$

It is shown by Zabrodsky [34] that for a laminar flow condition in the bed the second term in the denominator is much smaller than the first term, that is, $1400 \gg 5.22\sqrt{\mathrm{Ar}}$, and for the turbulent flow condition, the reverse is the case, that is $5.22\sqrt{\mathrm{Ar}} \gg 1400$.

It is widely known [5, 23, 35, 36] that the maximum Nusselt number, $\mathrm{Nu}_{\mathrm{max}}$, is a unique function of the Archimedes number, that is,

$$\mathrm{Nu}_{\mathrm{max}} = h_{\mathrm{w\,max}} d_{\mathrm{p}}/k_{\mathrm{g}} = f(\mathrm{Ar}) \tag{9}$$

It therefore follows that heat transfer behavior typified by the Nusselt number and the fluid flow conditions characterized by Reynolds number are related to each other through their common dependence on the Archimedes number. This, together with the fact that the Archimedes number, which is obtained from a force balance on a single particle in a flowing fluid field, effectively describes the fluid–particle interaction, constitutes the basis that led Saxena and Ganzha [14] to choose the Archimedes number as the characteristic number in their powder classification scheme. Subsequent efforts [37–40] to describe, understand, and interpret the heat transfer and hydrodynamic properties of fluidized beds on the basis of this scheme [14, 41] have met with good success.

In the relation of Eq. (8), it has been assumed that $\varepsilon_{\mathrm{mf}}$ has a constant numerical value of 0.4. Botterill *et al.* [32] and Lucas *et al.* [42] found from their measurements at high temperatures for fluidized beds of different size particles that $\varepsilon_{\mathrm{mf}}$ varies with temperature and that the nature of variation depends upon the size of particles. They found that for small particles $\varepsilon_{\mathrm{mf}}$ increases with temperature and tends to become constant at higher temperatures; for larger particles, $\varepsilon_{\mathrm{mf}}$ first decreases with increasing temperature, passes through a minimum, and then increases with a further increase in temperature; and, for still larger particles, $\varepsilon_{\mathrm{mf}}$ was found to be constant as the bed temperature increased. Saxena *et al.* [43] and Mathur *et al.* [44], from their measurements on sand particles with average diameters of 559, 571, 1225, and 3778 μm as a function of temperature in the range from 300 to 1250 K, established that $\varepsilon_{\mathrm{mf}}$ exhibits a characteristic variation with the corresponding Reynolds number, $\mathrm{Re}_{\mathrm{mf}}$, and the Archimedes number. These authors [43, 44] have shown that this variation is due to the changing fluid-flow field around the particles and hence to the interparticle forces. Further, it is shown that the qualitative variation of $\varepsilon_{\mathrm{mf}}$ with $\mathrm{Re}_{\mathrm{mf}}$ and Ar is in accord with the powder classification scheme of Saxena annd co-workers [14, 41]. The selection of the Archimedes number to develop the powder classification scheme is also substantiated by these investigations. We now present the details of this classification scheme.

The particle classification scheme of Saxena and Ganzha [14] is based on the hydrodynamic fact observed by many workers (including Aerov and Todes [45], Leva [46], and Geldart and Cranfield [16]) that the gas flow in packed beds is laminar as long as $\mathrm{Re} < 1$, but it becomes turbulent for $\mathrm{Re} > 200$.

Several authors [46, 11] have proposed that the flow through a packed bed is essentially laminar beyond the Reynolds number limit of 1 and up to 10. Consequently, for a bed of small particles, when the fluid flow is laminar around the particles [41] and there is no effective interaction between them, ε_{mf} will be expected to remain constant with increasing Re_{mf}. Based on Eq. (8), the upper limit for group I particles is $Ar \leq 21{,}700$ corresponding to $Re_{mf} \leq 10$. The dependence of h_w on d_p is well known (see Baskakov *et al.* [22]) and is displayed explicitly in Fig. 10.11 on p. 494 of Ref. 2, where h_w is plotted against d_p for a variety of particles. It will be noted that at first, as d_p increases beyond a characteristic particle diameter (such as 0.032 mm for corundum [21]), h_w decreases, then remains constant as d_p increases, and thereafter increases as d_p is further increased. This dependence of h_w on d_p is explained by the changing contribution of gas convection, $h_{w\,conv}$, to h_w as d_p is increased. Briefly, only the $h_{w\,conv}$ part of h_w is significant, that is, it is not negligible compared with the particle convection contribution, $h_{w\,cond}$, to h_w. The lower limit of Ar for group I powders is computed on the basis of the dependence of h_w on d_p [5, 22] under the requirement that h_w decreases as d_p increases. Thus, group I powders are characterized by

$$3.55 \leq Ar \leq 21{,}700 \tag{10}$$

$$1.5\left[\frac{\mu_g^2}{\rho_g g(\rho_s - \rho_g)}\right]^{1/3} \leq d_p \leq 27.9\left[\frac{\mu_g^2}{\rho_g g(\rho_s - \rho_g)}\right]^{1/3} \tag{11}$$

and physically for such powders ε_{mf} will remain constant with changing Re_{mf} and Ar, and the fluid flow around the particles will be laminar.

The transitional group II is defined by a wide range of Reynolds number, $10 \leq Re \leq 200$. In this range, the flow field around the particles becomes turbulent and gets almost fully developed around $Re = 40$ [5, 47]. For $Re < 40$, the laminar boundary layer around the particles becomes increasingly turbulent, and the active interphase contact surface area approaches the total surface area of the particles as Re is increased. For $Re > 40$, the active interphase contact surface becomes equal to the total surface of the particles in the bed, and heat transfer is mainly controlled by the turbulence in the thermal boundary layer and wake. For $Re > 200$, the thermal wakes of the particles are disrupted by the following particles, leading to a bed turbulence whose intensity increases with Re. Saxena and Ganzha [14] thus proposed groups IIA and IIB, defined by the limits $10 \leq Re \leq 40$ and $40 \leq Re \leq 200$, respectively, and group III, defined by $Re > 200$.

For particles of group IIA, defined such that

$$21{,}700 \leq Ar \leq 130{,}000 \tag{12}$$

$$27.9\left[\frac{\mu_g^2}{\rho_g g(\rho_s - \rho_g)}\right]^{1/3} \leq d_p \leq 50.7\left[\frac{\mu_g^2}{\rho_g g(\rho_s - \rho_g)}\right]^{1/3} \tag{13}$$

the laminar boundary layer around the particles becomes increasingly turbulent, and a wake is formed on the downstream side of the particles. The size of this wake increases with an increase in particle size and hence with $\mathrm{Re}_{\mathrm{mf}}$ and Ar. The low-pressure wake region will cause the neighboring particles to be drawn closer, thereby decreasing $\varepsilon_{\mathrm{mf}}$ as $\mathrm{Re}_{\mathrm{mf}}$ or Ar increases. From the heat transfer point of view, the $h_{\mathrm{w\,conv}}$ part of h_{w} is significant, that is, it is not negligible compared with the particle convection contribution to h_{w}, $h_{\mathrm{w\,cond}}$, for particles of group IIA.

Particles of group IIB are defined such that

$$130{,}000 \leq \mathrm{Ar} \leq 1.60 \times 10^{6} \tag{14}$$

$$50.7\left[\frac{\mu_{\mathrm{g}}^{2}}{\rho_{\mathrm{g}} g(\rho_{\mathrm{s}}-\rho_{\mathrm{g}})}\right]^{1/3} \leq d_{\mathrm{p}} \leq 117\left[\frac{\mu_{\mathrm{g}}^{2}}{\rho_{\mathrm{g}} g(\rho_{\mathrm{s}}-\rho_{\mathrm{g}})}\right]^{1/3} \tag{15}$$

the boundary layer around the particles becomes fully turbulent, the separation point having moved downstream of the equatorial plane of the particle causes the wake size to decrease. As a result, the interparticle forces decrease and $\varepsilon_{\mathrm{mf}}$ increases monotonically with an increase in Ar or $\mathrm{Re}_{\mathrm{mf}}$. For particles of group IIB, the two components, $h_{\mathrm{w\,conv}}$ and $h_{\mathrm{w\,cond}}$, are comparable with each other.

The group III particles are characterized by

$$\mathrm{Ar} \geq 1.60 \times 10^{6} \tag{16}$$

$$d_{\mathrm{p}} \geq 117\left[\frac{\mu_{\mathrm{g}}^{2}}{\rho_{\mathrm{g}} g(\rho_{\mathrm{s}}-\rho_{\mathrm{g}})}\right]^{1/3} \tag{17}$$

the gas flow through the bed is turbulent, and the interparticle forces do not play any role in influencing the particle arrangement in the bed as these are negligibly small compared with the fluid shear force. As a result, $\varepsilon_{\mathrm{mf}}$ does not change with changes in $\mathrm{Re}_{\mathrm{mf}}$ or Ar for particles of this group and remains effectively constant. From the heat transfer point of view, h_{w} is strongly controlled by the $h_{\mathrm{w\,conv}}$ part, which makes the dominant contribution. The $h_{\mathrm{w\,cond}}$ component is only a negligible part of h_{w}.

It is enlightening as well as interesting to examine the particle sizes for the previously mentioned three groups, I, II, and III, in relation to a particular gas–solid system. For this, we consider a bed of sand particles ($\rho_{\mathrm{s}} = 2500$ kg/m^{3}) at $T = 300$ K and $P = 0.1$ MPa as a typical example. Particles specified by the range $d_{\mathrm{p}} \leq 0.63$ mm obtained from Eq. (11) belong to group I and correspond to Geldart's group B particles and will be referred to here as small particles. The particles belonging to the transitional group II correspond to Geldart's group D particles, and for a sand–air system, the particle diameters for groups IIA and IIB are given by Eqs. (13) and (15) as $0.63 \leq d_{\mathrm{p}} \leq 1.1$ (mm) and $1.1 \leq d_p \leq 2.67$ (mm), respectively. For the proposed group III of large particles, the particle diameter is given by Eq. (17) and for the sand–air system under discussion, $d_p \geq 2.67$ mm. This group per se is not discussed by Geldart [9] but is included in his group D powders. At the present time, insufficient experimental data are available to characterize the reliable and accurate limits for the different groups as well as for the subgroups of group II. Availability of elaborate experimental data for particles of a wide variety, size range, and operating conditions will only enable to establish the demarcating limits for the different groups and subgroups. It will then be more meaningful to develop an elaborate particle classification system. Discussion of available hydrodynamic and heat transfer data as given later in this chapter supports and substantiates the powder classification scheme presented here.

III. Bed Hydrodynamic Behavior at High Temperatures and Pressures

In the previous section, we mentioned the hydrodynamic condition of the bed in the perspective of the proposed powder classification scheme. We will now describe the high-temperature and high-pressure hydrodynamic behavior of fluidized beds and examine the same in the light of the previously detailed powder classification. In particular, the available data on the minimum fluidization velocity, u_{mf}, and the bed voidage, ε_{mf}, will be interpreted.

A. BED VOIDAGE AT MINIMUM FLUIDIZATION

Botterill *et al.* [17, 32] pointed out that the bed voidage at incipient fluidization is not constant with temperature (very much in disagreement with the common belief), but instead its variation is nonmonotonic, exhibiting a characteristic minimum. They measured ε_{mf} for beds of silica sands of average particle diameters, d_p, 380, 460, 530, 660, 780, 890, 1280, and 2320 μm in the temperature range from ambient to about 1025 K. From these measurements, they inferred that ε_{mf} first decreased with increase in Ar, and thereafter ε_{mf} increased with further increases in Ar. The change in the trend of ε_{mf} variation occurred around Ar = 26,000 and $\mathrm{Re}_{mf} = 12.5$. Lucas *et al.* [42] investigated beds of silica sands with particles in the ranges of 177–250, 250–400, 400–500, and 500–700 μm in the temperature range from ambient to 1225 K. They concluded that ε_{mf} was approximately constant for $\mathrm{Re}_{mf} < 0.75$, decreased for $0.75 \leq \mathrm{Re}_{mf} \leq 2$, and remained constant for $\mathrm{Re}_{mf} \geq 2$.

Chitester *et al.* [48] observed that their measured bed voidage values at incipient fluidization, referring to ambient temperatures and at pressures in the range from 0.1 to 6.3 MPa, are feebly dependent on pressure. ε_{mf} increases slowly with an increase in pressure and reaches a constant value at higher pressures. Mathur and Saxena [41] have explained that these variations of ε_{mf} with temperature and pressure for particles of different sizes are essentially due to changing interparticle forces that are generated with changing fluid-flow patterns around the particles. It was, therefore, inferred [41, 43, 44] that the particle Reynolds number at incipient fluidization, or the corresponding Archimedes number, would be a more appropriate parameter to characterize a gas–solid system. The particle size appears in such considerations as a more involved parameter inasmuch as it will have to account for the particle shape, size range, surface morphology, etc. For a complete and unambiguous

TABLE I

SUMMARY OF EXPERIMENTAL RESULTS OF SAXENA AND CO-WORKERS[a]

d_p (μm)	T_b (K)	Ar	u_{mf} (m/s)	Re_{mf}	ε_{mf}	ϕ_s
559	1200	612	0.13	0.46	0.46	0.79
559	1050	812	0.14	0.59	0.45	0.82
559	725	1,910	0.15	1.22	0.46	0.83
559	525	3,990	0.17	2.35	0.47	0.84
559	375	9,130	0.20	4.81	0.46	0.85
559	300	15,700	0.22	6.39	0.45	0.85
751	1150	1,633	0.70	3.5	0.58	0.70
751	925	2,672	0.56	4.0	0.57	0.75
751	725	4,618	0.47	5.1	0.53	0.72
751	500	10,886	0.45	6.8	0.47	0.73
751	385	20,283	0.40	7.2	0.40	0.74
751	300	38,013	0.30	15.0	0.43	0.72
1225	1200	6,400	0.96	7.3	0.53	0.78
1225	1000	9,728	0.75	7.8	0.48	0.80
1225	750	18,534	0.62	10.2	0.44	0.80
1225	500	47,258	0.82	26.4	0.47	0.79
1225	385	88,060	0.94	47.6	0.49	0.81
1225	300	165,050	0.96	74.7	0.50	0.81
3788	1250	1.68×10^5	1.12	25	0.44	0.66
3788	1200	1.89×10^5	1.18	28	0.44	0.64
3788	1000	2.86×10^5	1.26	40	0.45	0.63
3788	950	3.20×10^5	1.30	45	0.46	0.65
3788	825	4.41×10^5	1.34	58	0.46	0.62
3788	750	5.44×10^5	1.38	70	0.47	0.61
3788	500	1.39×10^6	1.32	132	0.47	0.63
3788	385	2.59×10^6	1.26	195	0.46	0.62
3788	300	4.85×10^6	1.24	298	0.47	0.60

[a] From Saxena *et al.* [43] and Mathur *et al.* [44].

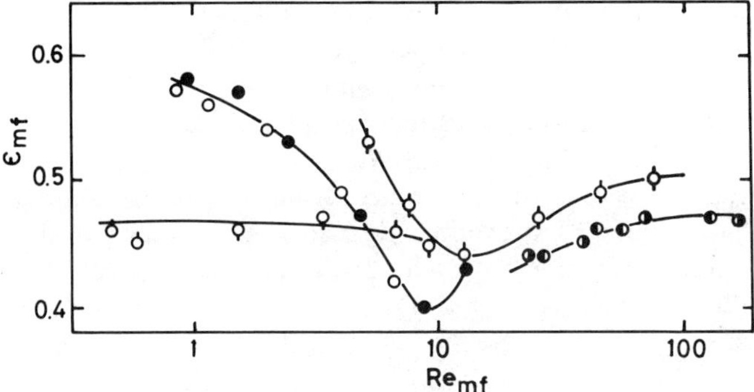

FIG. 1. Variation of incipient bed voidage, ε_{mf}, with Reynolds number at minimum fluidization, Re_{mf}. (◐ —— ◑), $d_p = 3788$ μm; (ϙ), $d_p = 1225$ μm; (○ —— ●), $d_p = 751$ μm; (ϙ̇ —— ϙ), $d_p = 559$ μm (from Mathur *et al.* [44]).

understanding of this phenomenon, Saxena *et al.* [43, 44] conducted very careful and controlled experiments over a wide range of average bed particle sizes and temperatures, spanning over several orders of magnitude of Re_{mf} values. The results of these workers [43, 44] are reproduced in Table I, and Figs. 1 and 2. These data established a characteristic qualitative variation of ε_{mf} with Re_{mf} and Ar, which is in agreemennt with all earlier workers [17, 32, 42] as well as with the powder classification scheme of Saxena and Ganzha [14, 41]. The bed voidage, ε_{mf}, is constant for small values of Re_{mf} or Ar, decreases as Re_{mf} (or Ar) is increased, goes through a minimum, increases with a further increase of Re_{mf} (or Ar), and finally becomes constant as Re_{mf} (or Ar)

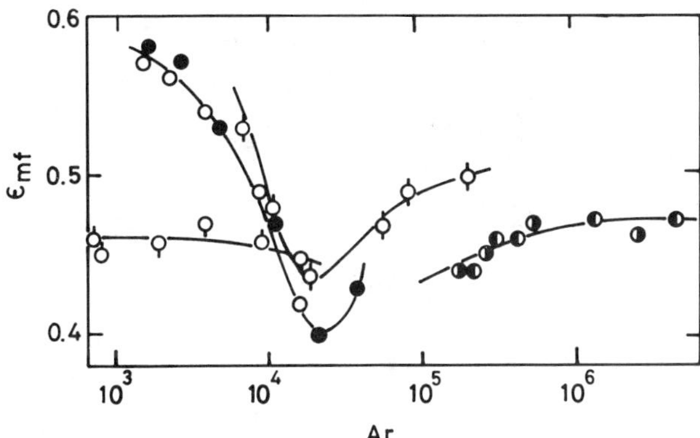

FIG. 2. Variation of incipient bed voidage, ε_{mf}, with Archimedes number, Ar. Symbols represent different values for d_p as defined in Fig. 1 legend (from Mathur *et al.* [44]).

is further increased. The minimum occurred for Re_{mf} values in the range from 9 to 13 and for the Archimedes number in the range from 20,000 to 25,000. These values are in good agreement with the similar findings of Botterill *et al.* [17].

Mathur *et al.* [44] and Saxena and Mathur [49] have shown that the experimental findings of Pattipati and Wen [50], concerning the constancy of ε_{mf} with Re_{mf}, are neither wrong nor inconsistent with the findings of Botterill *et al.* [17], who found ε_{mf} to exhibit a characteristic variation with an increase in Re_{mf}. The apparent different qualitative variations of ε_{mf} observed by these workers [50, 17] with an increase in Re_{mf} are due to their data being in different ranges of Re_{mf}, where the interparticle forces are different and hence the observed different variations in ε_{mf}. It should be stressed that all these variations are completely consistent with the powder classification scheme of Saxena and Ganzha [14, 41].

B. Minimum Fluidization Velocity

Bed pressure drop data as a function of superficial fluidizing velocity have been routinely analyzed to establish the minimum fluidizing velocity at specified values of temperature and pressure [10, 18]. Some workers have considered beds of either small or large particles at ambient pressures but over a wide temperature range [17–19, 32, 42, 51–54], whereas others have considered beds at ambient temperatures but over a wide pressure range [7, 18, 48, 55, 56]. These minimum fluidizing velocities have been correlated, and a large number of predictive equations have been proposed. Three of the most recent compilations of such correlations are by Babu *et al.* [57], Grewal and Saxena [58], and Thonglimp *et al.* [59]. The positive root of Eq. (2), giving the value of Re_{mf}, may be expressed as

$$\mathrm{Re}_{mf} = \sqrt{C_1^2 + C_2\,\mathrm{Ar}} - C_1 \tag{18}$$

where

$$C_1 = \frac{85.71(1 - \varepsilon_{mf})}{\Phi_s} \tag{19}$$

and

$$C_2 = \frac{\phi_s \varepsilon_{mf}^3}{1.75} \tag{20}$$

Working on the assumptions that the particles constituting the bed can be approximated by a constant value of ϕ_s and that ε_{mf} remains constant over the entire range of operating conditions of temperature and pressure, various investigators have proposed values of C_1 and C_2 on the basis of experimental data. These are listed in Table II. Comparison of the experimental data of

TABLE II

VALUES OF THE CONSTANTS C_1 AND C_2 IN EQ. (18) OF DIFFERENT INVESTIGATORS[a]

Investigators	C_1	C_2
Wen and Yu [60]	33.7	0.0408
Bourgeois and Grenier [61]	25.46	0.0384
Richardson and Jeronimo [62]	25.7	0.0365
Saxena and Vogel [18]	25.28	0.0571
Babu *et al.* [57]	25.25	0.0651
Thonglimp *et al.* [59]	19.9	0.0320
Grace [12]	27.2	0.0408
Zheng *et al.* [63]	18.75	0.0313

[a] From Mathur *et al.* [44].

Saxena and co-workers [43, 44] with the predictions of Eq. (18) with the values of C_1 and C_2 listed in Table II is shown in Fig. 3 as a plot of Re_{mf} versus Ar. It is clear that the four experimental data sets cannot be represented by a universal curve. Two symbols are used to represent data sets belonging to different runs for a given data set. Further, none of the theoretically computed curves can represent all of the data adequately. The disagreement between the computed and experimental values is to be ascribed to the two assumptions listed

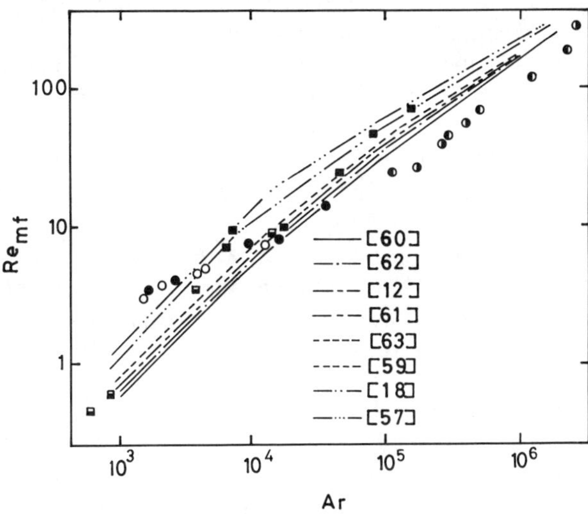

FIG. 3. Comparison of the variation of experimental Re_{mf} with Ar and the predictions based on the correlations of eight different studies (references indicated in brackets). (◐ ◑), $d_p = 3788\ \mu m$; (■), $d_p = 1225\ \mu m$; (● ○), $d_p = 751\ \mu m$; (⬓ ⬒), $d_p = 559\ \mu m$ (from Mathur *et al.* [44]).

previously. Mathur *et al.* [44] resolved the assumption of a constant ϕ_s value by determining it from Eq. (4), using all other quantities as obtained from experiments. The ϕ_s values so determined are listed in the last column of Table 1. They found that for each bed an average value for the entire temperature range can be ascribed with a high degree of reliability. The scatter from the average value of individual values at different temperatures can certainly be explained by the uncertainties in the experimental data. Hence, it was inferred that the assumption of a constant ϕ_s value for a system seems quite reasonable, and the inability of Eq. (18) to represent experimental data with a higher degree of agreement must be ascribed to the assumption of constancy of ε_{mf} values with temperature. As discussed and explicitly shown in Figs. 1 and 2, ε_{mf} varies with temperature, and this characteristic variation is physically plausible and can be understood in terms of the interparticle forces [14, 41].

In view of the previous discussion, it is imperative that any predictive equation for u_{mf} must be assessed in conjunction with appropriate values of ε_{mf} (Shrivastava *et al.* [64]). Also, in predicting behavior of the bed in a particular regime from data taken in another regime, caution must be exercised in choosing the appropriate values of ε_{mf} [65, 66]. In summary, the hydrodynamic behavior of fluidized beds requires a very careful determination of ε_{mf} for the specific operating conditions under discussion, and such values are dependent on the system Reynolds number (or Archimedes number) characterizing the fluid flow around the bed particles.

The success of the Ergun correlation, Eq. (4), or its simplified version, Eq. (8), in representing experimental data at ambient pressures and at ambient, moderate and high temperatures was previously discussed. Some remarks will now be made in relation to operations at moderate and high pressures following the work of Saxena and Ganzha [37]. These authors [37] have analyzed the available experimental values of Re_{mf} [18, 67–72] at high pressures as a function of Ar. In Fig. 4, these data are displayed and also compared with the predictions of Eq. (8). It is to be noted that the experimental and predicted values are in fair agreement with each other. This provides a rationale for the use of Eq. (8) in developing the powder classification scheme [14], as well as for using Ar as a characteristic parameter for fluidized-bed operations at high pressures. It follows from Eq. (8) that for group I particles $\mathrm{Re}_{mf} \leq 10$, the second term in the denominator is negligible compared to the first term, and Re_{mf} will be proportional to Ar. As a result, u_{mf} will be independent of ρ_g or P. On the other hand, for group III particles, $\mathrm{Re}_{mf} \leq 200$, the first term in the denominator of Eq. (8) is negligible compared to the second term, and Re_{mf} is proportional to Ar. Consequently, u_{mf} will vary as $\rho_g^{0.5}$ or $P^{0.5}$. For group II particles, the dependence of u_{mf} on ρ_g or P will be more complicated and will be controlled by the nature and extent of turbulence in the bed.

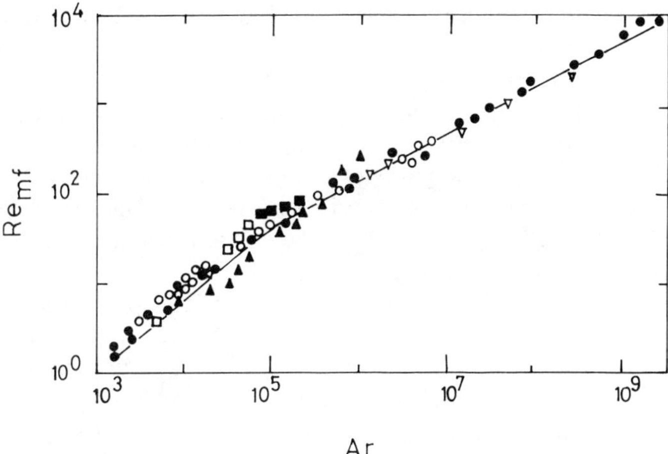

FIG. 4. Comparison of the experimental values of Re_{mf} with predictions based on Eq. (8) at different values of Ar.

A critical examination of the works of Botterill and Desai [68], Borodulya *et al.* [28], Chesnokov *et al.* [70], and Varadi and Grace [73] reveals that u_{mf} decreases with increasing pressure in all cases. The magnitude of variation seems to vary with the investigation and depends on the system characteristics and operating conditions. Altshuler and Sechenov [74] and Boguslavskiy and Melik-Akhnazarov [75] found that u_{mf} does not vary with pressure even for particles of diameter 0.3 and 0.45 mm. It would thus appear that there is a contradiction on the probable dependence of u_{mf} on pressure, and it remains to be resolved. However, for small particles ($Ar \leq 130{,}000$), u_{mf} should be almost independent of pressure. For gas–solid systems characterized by $130{,}000 \leq Ar \leq 1.6 \times 10^6$, u_{mf} is dependent on pressure. For systems characterized by $Ar > 1.6 \times 10^6$, u_{mf} becomes proportional to $P^{0.5}$.

IV. Heat Transfer to and from Immersed Surfaces

Several reviews [1–5, 37–40, 76, 77] of varying scope describing the heat transfer process to or from an immersed surface in a gas-fluidized bed have appeared from time to time. The value of the heat transfer coefficient will undoubtedly depend on the fluid mechanics of the individual particles and of the bed as a whole. As a result, the previously described and discussed powder classification scheme [14, 41] provides a very valid basis for assessing and classifying the different correlations, a majority of them are developed and tested from a limited amount of experimental data only. Efforts made to model heat transfer processes or correlate the experimental data will be examined in the following in this perspective.

Fluidized-bed systems belonging to group I are characterized by bubble coalescence, good solids mixing, and laminar gas flow around the particles. The random motion of the particles brings them in contact with the heat transfer surface, where the heat transfer process occurs before the particles move back to the bulk of the bed. The particle-surface heat transfer process for such systems is regarded as an unsteady-state heat conduction through the gas lens between the particle and the surface [1]. Direct particle-surface conduction at the contact points between the particle and the surface is negligible because of the smallness of the contact area [78]. As a result, heat transfer for such systems is controlled mainly by the thermal conductivity of the gas, k_g, and the volumetric heat capacity of the particles, $\rho_s C_{ps}$ [3]. The dependence of h_w on k_g becomes increasingly more important as the temperature is raised. The increase in the size of the particles decreases the heat transfer coefficient. This is because for larger particles, the effective thickness of the gas lens increases while the surface to volume ratio decreases. The residence time, τ, of the particles at the heat transfer surface depends on the fluidizing velocity, u, among other factors, and increasing u causes τ to decrease. Further, increases in u cause the bed voidage, ε, to increase and the particle concentration, $1 - \varepsilon$, to decrease. These two quantities, τ and $(1 - \varepsilon)$, which decrease with an increase in u, cause the heat transfer coefficient, h_w, to increase initially with an increase in u to a maximum value, $h_{w\,max}$, at a characteristic optimum fluidizing velocity, and then to decrease with a further increase in u [79, 80]. It should be emphasized that this type of qualitative variation of h_w with u is characteristic of fluid-particle systems belonging to groups I and IIA.

The orientation of the heat transfer surface in the fluidized bed influences the value of h_w for systems belonging to these two groups. This is because the surface orientation influences the bed voidage. This has been experimentally confirmed by measurement of h_w for horizontal and vertical tubes immersed in gas-fluidized beds by Verma and Saxena [81]. Horizontal tubes have been found to have a defluidized cap of solids on the downstream side and a gas bubble trapped on the upstream side [82]. Saxena *et al.* [83] have developed an image-carrying, fiber-optic probe technique for the direct measurement of voidage or porosity at the immersed surface as a function of fluidizing velocity and at different angular positions of an immersed tube. They [83] found that for a horizontal tube the voidage values were smallest at the top or downstream side of the tube, were largest and identical at the two lateral sides of the tube, and intermediate at the bottom or upstream side of the tube. At each of the four positions, the voidage increased with an increase in fluidizing velocity above u_{mf}, though the rate of increase was different at different positions. The bulk bed voidage values were sufficiently smaller than these surface voltage values. These investigations, conducted in a bed of spherical glass beads of diameter 1.4 mm and a horizontal glass tube of outside diameter 11 mm, have also been extended to tubes of diameter 28 mm [84] and 50.8 mm [85]. In Fig. 5, the typical results for a tube with a 50.8-mm outside diameter of surface voidage ε_s, at the angular positions of 0°, 60°, 180°, and 300°, as well as of the buld bed voidage, ε_b, as a function of u/u_{mf} are shown. The mean of the surface voidage values (ε_s) at the four positions, $\bar{\varepsilon}_s$, is also shown for comparison with ε_b to get a gross idea of differences in the two types of voidage values, namely, surface and bulk. These results have an important bearing on the dependence of h_w for a horizontal tube on tube diameter as observed by different workers [79, 80, 86–93]. This feature of h_w variation with tube diameter will be discussed later in this chapter.

For fluid particle systems belonging to subgroup IIA, the contribution of gas convection to h_w, $h_{w\,conv}$, is significant and is not negligible compared to the particle convection contribution to h_w, $h_{w\,cond}$. For powders of subgroup

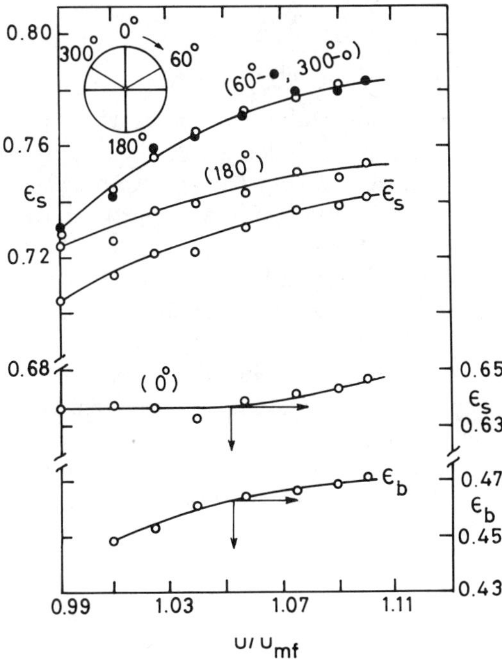

FIG. 5. Variation of ε_s at four peripheral positions (0°, 60°, 180°, and 300°) of the 50.8-mm diameter glass tube, and ε_b as a function of u/u_{mf} (from Saxena and Patel [85]).

IIB, the two components $h_{w\,conv}$ and $h_{w\,cond}$ of h_w are comparable to each other. In this perspective for subgroup IIA powders, only the packet model cannot satisfactorily predict the heat transfer coefficient between the immersed surface and the bed. An adequate model for subgroup IIA should express h_w as the sum of two terms, namely $h_{w\,cond}$ and $h_{w\,conv}$. The former must be computed very accurately, while the latter can be relatively simpler representation of the gas convection heat transfer process. It may be mentioned that to accomplish the latter, a laminar boundary layer at turbulent flow along the surface may be assumed so that this term should contain $Re^{0.5}$.

For subgroup IIB powders, the same two-term expression for h_w would be essential, except that now, in addition to the accurate calculation of the $h_{w\,cond}$ term, the $h_{w\,conv}$ term should also be computed with a high degree of accuracy. In this region, h_w is independent of d_p, and the $h_{w\,conv}$ can be calculated from the similarity of heat transfer between a plate and a gas flow in the transition regime so that the term should involve $Re^{0.67}$. For the group III powders, h_w is strongly controlled by the $h_{w\,conv}$ part, which makes the dominant contribution. The $h_{w\,cond}$ component is only a small part of h_w. Consequently, for a precise prediction of h_w, the $h_{w\,conv}$ part must be formulated with a high degree of accuracy.

A. MAXIMUM HEAT TRANSFER COEFFICIENT

We will now discuss the applicability of different heat transfer models and various proposed correlations in reference to their appropriateness for different groups of the Saxena and Ganzha [14] powder classification scheme. However, the models and correlations developed for the maximum heat transfer coefficient, $h_{w\,max}$ or Nu_{max}, will be discussed first. In view of their relevance in the design of fluidized-bed coal combustors, the high-pressure measurements are described in relatively greater detail from the analysis of Saxena and Ganzha [37].

A little history of the measurement of h_w as a function of pressure is interesting to bring out its dependence on gas density or operating gas pressure and particle diameter. Traber *et al.* [94] measured h_w for a 30-mm-diameter copper coil with outer and inner diameters of 6 and 4 mm, respectively, immersed in beds of different sized particles ($d_p = 0.38$, 0.75, and 1.5 mm) in the pressure range from 0.1 to 23 MPa at 423 K, fluidized by a gas mixture containing 80% H_2, 10% N_2, 7% CO, 2% CH_4, and 1% CO_2. They found that h_w increases with increasing pressure, and they correlated their data by

$$Nu_{max} = 0.021\, Ar_b^{0.4}\, Pr^{0.33} (D_b/d_p)^{0.13} (H_{mf}/d_p)^{0.16} \tag{21}$$

where

$$Ar_b = g d_p^3 \rho_g \rho_b / \mu_g^2 \tag{22}$$

and

$$\rho_b = \rho_s(1 - \varepsilon_{mf}) \tag{23}$$

Rabinovich and Sechenov [95] measured h_w between the wall of a 44-mm-diameter cylindrical column and fluidized beds of particles of diameters in the range from 0.40 to 0.63 mm and 0.63–1.0 mm as a function of pressure in the range from 0.1 to 3.1 MPa at temperatures in the range from 423 to 1273 K, fluidized by different gases. They correlated their data by

$$\mathrm{Nu}_{D_b} = 42.17\,\mathrm{Re}_{D_b}^{0.9}(T_b - 273)/250 \tag{24}$$

and Borodulya *et al.* [67] have recorrelated these data by

$$\mathrm{Nu}_{D_b} = 25.18\,\mathrm{Re}_{D_b}^{0.9}[(T_b - 273)/250]^{0.41} \tag{25}$$

Botterill and Desai [68] measured h_w for fluidized beds of copper shots (150 and 625 μm), sands (160, 800, and 2740 μm) and coal (1430 μm) at 0.1 and 1.13 MPa at ambient temperature. They concluded that h_w is almost independent of pressure for small particles but increases appreciably for large particles as the pressure is increased. The distinction between small and large particles was made somewhat arbitrarily at about 500 μm. They also commented that "the quality of fluidization" improves for dense, large particles. The vertical heat transfer surface in these experiments was made as small as possible so that the particle residence times are small and comparable with the shortest times achievable in associated flowing packed bed experiments.

In a later effort, Denloye and Botterill [24] determined h_w for a heat transfer surface 120 mm in length and 12.7 mm in diameter immersed in beds of copper shot (160, 340, and 620 μm), sand (160, 590, 1020, and 2370 μm), and soda glass (415 μm) fluidized by air, argon, carbon dioxide, and freon in the pressure range from 0.1 to 1 MPa. Based on these data, they proposed that for $10^3 < \mathrm{Ar} < 2 \times 10^6$,

$$\mathrm{Nu}_{max} = 0.843\,\mathrm{Ar}^{0.15} + 0.86\,\mathrm{Ar}^{0.39}\,d_p^{0.5} \tag{26}$$

the two additive components represents the particle and gas convective components of heat transfer, respectively.

Borodulya *et al.* [28] measured h_w for a vertical cylindrical surface 60 mm in length and 13 mm in diameter immersed in fluidized beds of sands (126, 250, 800, and 1220 μm) and glass beads (450, 950, and 3100 μm) in the pressure range from 0.6 to 8.1 MPa. The experimental results are correlated by

$$\mathrm{Nu}_{max} = 0.064\,\mathrm{Ar}^{0.4} \tag{27}$$

Xavier *et al.* [96] concluded that the influence of pressure on h_w is insignificant for small particles, while for large particles it increased approximately proportional to the square root of pressure. This was inferred from their work on heat transfer between a vertical plate and fluidized beds of glass beads (61, 475, and 615 μm) and polymer particles (688 μm) at pressures to about 2.5 MPa at ambient temperatures with fluidizing gases, nitrogen, and carbon dioxide.

This review clearly shows that h_w generally increases with an increase in pressure. Later workers [22, 24, 28, 96] following Botterill and Desai [68] have emphasized that for small particles h_w is almost independent of pressure and for large particles h_w increases with pressure. It is further suggested that the predominant contribution to h_w for small particles comes from particle convection, $h_{w\,cond}$, and this is not significantly influenced by changes in pressure [5]. However, it is also suggested [28] that the increase in pressure improves the hydrodynamic behavior of the bed and hence may give rise to an increase in the value of $h_{w\,cond}$. On the other hand, for large dense particles, $h_{w\,cond}$ makes only a small contribution to h_w, and gas convection is the important mode of heat transfer, $h_{w\,conv}$. The latter heat transfer mechanism is sensitively dependent on gas density and hence on pressure.

For small particle systems characterized by $Re_{mf} < 40$ and $Ar < 130{,}000$ (groups I and IIA), Saxena and Ganzha [37] concluded that the packet model or one of its modified forms [5] would be quite successful in predicting the heat transfer coefficient. The results of these formulations do not express any explicit dependence of h_w on ρ_g or P. However, a small pressure dependence can arise through improved bed hydrodynamics at high pressures when the packet residence time at the heat transfer surface may alter, or the effective thermal conductivity of the bed may change through a change in gas conductivity at higher pressures. The effect of these changes on h_w are likely to be small and more so if the pressure increase is only modest. As a result, it is reasonable to conclude that $h_{w\,max}$ or Nu_{max} will have only a weak dependence on ρ_g or P, and consequently, it will not be appropriate to regard Nu_{max} as a function of the Archimedes number only because the latter group contains ρ_g explicitly and will consequently exhibit a pronounced dependence on P. For such small particles, it will be appropriate to express $h_{w\,max}$ by the expression from Zabrodsky [1, 36]:

$$h_{w\,max} = 35.7 k_g^{0.6} \rho_s^{0.2} d_p^{-0.36} \tag{28}$$

No limits of applicability for this correlation in terms of d_p are given.

For large particles, defined as those where the gas convective contribution is significant compared to the particle convective contribution, groups II and III, the $h_{w\,max}$ or Nu_{max} will depend upon gas density and will increase as the pressure increases. Hence, for large particles, Nu_{max} may be regarded as a function of the Archimedes number. Saxena and Ganzha [37] have therefore examined such correlations in conjunction with experimental data [67, 68] to develop specific recommendations. Their findings will now be summarized. The correlations considered by them [37] include the following equations.

Varygin and Martyushin [1, 97]:

$$Nu_{max} = 0.86\,Ar^{0.2} \qquad 30 < 1.35 \times 10^5 \tag{29}$$

Zabrodsky *et al.* [36]:

$$Nu_{max} = 0.88\,Ar^{0.213} \qquad 10^2 < Ar < 2 \times 10^5 \tag{30}$$

Grewal and Saxena [80, 98]:

$$Nu_{max} = 0.9(Ar\,D_{12.7}/D_T)^{0.21}(C_{ps}/C_{pg})^{0.2} \qquad 75 < Ar < 20{,}000 \tag{31}$$

Borodulya *et al.* [28]:

$$Nu_{max} = 0.064\,Ar^{0.4} \qquad 5 \times 10^4 < Ar < 5 \times 10^5 \tag{32}$$

Maskaev and Baskakov [21]:

$$Nu_{max} = 0.21\,Ar^{0.32} \qquad 1.4 \times 10^5 < Ar < 3 \times 10^8 \tag{33}$$

Denloye and Botterill [24]:

$$Nu_{max} = 0.843\,Ar^{0.15} + 0.86\,Ar^{0.39}\,d_p^{0.5} \qquad 10^3 < Ar < 2 \times 10^6 \tag{26}$$

For small particles, workers have correlated their heat transfer data at ambient pressure with a functional form of the type $Nu_{max} = f(Ar)$. As the pressure is increased, these correlations do not succeed in reproducing the experimental Nu_{max} data, and this fact is explicitly demonstrated by the work of Saxena and Ganzha [37] on the basis of the experimental data of Borodulya *et al.* [67] and Botterill and Desai [68] for sand particles of diameter 0.126, 0.26, 0.794, and 1.225 mm and glass beads of diameter 0.45 and 3.1 mm, at pressures in the range from 0.1 to 8.1 MPa. The calculations [37] revealed that the agreement between the experimental and predicted values based on the correlations of Eq. (29), (30), and (31) for particles of $d_p = 0.126$, 0.26, 0.45, and 0.794 mm at ambient pressure (0.1 MPa) is good, always within 15% . It was thus confirmed that for such systems $Nu_{max} = f(Ar)$. However, as the pressure was increased, the agreement between experimental and calculated values became worse and the functional dependence of Nu_{max} on Ar only appeared to be invalid. On the other hand, Eq. (28) was found to reproduce the experimental data for $h_{w\,max}$ much better. Denloye and Botterill [24], while correlating their data for ambient and higher pressures, found it necessary to express Nu_{max} as the sum of two terms, one involving only Ar and the other involving d_p in addition to Ar [Eq. (26)]. Even then it was noticed by Saxena and Ganzha [37] that the agreement between experimental and calculated values was not satisfactory. It should also be remembered that the experimental data of Botterill and Desai [68] was generated by employing a very short heat transfer probe and therefore may be somewhat larger than those obtained under similar conditions but using normal size probes.

From the experimental data, it was also noticed [37] that $h_{w\,max}$ and Nu_{max} for the same particle size bed increased with an increase in pressure. For $Re_{mf} < 40$, that is, for particles of groups I and IIA, the increasing pressure increased to some extent the gas thermal conductivity and also improved the bed hydrodynamics [28, 68, 73], which caused the value of the conductive component of h_w to increase slightly. For systems of group IIB, $40 < Re_{mf} < 200$, the convective component of h_w is important, and it increases with an increase in pressure, and hence the dependence on the heat transfer rate on pressure becomes significant. As a result, the correlations developed for large particles at ambient pressures [21] become appropriate for small particles at high pressures. For large particles ($Re_{mf} > 200$, group III), the correlations of Eqs. (32) and (33) are applicable and Saxena and Ganzha [37] found the experimental data to agree with the calculated values within 25%. Equation (32) overestimated the experimental values whereas they were underestimated by Eq. (33). The reason for this was found in the explicit dependence of h_w on d_p. According to Eqs. (32) and (33), h_w varies with $d_p^{0.2}$ and $d_p^{-0.04}$, respectively, so that Eq. (32) will give an increasing value of h_w with d_p, whereas Eq. (33) will exhibit the opposite trend but much less significantly in magnitude. For systems belonging to groups IIB and III, Saxena and Ganzha [37] have recommended correlations of Eq. (33) and the following alternative form of Eq. (27), also given by Borodulya *et al.* [28], respectively:

$$Nu_{max} = 0.116\,Ar^{0.3}\,Pr^{0.33} + 0.0175\,Ar^{0.46}\,Pr^{0.33} \tag{34}$$

The Maskaev and Baskakov [21] correlation of Eq. (33) predicts that $h_{w\,max}$ is proportional to $d_p^{-0.04}$ or almost independent of d_p. It should, therefore, be most appropriate for particles belonging to subgroup IIB. On the other hand, the Borodulya *et al.* [28] correlation suggests that $h_{w\,max}$ should increase with $d_p^{0.20}$, and therefore it is most appropriate for particles of group III. Also, $h_{w\,max}$ is proportional to $d_p^{-0.40}$, $d_p^{-0.36}$, and $d_p^{-0.37}$ according to Eqs. (29), (30), and (31), respectively, and this implies that $h_{w\,max}$ will decrease rapidly with increasing d_p. The particles of group I and, to a large extent, of subgroup IIA will satisfy this type of dependence of $h_{w\,max}$ on d_p. Saxena and Ganzha [14] have, therefore, proposed that for these three correlations the upper limit of validity be Ar of about 130,000. In this light, the applicability of the Grewal correlation [98] appears to be valid up to much higher values of Ar than that of 20,000 assigned by him.

B. Heat Transfer Coefficient

The heat transfer coefficient, h_w, for an immersed surface in a gas-fluidized bed of small particles, or more specifically systems belonging to groups I and IIA, will depend upon the orientation of the surface with the flow of gas in the fluidized bed. We will first consider the smooth horizontal tubes and then the vertical tubes. Tubes with rough surfaces or finned tubes, as well as bundles of smooth and rough tubes, will also be discussed later in this chapter.

1. *Smooth Horizontal Tubes and Fluidized Beds of Small Particles*

Based on the measurement of the heat transfer rate between a horizontal tube and fluidized beds, several correlations have been proposed for the total average heat transfer coefficient, h_w, for small particles. Some of these are listed here.

Vreedenberg [88]:

$$\mathrm{Nu_T} = 420\left[\left(\frac{GD_T\rho_g}{\rho_g\mu_g}\right)\left(\frac{\mu_g^2}{d_p^3\rho_s^2 g}\right)\right]^{0.3}\mathrm{Pr}^{0.3} \tag{34a}$$

Andeen and Glicksman [99]:

$$\mathrm{Nu_T} = 900(1-\varepsilon)\left[\left(\frac{GD_T\rho_s}{\rho_g\mu_g}\right)\left(\frac{\mu_g^2}{d_p^3\rho_s^2 g}\right)\right]^{0.326}\mathrm{Pr}^{0.3} \tag{35}$$

Grewal and Saxena [79]:

$$\mathrm{Nu_T} = 47(1-\varepsilon)\left[\left(\frac{GD_T\rho_s}{\rho_g\mu_g}\right)\left(\frac{\mu_g^2}{d_p^3\rho_s^2 g}\right)\right]^{0.325}\left[\frac{\rho_s C_{ps} D_T^{3/2} g^{1/2}}{k_g}\right]^{0.23}\mathrm{Pr}^{0.30} \tag{36}$$

where the bulk bed porosity, ε, is given by [79, 104]

$$\varepsilon = \frac{1}{2.1}\left[0.4 + \left\{4\left[\frac{\mu_g G}{d_p^2(\rho_g(\rho_s-\rho_g))\phi_s^2 g}\right]^{0.43}\right\}^{1/3}\right] \tag{37}$$

Goel and Saxena [91]:

$$\mathrm{Nu_T} = 17.9(1-\varepsilon)\left[\left(\frac{GD_T\rho_s}{\rho_g\mu_g}\right)\left(\frac{\mu_g^2}{d_p^3\rho_s^2 g}\right)\right]^{0.325}\left[\frac{\rho_s C_{ps} D_T^{3/2} g^{1/2}}{k_g}\right]^{0.23}$$
$$\times\left[\frac{D_T G^2}{g d_p^2 \rho_g^2}\right]^{0.147}\mathrm{Pr}^{0.30} \tag{38}$$

Petrie *et al.* [89]:

$$\mathrm{Nu_T} = 14(G/G_{mf})^{1/3}\,\mathrm{Pr}^{1/3}(D_T/d_p)^{2/3} \tag{39}$$

Ainshtein [100]:

$$\mathrm{Nu_T} = 5.76(1-\varepsilon)(Gd_p/\mu_g\varepsilon)^{0.34}\,\mathrm{Pr}^{0.33}(H_s/D_b)^{0.16}(D_T/d_p) \tag{40}$$

Gelperin *et al.* [101]:

$$\mathrm{Nu_T} = 4.38\left[\frac{1}{6(1-\varepsilon)}\left(\frac{Gd_p}{\mu_g}\right)\right]^{0.32}\left(\frac{1-\varepsilon}{\varepsilon}\right)\frac{D_T}{d_p} \tag{41}$$

Genetti *et al.* [102]:

$$\mathrm{Nu_T} = \frac{11(1-\varepsilon)^{0.5}}{\left[1 + \dfrac{0.2512}{\left(\dfrac{Gd_p}{\mu_g}\right)^{0.34}\left(\dfrac{d_p}{0.000203}\right)^2}\right]^2}\left(\frac{D_T}{d_p}\right) \tag{42}$$

Ternovskaya and Korenberg [103]:

$$\mathrm{Nu_T} = 2.9\left[\left(\frac{1-\varepsilon}{\varepsilon}\right)\left(\frac{G\,d_p}{\mu_g}\right)\right]^{0.4}\left(\frac{D_T}{d_p}\right)\mathrm{Pr}^{0.33} \tag{43}$$

Martin [105, 106]:

$$\mathrm{Nu} = Z(1-\varepsilon)\{1 - \exp(-N)\} \tag{44}$$

$$Z = \frac{\rho_s C_{ps}}{6k_g}\left\{\frac{g\,d_p^3(\varepsilon - \varepsilon_{mf})}{5(1-\varepsilon_{mf})(1-\varepsilon)}\right\}^{1/2} \tag{45}$$

$$N = \frac{1}{CZ}\left[\frac{1}{\mathrm{Nu_{max}}} + \frac{k_g}{4k_s(1 + \sqrt{(3C/2\pi)(k_g/k_s)Z})}\right]^{-1} \tag{46}$$

$$\mathrm{Nu_{max}} = 4\left[(1 + \mathrm{Kn})\ln\left(1 + \frac{1}{\mathrm{Kn}}\right) - 1\right] \tag{47}$$

and

$$\mathrm{Kn} = \frac{2l_1}{d_p} = \frac{4}{d_p}\left(\frac{2}{\sigma} - 1\right)\frac{k_g\sqrt{2\pi R(T/M)}}{p[2C_{pg} - (R/M)]} \tag{48}$$

Grewal and Saxena [79] have examined all these correlations, except the very last one [105, 106], on the basis of available data in the literature and their own data on small particles ($>504\ \mu$m) of alumina, dolomite, glass beads, silica, silicon carbide, and lead glass at ambient temperatures and pressures. They found that the correlation of Eq. (36) is the only one that can successfully represent all the available data in the literature not employed in its development within an uncertainty of $\pm 25\%$. We, therefore, recommend it for the calculation of h_w for systems belonging to group I and for the calculation of the particle convection heat transfer coefficient for systems of group IIA.

Goel and Saxena [91] observed that for horizontal tubes of diameters equal to or greater than 50.88 mm and for particles of wide size range at low gas fluidizing velocities, the correlation of Eqs. (36) is inadequate to represent the dependence of h_w on D_T. They [91] correlated their data within an uncertainty of $\pm 20\%$ by the relation of Eq. (38). Saxena and co-workers have further investigated this phenomenon by performing systematic experiments with tubes with diameters of 28.6, 50.8, and 76.2 mm immersed in fluidized beds of silica sands with average particle diameters of 497 and 773 μm and glass beads with an average particle diameter of 270 μm as a function of fluidizing air velocity at ambient temperature and pressure. It was noticed by them [107] that h_w first decreases as D_T is increased, then increases as D_T is further increased. This observed minimum in h_w with respect to D_T is explained by the detailed interpretation of the changing hydrodynamic condition of the bed, particularly in relation to solids motion and related bubble dynamics.

Doherty *et al.* [107] explained the decrease in h_w as the tube diameter was increased from 28.6 to 50.8 mm on the basis of an increase in particle residence time with the increase in tube diameter, and as a result the temperature difference between the tube and the particle decreases. This reduction in driving force brings about a reduction in the heat transfer coefficient. They also emphasized that the gas trapped on the upstream side of the tube played an important role in establishing the value of h_w, that its size increased with the tube diameter, and that it had a complicated dependence on gas velocity. The subsequent increase of h_w with D_T was explained on the basis of a greater solids renewal rate at the downstream side of the tube in the defluidized cap and the subsequent transport of these solids by the bubble arising from the gas pocket located in the upstream side of the tube. Further, the increase was more for smaller particles because of the greater ability of the bubbles to sweep them away from the region close to the tube to farther regions of the bed. They have also remarked that the quantitative correlation of h_w with D_T stipulating a minimum is a complicated matter in view of the fact that its origin lies in local solids motion and bubble dynamics around the tube and is therefore dependent on all possible operating and system parameters. Cherrington *et al.* [92] observed the increase in the average value of the heat transfer coefficient to be proportional to $D_T^{0.2}$. Doherty *et al.* [107] found their $h_{w\,max}$ values to follow this dependence on D_T. Information of this nature, along with the knowledge of the detailed behavior of the gas pocket shrouding the upstream side of the tube and defluidized cap of solids on the downstream side of the tube, is very useful from the practical viewpoint of designing heat transfer tubes for a particular application and corresponding operating conditions.

Grewal *et al.* [76] tested the model of Martin [105, 106] on the basis of heat transfer data generated in an atmospheric fluidized-bed combustor with horizontal tubes 25.4 mm in diameter and burning, low-ranking coals in fluidized beds of silica sand (888–1484 μm) and limestone (716–1895 μm) at average bed temperatures ranging from 1047 to 1125 K and a superficial fluidizing velocity of 1.66–2.04 m/sec. They found that the Martin's model [108, 109] underpredicts the data by as much as 36%. The only adjustable parameter, C, in this model was given the value of 3.0. In a later effort [106], this value was revised to 2.6 and the earlier work [108] refined by including the internal transient conduction resistance of the solid particle to the dominant resistance of the gas filled wedge or lens between the wall and the particle. The Grewal *et al.* [76] assessment is based on the earlier formulation of Martin [108, 109], and detailed examination of the recent formulation of Martin's model for heat conduction [105, 106] on the basis of commonly employed heat transfer data [79] has not been yet undertaken. It is partly because this theory expresses h_w in terms of ε and not u and also employs a number of parameters that are not readily available in relation to many experimental investigations.

2. *Smooth Vertical Tubes and Fluidized Beds of Small Particles*

For small particles (groups I and IIA), there have been only limited efforts to investigate the dependence of h_w on D_T for vertical tubes (Verma and Saxena [81], Mathur *et al.* [110], and White *et al.* [111]). The symmetrical flow pattern around a vertical tube precludes the possibility of an angular variation of voidage around the tube circumference and hence suggests h_w to be likely independent of D_T, at least at first glance. Chen and Withers [112] reported results of h_w as a function of mass fluidizing velocity for electrically heated vertical tubes 12.7 and 22 mm in diameter immersed in beds of glass beads (127, 254, and 609 μm) at ambient conditions. Their results indicate a very small decrease in h_w with an increase in D_T for all three particles, but they preferred to represent their results by a single, averaged curve, thereby implying a negligible dependence of h_w on D_T. Antonishin [113] carried out experiments with electrically heated tubes 14, 21, and 28 mm in diameter in a bed of sand (268 μm) and observed a consistent decrease in h_w with an increase in D_T. It should be noted that the magnitude of the deviation of h_w values for the 14- and 21-mm-diameter tubes is of the same order, about 10%, as that observed by Chen and Withers [112] for tubes 12.7 and 22 mm in diameter in beds of glass beads of three different sizes.

Verma and Saxena [81] performed experiments with electrically heated vertical tubes of 12.7, 28.6, and 50.8 mm diameter in beds of glass beads (427 μm) and silica sands (167, 499, and 745 μm) at ambient conditions. Their qualitative inference was that for all the particles, h_w initially increased as the tube diameter increased (12.7–28.6 mm) but decreased with further increase in D_T to 50.8 mm, except for 488-μm silica sand, where it continued to increase. White *et al.* [111] examined the experimental data of Verma and Saxena [81], Baerg *et al.* [114], and Gennetti *et al.* [102] on glass beads of approximately the same average size (427–470 μm) and for tubes of diameters 12.7, 15.9, 28.6, 31.8, and 50.8 mm at ambient conditions and inferred a probable trend that h_w decreases as D_T increases. For data on silica sand beds of 503 μm and tubes of diameter 31.8 and 50.8 mm from Verma and Saxena [81] and Baerg *et al.*

[114], White *et al.* [111] concluded a trend for h_w to decrease with an increase in tube diameter under identical fluidizing conditions. The same trend was confirmed by the analysis of the data from Baerg *et al.* [114] for $D_T = 31.8$ mm and Mickley *et al.* [115, 116] for $D_T = 6.3$ mm by White *et al.* [111].

White *et al.* [111] measured h_w in 15.2-cm internal diameter air-fluidized beds of glass beads (215 and 440 μm), silica sand (470, 552, and 860 μm), and silicon carbide (362 μm) and heat transfer tubes of diameters 25.4 and 60.3 mm as a function of fluidizing velocity at 383 K. The h_w values were found to increase with increasing tube diameter over the entire fluidization velocity range. Mathur *et al.* [110] measured h_w for vertical tubes of diameters 12.7, 28.6, and 50.8 mm immersed in beds of glass beads (275 and 410 μm) and silica sands (500, 665, and 803 μm) at ambient conditions as a function of fluidizing velocity. They found a clear trend in $h_{w\,max}$ values, which consistently increased as the tube diameter decreased. To understand these qualitative trends of the variation of h_w and $h_{w\,max}$ with D_T, it is necessary to consider the bed hydrodynamics in terms of solids movement and associated bubble dynamics in the context of distributor design and bed geometry for the specific conditions. In the following, we first present the general information that can help in establishing the bed hydrodynamics and then consider the specific investigations for which enough details have been reported.

The maldistribution of flow at the distributor plate and the resulting bubble flow pattern and movement of solids in the unbaffled beds have been investigated by many workers. Whitehead *et al.* [117–120], based on their work in square beds of sides 0.61, 1.22, and 2.44 m, concluded from the bubble eruption patterns at the bed surface that bubbles were formed at the four corner regions of the bed. This pattern was more typical for the two larger beds, particularly at high fluidizing velocities. These workers noted the presence of a central downflow track of solids in shallow beds ($H/D_b \simeq 1$). As the bed height was increased, the four bubble tracks moved toward the bed center and a single central bubble track was observed in beds with an H/D_b ratio of about 2. Park *et al.* [121] confirmed the same qualitative bubble flow pattern from their experiments in cylindrical beds of 0.1-m diameter, using an electroresistivity probe. Werther and co-workers [122–124] further confirmed this qualitative picture based on their experiments with beds of diameters 0.1, 0.2, 0.45, and 1 m, using a capacitance probe. They found an annular zone near the bed wall of high bubble activity. It moved toward the bed center and finally merged at the vessel centerline at a bed height of about twice the bed diameter. The solids carried in the wakes of these bubbles descended down the bed in a single stream in the central region, as shown in Fig. 6. In tall beds, $H/H_b > 2$, the converging bubble streams merged into a single stream of rising bubbles at the vessel centerline at a bed height of about two bed diameters, as shown in Fig. 7. The solids in such tall beds descended in

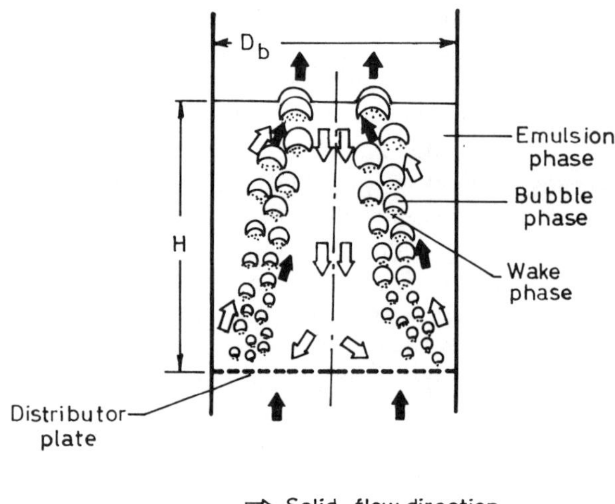

FIG. 6. Bubble and solids flow patterns in shallow beds, $(H/D_b) < 2$ (from Mathur *et al.* [110]).

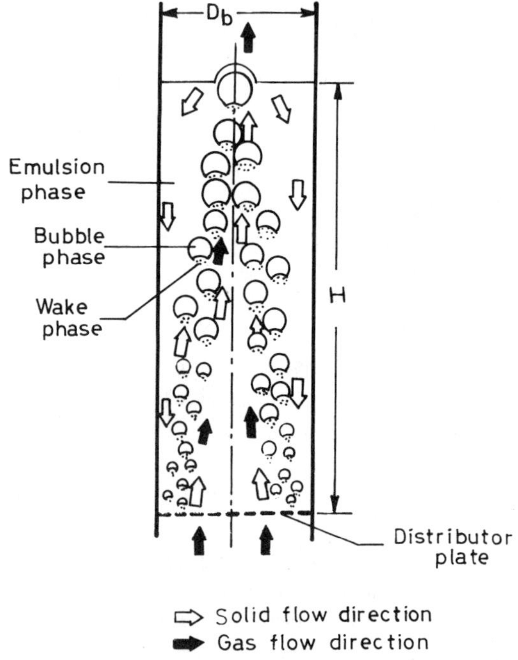

FIG. 7. Bubble and solids flow patterns in tall beds, $(H/D_b) > 2$ (from Mathur *et al.* [110]).

the outer region of the bed near the bed wall. Whitehead *et al.* [117–120] employed multituyere with a typical bubble-cap design distributor plate, whereas the two other groups [121–124] have used distributors typical of a porous plate type. These investigations clearly bring out a common pattern of bubble flow and movement of solids for beds of a wide size range and fitted with different design distributor plates for the majority of the bed height, except for a small region near the distributor plate. It is therefore reasonable to conclude that the bed size and distributor design will not influence the general hydrodynamics of the bed above the grid region as long as the bed dimensions and distributor design conform to the general designs previously mentioned.

The pattern of bubble flow and associated solids movement in fluidized beds is also reported by Chen *et al.* [125] and Lin *et al.* [126]. These investigators measured the solids velocity profile and inferred the bubble flow pattern therefrom. The measurements were taken in two 14-cm-diameter beds of glass beads of average diameters ranging between 420 and 800 μm. The solids velocity profiles were obtained by tracking the instantaneous position of a radioactive tracer particle by 16 strategically located scintillation counters connected to a minicomputer for real-time analysis. The flow pattern in the bed changed with the fluidizing velocity. They found at low velocities an ascending wall descending center (AWDC) vortex, which implied that the bubbles existed predominantly near the bed walls and moved very little toward the vessel centerline during their rise in the bed. As a result, solids were observed to rise near the walls and descend in the central region of the bed. As the fluidizing velocity was increased, a second countervortex, ascending center descending wall (ACDW), was observed in the upper region of the bed near the walls. This implied that the bubbles moved more toward the vessel centerline and solids descended both in the central region of the bed and along the bed walls in the upper region of the bed, and these were carried upward by the rising bubbles, leading to the formation of the ACDW vortex. Further increases in the fluidizing velocity caused the ACDW vortex to increase its size in the bed at the expense of the AWDC vortex. At sufficiently high velocities ($u/u_{mf} > \sim 4.6$), the AWDC vortex disappeared completely.

The influence of internals on the hydrodynamics of bubbling beds is found to be dependent on the shape and size of the immersed surface. We will briefly mention some other works that are relevant to immersed vertical single tubes. Using X-ray cinephotography, Rowe and Everett [127] found that the motion of single bubbles injected into cylindrical beds of 254-mm diameter was not influenced by vertical rods of 3.2- and 6.4-mm diameter, unless they were formed directly beneath the rod. In such a case, the bubble tended to stabilize on the rod and hindered bubble coalescence. Rowe and Masson [128, 129] found that slender vertical probes caused acceleration and elongation of the bubbles. A relatively massive vertical probe, about 40 mm in diameter, only slowed down the bubbles somewhat. Only a multipronged vertical probe promoted splitting of bubbles larger than 30 mm. Volk *et al.* [130] investigated the effect of vertical internals on bubbles, growth, and the mixing of solids indirectly by measuring the conversion of a catalyzed CO reaction. The equivalent diameter of a 610-mm bed was varied by changing the number of vertical tubes immersed in the bed, and then the bed pressure fluctuations and CO conversion levels were observed. The measurements suggested that bubble growth is checked in the baffled beds and better gas–solid contacting occurs.

From this discussion, it is clear that the change in the bubble flow profile with the change in the aspect ratio and the fluidizing velocity has a pronounced effect on the nature of the descending solids stream and hence on the nature of the solids mixing pattern in the bed. It will, therefore, appear that the value of h_w for an immersed surface will characteristically depend upon its location in the bed, its size, and its size relative to the bed diameter. In the case of shallow beds, the heat transfer surface located in the central region of the bed is continuously exposed to a descending stream of solids. The extent of bubbles reaching it would be minimal, unless the tube diameter is large compared to the bed diameter. One would therefore expect the heat transfer coefficient to show a weak dependence on the tube diameter for a wide range of D_T values provided the bed diameter is sufficiently wide in comparison to D_T.

On the other hand, if the bed height is sufficiently greater than $2D_b$ the heat transfer tube located in the central region of the bed will be exposed to a continuously rising stream of bubbles. In case the tube diameter is quite large as compared to the bed diameter, so that the annulus, $(D_b - D_T)/2$, is narrow, the radial intermixing between the solids in the rising bubble, the descending solids streams, and the enhanced turbulence will reduce the solids residence time on the heat transfer surface. A dependence of h_w on D_T would, therefore, be observed, with h_w being larger for tubes having larger values of D_T. This has been observed by Baskakov *et al.* [22], who measured the solids residence time on tubes of 15- and 30-mm diameter in beds of 92- and 98-mm diameter. They found the dependence of the solids residence time on the tube diameter to vary as $D_T^{-0.225}$. White *et al.* [111] have also observed a similar trend in a cylindrical fluidized bed of 0.152-m diameter for tubes of 25.4- and 60.3-mm diameter. They found that h_w values for the wider tube to be greater than those for the narrower tube over the entire range of fluidizing velocities in six different beds with an aspect ratio of 4.5 ± 0.2. The width of the annulus in the two cases is 63.3 and 45.9 mm. An approximate quantitative criterion for the annulus width to be characteristically narrow or wide can be developed by its analogy with heat transfer measurements using tube bundles immersed in a fluidized bed. The analog of the annulus width will be the pitch of the tube bundle, the latter being defined as the distance between the centers of the two adjacent tubes. Borodulya *et al.* [131] found that the characteristics of heat transfer from a single tube in a bundle are not influenced by the presence of other tubes in the bundle as long as the relative pitch is greater than three. The relative pitch is defined as the ratio of pitch to the tube diameter. This will suggest that as long as the annulus is less than twice the tube diameter, good mixing between the wake solids and solids in the descending stream may be expected. Pending the availability of more elaborate experimental data, this qualitative rule may be helpful in design work.

For small particles, the predominant heat transfer between an immersed surface and a fluidized bed takes place via gas film conduction and particle heat absorption at the surface. The latter is controlled by the particle residence time and the particle concentration near the heat transfer surface. A small dependence of h_w on D_T comes through these factors, and these will be described briefly here following the elaboration by White *et al.* [111]. Baskakov *et al.* [22] experimentally measured the packet residence time, τ, and the fraction of time an immersed surface is covered by bubbles, f_o, using low thermal capacity foils mounted on the surface of heat transfer probes. Two vertical tubes with diameters of 15 and 30 mm were employed as the experimental probes in beds of corundum particles and slag beads at temperatures up to 823 K. The beds were 98 and 92 mm in diameter and equipped with porous tile distributors. The temperature fluctuations of the foils were employed to determine the bubble and packet contact times, the bubble and packet frequencies, and the fraction of time the foil is covered by bubbles. They correlated their results by the empirical relation

$$\tau = 0.44\left[\frac{d_p g}{u_{mf}^2(u - A)^2}\right]^{0.14}\left(\frac{d_p}{D_T}\right)^{0.225} \tag{49}$$

Here A is an empirical constant that is found to decrease as D_T increases and as the sphericity of the particles approaches unity. This relation suggests that τ decreases as D_T increases, which implies an increase in h_w with increase in D_T.

Denloye [132] developed the following relationship for ε_s of a quiescent bed based on the correlation of Kimura *et al.* [133] in a packed bed and a material balance of solids in the vicinity of the heat transfer tube:

$$\varepsilon_s = 1 - \frac{(1 - \varepsilon)\{0.7293 + 0.5139(d_p/D_T)\}}{1 + (d_p/D_T)} \tag{50}$$

Since Eq. (50) is developed for quiescent beds, ε_s and ε may be considered as the values of the bed voidage at minimum fluidizing conditions near the tube and in the bulk of the bed, respectively. Equation (50) indicates that at minimum fluidization, the tube diameter has the effect of reducing ε_s as D_T increases. The particle–tube contact geometry suggests that this cause–effect relationship, that is, a decrease in ε_s because of an increase in D_T, will also be valid at velocities above u_{mf}.

The packet model [5] leads to this expression for h_w:

$$h_w = \frac{1 - f_o}{R_t + 0.5R_p} \tag{51}$$

where f_o is the fraction of time the tube is covered by bubbles, and R_t and R_p are the thermal and packet contact resistances, respectively. The packet contact resistance R_p is dependent on both the particle concentration $(1 - \varepsilon)$ and the particle residence time τ and is given by

$$R_p = \left[\frac{\pi\tau}{k_{eff}\rho_s C_{ps}(1 - \varepsilon)}\right] \tag{52}$$

Thus, an increase in $(1 - \varepsilon)$ and a decrease in τ (both experimentally shown to accompany an increase in D_T) lead to a smaller packet contact resistance and hence a higher value of h_w.

Also, f_o has been correlated by Baskakov *et al.* [22] as

$$f_o = 0.33\left[\frac{u_{mf}^2(u - A)^2}{d_p g}\right]^{0.14} \tag{53}$$

The constant A is the same as in Eq. (49). The thermal contact resistance R_t is approximated [133] as

$$R_t = \frac{d_p}{2k_{eff}} \tag{54}$$

White *et al.* [111] employed Eqs. (49)–(54) to compute $h_{w\,max}$ and found that the packet-model does predict an increase in h_w as D_T is increased from 25.4 to 60.3 mm. However, the magnitude was considerably less than that experimentally observed for all the cases. They [111] also pointed out that the correlations of Baskakov *et al.* [22], Eqs. (49)–(54), are based on data collected on a single experimental unit and hence extrapolation to other conditions must be considered with some caution. In the experimental arrangement of White *et al.* [111], there was a larger free bed cross-sectional area than that of Baskakov *et al.* [22], and similarly the maximum fluidizing velocity was much higher in the former investigations than in the latter. From an analysis of this nature, White *et al.* [111] concluded that the available experimental and theoretical findings substantiate their finding, namely, the increase in the value of h_w with an increase in tube diameter.

In conclusion, we must emphasize that the dependence of h_w on the tube diameter is a complicated matter and each specific situation must be thoroughly investigated for proper understanding in terms of system geometry, operating conditions, and gas–solid system properties. These detailed and varied considerations will facilitate development of proper interpretation and hence an adequate explanation of the dependence of h_w on D_T, which could be quite involved.

Experimental investigations of h_w for vertical tubes are relatively more scarce than for horizontal tubes; here, a brief discussion will be presented from the recent work of Mathur *et al.* [110]. They have examined the correlations of Nu_{max} of Zabrodsky *et al.* [36], Eq. (30); Varygin and Martyushin [1, 97], Eq. (29); Borodulya *et al.* [28], Eq. (32); Grewal [98], Eq. (31); and Denloye and Botterill [24], Eq. (26). The experimental data of 13 different workers comprising 86 data points were considered by them [110], and some representative details of these works are listed in Table III. Some comments made and inferences drawn by Mathur *et al.* [110] on the basis of the comparison of these experimental data and predictions of these correlations are reproduced later.

The percentage deviations between experimental and predicted values, based on the correlation of Zabrodsky *et al.* (36), range between the limits of -34.5 and 59.6% with a root-mean-square deviation of 25.1%. Most of the

TABLE III

HEAT TRANSFER INVESTIGATIONS ON A SINGLE SMOOTH VERTICAL TUBE

Material	d_p[a] (μm)	ρ_s (kg/m^3)	C_{ps} (J/kg K)	D_T (mm)	Reference
Sand	268 (1)	2700	800	14, 21, 28	113
Iron powder	195–375 (3)	6700	574	31, 75	114
Sand	128–878 (8)	2300	800	31, 75	114
Glass beads	158–600 (3)	2400	754	31, 75	114
Alumina	136 (1)	2470	766	31, 75	114
Sand	800 (1)	2480	800	13	28
Glass beads	127–610 (3)	2500	754	12, 7, 22	112
Sand	250, 352 (2)	2700	800	20	5
Glass beads	114–270 (3)	2500	754	15, 9	102
Glass beads	104–430 (4)	2464	754	6, 35	116
Glass beads	136–610 (4)	2470	754	12.7	134
Glass beads	106–848 (4)	2700	800	12.7	135
Sand	167–745 (4)	2700	800	12.7, 28.6, 50.8	81
Glass beads	427 (1)	2470	754	12.7, 28.6, 50.8	81
Glass beads	100–770 (9)	2500	754	30, 40	136
Bronze	108, 150 (2)	8620	343	40	136
Aluminium	900 (1)	2730	904	40	136
Lead	470 (1)	11,180	128	40	136
Polystyrene	1050 (1)	1050	1314	40	136
Nickel	256 (1)	8750	528	12.7	137
Sand	500–803 (3)	2623	800	12.7, 28.6, 50.8	110
Glass beads	275, 405 (2)	2364	754	12.7, 28.6, 50.8	110

[a] Numbers in parentheses denote the number of average particle diameters in the size range.

data corresponding to large particles or the Archimedes number are overpredicted by Eq. (30). Thirty-seven percent of the data considered were overpredicted by more than 25%. The data that were underpredicted by more than 25% are those of Genetti *et al.* [102], some of the data points of Mickley *et al.* [116], and one of the three data points of Mathur *et al.* [110] for 275-μm glass beads. A close examination of the work of Gennetti *et al.* [102] suggested that the single thermocouple measuring the bed temperature and positioned upstream of the tube could have registered erroneous values of the bed temperature and thereby yielded higher values of h_w in relation to that given by Eq. (30). The bed charge corresponding to the 275-μm average diameter glass beads in the work of Mathur *et al.* [110] was of a wide size range having a preponderance of fines, and this could have caused the measured h_w values to be higher than for a bed charge of narrow size range having the same mean diameter. In heat transfer work, in a regime controlled by particle convection, the presence of fines contributes significantly to augment h_w, but their influence on d_p computation through the relation

$$d_p = 1 \Big/ \sum_i (x_i/d_{pi}) \tag{55}$$

is relatively insignificant. Equation (30) predicts 90% of the data points within ±35%. The overprediction by Eq. (30) of $h_{w\,max}$ for large particles implies that the dependence of Nu_{max} on d_p is steeper than the implicit variation of h_w as $d_p^{-0.361}$ in Eq. (30).

Varygin and Martyushin [1, 97] proposed the correlation of Eq. (29) on the basis of heat transfer data for preheated silver spheres dropped in beds of sands and glass beads ranging in size from 82 to 1160 μm. Most of these 86 data points are underpredicted by Eq. (29). The deviations range between −46.9% and 12.6% with a root-mean-square deviation of 19.4%.

The correlation of Borodulya *et al.* [28] of Eq. (32) grossly underestimates the experimentally observed values in the Ar range for which it was developed. The smallest deviation is 27%, and in most cases, it is much more pronounced. This is because of the fact that this correlation was developed from data taken in a fluidization regime in which the heat transfer is predominantly by gas convection; that is why the exponent of 0,4 on Ar in Eq. (32) is relatively higher. The data considered by Mathur *et al.* [110] for a current assessment of Eq. (32) refer to a heat transfer regime dominated by particle convection that is encountered at atmospheric pressure operations with relatively smaller particles.

Grewal's correlation of Eq. (31) underpredicts all 86 data points; the deviations range from about −40% to −80%. This clearly demonstrates the inapplicability of this correlation for predicting $h_{w\,max}$ values for vertical tubes. It is considered to be a reliable and accurate correlation for horizontal tubes, and this comparison may be taken to infer that $h_{w\,max}$ values for vertical tubes are larger than for the same tube in the horizontal configuration under otherwise identical conditions.

The correlation of Eq. (26) from Denloye and Botterill [24] was also found to underpredict all the data points; the deviations are about -10% to -55% for $Ar > 1000$, where this relation is strictly valid. However, even for $Ar < 1000$, the deviations are of the same order, and this comparison simply illustrates the incapability of Eq. (26) to predict Nu_{max} for vertical tubes.

Mathur *et al.* [110] proposed the correlation

$$Nu_{max} = 209(Ar\, C_{ps}/C_{pg})^{0.145}(d_p/D_T)^{0.065} \tag{56}$$

and its predictions are compared with the 86 data points in Fig. 8. An uncertainty of $\pm 25\%$ is assigned to this correlation. The two data points showing a larger deviation are from Gennetti *et al.* [102], and as explained earlier, these involve an experimental error that is responsible for these larger deviations. The adequacy of the proposed correlation in reproducing the dependence of Nu_{max} on the tube diameter is displayed in Fig. 9. The data of Verma and Saxena [81], Antonishin [113], Chen and Withers [112], and Mathur *et al.* [110], all reporting Nu_{max} data for operations under identical

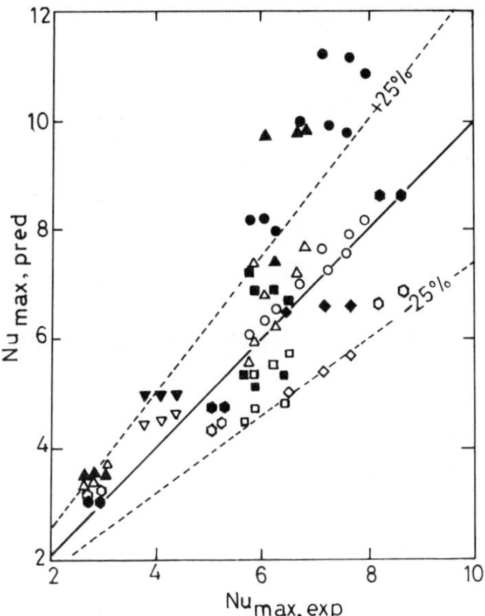

FIG. 8. Comparison of the experimental values with the predictions based on Eq. (30) (filled symbols) and Eq. (56) (open symbols) (from Mathur *et al.* [110]).

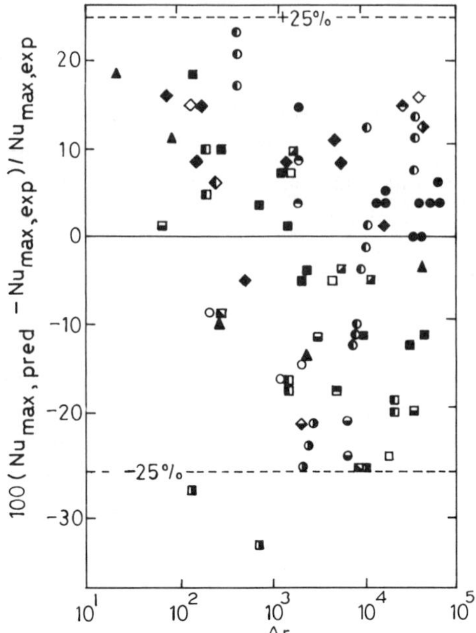

FIG. 9. Percentage deviations of the Nu_{max} values as predicted by Eq. (56) from the corresponding experimental values (from Mathur *et al.* [110]).

conditions but tubes of different diameters, have been considered. The filled symbols are based on Eq. (56). It is to be noticed that the predictions based on the Zabrodsky *et al.* [36] correlation are, in general, in poorer agreement with the experimental data than are the predictions based on the presently the experimental data than are the predictions based on the presently developed correlation of Eq. (56). It is important to note here that the experimental trend of the Nu_{max} dependence on D_T for a particular investigation is much better reproduced by the correlation of Eq. (56). The uncertainty of $\pm 25\%$ assigned to this correlation is not considered large when one carefully examines the scatter in the data of the same investigator or different investigators for comparable systems under identical conditions. The earlier discussion of bed hydrodynamics defines the system and operating parameters in conjunction with bed geometry to which the proposed correlation of Eq. (56) applies.

A number of correlations have been proposed to represent h_w data for vertical tubes by individual investigators, and these will be mentioned here following the review of Verma and Saxena [81]. Mickley and Trilling [115] fluidized glass beads (41–460 μm) in a 3.3-cm-diameter bed by air and examined the effect of the size and concentration of the solids on heat transfer. The air velocity was so high that a considerable portion of the bed was lifted above the vertical tube so that their results really correspond to heat transfer in a lean phase. They proposed the correlation

$$h_w = 0.028(\rho_b^2/d_p^3)^{0.238} \tag{57}$$

Here, ρ_b and d_p are to be expressed in S.I. units. The correlation was found to be inadequate for fine particles, and this was attributed to particle agglomeration. Present data are overestimated by this correlation and the computed values are several orders of magnitude greater than the experimental values. This correlation does not contain any thermal properties of the gas or solid particles, bed voidage, particles residence time, tube geometry, etc., and is therefore, inadequate for calculating h_w.

Baerg *et al.* [114] conducted experiments to demonstrate the influence of particle size, density, and fluidization air velocity on h_w. The heat transfer unit consisted essentially of a heavy walled brass cyclinder with a 14.0-cm internal diameter and 55.9-cm height, and particles of iron powder, round sand, foundry sand, jagged silica, scotchite beads, cracking catalyst, and alumina filled to about 25.4 cm constituted the bed. A 3.2-cm outer diameter copper tube 10.2 cm long was employed as the heat exchanger and air was used as the fluidizing gas. Then h_w was expressed by the dimensionally inconsistent relation

$$h_w = h_{w\max} - 312\exp[-8.85(G - 6 \times 10^{-5}\rho_b)] \tag{58}$$

where

$$h_{w\max} = 120.8\ln(7.05 \times 10^{-6}\rho_b/d_p)$$

(all quantities expressed in S.I. units). Here, d_p and ρ_b are the only properties employed in expressing h_w, and, therefore, one would consider this correlation to be inadequate for h_w prediction. Verma and Saxena [81] found on the basis of their experimental data that the calculated values are in fair agreement with the experimental results but the dependence of h_w on G is poorly reproduced, and this correlation is considered inadequate for h_w estimation.

Miller and Logwinuk [138] measured h_w in 5.1-cm-diameter fluidized beds of alumina, carborundum, and silica gel particles. Particle size and operating temperatures were varied and the following correlation was proposed.

$$h_w = 7.1 \times 10^3 G^{0.2} k_s^{0.045} k_g^{1.5}/d_p^{0.6} C_{pg} \tag{59}$$

This correlation, which is dimensionally inconsistent, predicted h_w values that were about 60% greater than the experimental values [81].

Vreedenberg [139, 140] conducted detailed experimental work with a vertical tube and beds of various mean size sands in the range from 70 to

600 μm for fluidization velocities lying between 0.9 and 1.6 kg/m^2 s and bed temperatures between 293 and 493 K. Here, h_w was found to increase with bed temperature as well as with G. The former result was attributed to better solids mixing resulting from an increase in the kinematic viscosity and to a reduction in the segregation of particles. The recommended correlations are reproduced in the following. For small particles, $(Gd_p\rho_s/\rho_g\mu_g) < 2050$:

$$\frac{h_w(D_b - d_o)}{k_g}\left(\frac{d_o}{D_b}\right)^{1/3}\left(\frac{k_g}{C_{pg}\mu_g}\right)^{1/2} = 0.027 \times 10^{-15}\left(\frac{G(D_b - d_o)\rho_s}{\rho_g\mu_g}\right)^{3.4} \tag{60}$$

for $G(D_b - d_o)\rho_s/\rho_g\mu_g \leq 0.237 \times 10^6$; and

$$\frac{h_w(D_b - d_o)}{k_g}\left(\frac{d_o}{D_b}\right)^{1/3}\left(\frac{k_g}{C_{pg}\mu_g}\right)^{1/2} = 2.2\left(\frac{G(D_b - d_o)\rho_s}{\rho_g\mu_g}\right)^{0.44} \tag{61}$$

for $G(D_b - d_o)\rho_s/\rho_g\mu_g > 0.237 \times 10^6$.

For coarser and heavier particles, $(Gd_p\rho_s/\rho_g\mu_g) > 2.5 \times 10^3$:

$$\frac{h_w(D_b - d_o)}{k_g}\left(\frac{d_o}{D_b}\frac{d_p}{(D_b - d_o)}\frac{k_g}{\mu_g C_{pg}}\right)^{1/3} = 0.105 \times 10^{-3}\left(\frac{G(D_b - d_o)}{\rho_g(d_p^3 g)^{1/2}}\right)^2 \tag{62}$$

for $G(D_b - d_o)/\rho_g d_p^{3/2} g^{1/2} < 1070$; and

$$\frac{h_w(D_b - d_o)}{k_g}\left(\frac{d_o}{D_b}\frac{d_p}{(D_b - d_o)}\frac{k_g}{\mu_g C_{pg}}\right)^{1/3} = 240\left(\frac{G(D_b - d_o}{\rho_g(d_p^3 g)^{1/2}}\right)^{-0.10} \tag{63}$$

for $G(D_b - d_o)/\rho_g d_p^{3/2} g^{1/2} \geq 1070$.

These correlations reproduced the available experimental data within $\pm 100\%$. These do not predict a maximum in the plot of h_w versus G as found in the experiments. This deficiency of the correlations is probably because of the absence of a particle concentration term, $(1 - \varepsilon)$. The correlations substantially underestimate the data of Verma and Saxena [81], the experimental values being about three times the calculated values.

Wender and Cooper [141] correlated the data of Mickley and Trilling [115], Baerg *et al.* [114], Olin and Dean [142], Toomey and Johnstone [135], Mickley and Fairbanks [143], and their own from a commercial Kellog catalyst-regenerator bed and proposed

$$\frac{\mathrm{Nu}_T}{(1 - \varepsilon)}\left(\frac{k_g}{C_{pg}\rho_g}\right)^{0.43} = 3.5 \times 10^{-4} C_R\left(\frac{d_p G}{\mu_g}\right)^{0.23}\left(\frac{C_{ps}}{C_{pg}}\right)^{0.8}\left(\frac{\rho_s}{\rho_g}\right)^{0.66} \tag{64}$$

for $0.01 < (d_p G/\mu_g) < 100$; C_R is a correlation factor for the nonaxial tube location as given by Vreedenberg [139]. An average deviation of $\pm 20\%$ was found for the 323 data points used to develop this correlation. However, the data of Baerg *et al.* [114] deviated significantly from the correlation at low gas velocities. Verma and Saxena [81] found this correlation to reproduce their data best. The average absolute deviations for sand are 3.4% for $d_p = 745\ \mu m$, 2.6% for $d_p = 488\ \mu m$, 45% for $d_p = 167\ \mu m$, and 15% for glass beads ($d_p = 427\ \mu m$). The worst agreement for the smallest size sand of $d_p = 167\ \mu m$ may be attributed to particle agglomeration.

Noe and Knudsen [236] correlated their own data as well as those considered by Wender and Cooper [141] and put $C_R = 2$ in Eq. (64). The accuracy of this correlation was judged to be within $\pm 50\%$ and the experimental data of Verma and Saxena [81] were reproduced to within $\pm 60\%$.

Borodulya *et al.* [28] reported h_w data for an 18-mm vertical heat transfer probe immersed in 45-cm deep and 10.5-cm diameter fluidized beds of quartz sands ($d_p = 126$–$1220\ \mu m$) and glass balls ($d_p = 950$–$3100\ \mu m$) at pressures up to 8.1 MPa. Here, h_w was found to increase with pressure and was attributed to better hydrodynamic conditions at the heat transfer surface for small particles and to the increase on the gas convective component for large particles. They [28] proposed

$$\mathrm{Nu} = 0.37\,\mathrm{Re}^{0.71}\,\mathrm{Pr}^{0.31} \tag{65}$$

for $20 < \mathrm{Re} < 5 \times 10^3$. The experimental data of Verma and Saxena [81] were not satisfactorily reproduced by Eq. (65) over the entire range of the Reynolds number. The agreement between theory and experiment was better for $\mathrm{Re} > 20$, and the data points for $\mathrm{Re} < 20$ were better reproduced by the correlation

$$\mathrm{Nu}_p = 0.96\,\mathrm{Re}^{0.71}\,\mathrm{Pr}^{0.31} \tag{66}$$

In general, the agreement of experimental points with the predictions of Eqs. (65) and (66) was considered to be inadequate by Verma and Saxena [81] because the form of these equations does not include all essential parameters, and, consequently, they even fail to reproduce the qualitative trends of the experimental data.

From the above discussion of the existing correlations of h_w for vertical tubes, it is clear that none of these are capable of representing the experimental data corresponding to a wide range of operating conditions with sufficient accuracy. The correlation of Eq. (64) is about the best. However, it will be seen later that some of the mechanistic models are probably more reliable for prediction and estimation of h_w for design work.

3. *Smooth Surfaces and Fluidized Beds of Large Particles*

In the above, we have discussed the heat transfer correlations for horizontal and vertical smooth tubes immersed in fluidized beds of small particles belonging to groups I and IIA of the powder classification scheme of Saxena and Ganzha [14]. Now, we will discuss similar correlations for large particles belonging to groups IIB and III of this scheme [14] characterized by Ar > 130,000 following the work of Mathur and Saxena [40]. These authors [40] assessed the appropriateness of correlations of Borodulya *et al.* [28], Catipovic *et al.* [30], Glicksman and Decker [27], and Zabrodsky *et al.* [29] and proposed a new one on the basis of available data comprising 366 data points at ambient temperatures but pressures ranging from ambient to 8.1 MPa and Reynolds numbers ranging up to 5000. A summary of these data with some details of the experimental arrangement and operating conditions are listed in Table IV. These data are from 8 different investigations on 5 bed materials comprising 13 different beds and 7 different heat transfer tubes. The data belong to 59 pressure levels and in each case there are several velocities. There are 119 data points that belong to group IIB, while the remaining 217 fall in group III. The ranges of Archimedes, Reynolds, and Prandtl numbers for the 19 data sets are also given in Table IV.

The correlation of Borodulya *et al.* [28], given by Eq. (65) and claimed to have an accuracy of $\pm 25\%$, is tested on the basis of the data in Table IV by Mathur and Saxena [40]. They found that the deviations between experimental values and theoretical predictions range from -65 to 340. However, most of the deviations beyond about $\pm 50\%$ are for the data points belonging to investigations at 0.1 MPa. This is not surprising if one examines the form of Eq. (65) and the fact that this correlation was obtained on the basis of numerical regression analysis of their own experimental data in the range from 1.1 to 8.1 MPa. Their study [40] also revealed that the uncertainty of their correlation judged by them as $\pm 25\%$ is quite an underestimate. They [40] proposed on the basis of their analysis that the majority of the high-pressure data (1.1–8.1 MPa) can be correlated by Eq. (65) within an uncertainty of $\pm 50\%$. The correlation of Eq. (65) is judged to be inadequate for pressures around ambient pressure.

Catipovic *et al.* [30] evaluated the total heat transfer coefficient as the sum of three different contributions, namely, from solids, bubbles, and the emulsion gas. The solids contribution was estimated by treating it as steady-state conduction, the bubble fraction contribution was obtained by regarding it to occur by conduction through a gas film, and the emulsion gas contribution was regarded as constant over the entire range of fluidizing velocities, being equal to that at minimum fluidization. Their final expression is

$$\begin{aligned} \mathrm{Nu} = {} & 6(1 - \beta_1) + (0.0175\,\mathrm{Ar}^{0.46}\,\mathrm{Pr}^{0.33})(1 - \beta_1) \\ & + \beta_1 (d_p/D_T)(0.88\,\mathrm{Re}_{mf}^{0.5} + 0.0042\,\mathrm{Re}_{mf})\,\mathrm{Pr}^{0.33} \end{aligned} \tag{67}$$

where β_1, the time fraction that the tube is in contact with the bubbles, is

TABLE IV

SUMMARY OF EXPERIMENTAL HEAT TRANSFER DATA FOR LARGE PARTICLES

Investigators	Bed dimensions (m)	D_T (m)	Bed material	d_p (mm)	P (MPa)	Ar ($\times 10^{-6}$)	Re	Pr
Borodulya *et al.* [28]	Cylindrical, 0.015 i.d.	0.018	Sand	0.794	1.1–8.1	0.6–3.3	98–610	0.94–2.4
			Glass beads	3.1	0.1–8.1	3.3–204	330–5000	0.71–2.4
Borodulya *et al.* [131]	Cylindrical, 0.105 i.d.	0.013	Sand	0.794	1.1–8.1	0.6–3.4	97–481	0.94–2.4
				1.225	0.1–8.1	0.2–11.6	208–965	0.7–2.4
			Glass beads	1.25	0.1–8.1	0.2–13.1	45–1150	0.71–2.4
				3.1	0.1–8.1	3.3–199	253–4052	0.71–2.4
Borodulya *et al.* [144]	Cylindrical, 0.015 i.d.	0.013	Sand	0.794	1.1–8.1	0.6–3.2	95–533	0.94–2.4
				1.225	0.1–8.1	0.2–11.6	171–1072	0.71–2.4
			Glass beads	1.25	0.1–8.1	0.2–12.7	56–1345	0.71–2.4
				3.1	0.1–8.1	3.9–231	258–4735	0.71–2.4
Botterill and Denloye [25]	Cylindrical, 0.114 i.d.	0.013	Copper shot	0.62	0.4–0.9	0.3–0.8	94–221	0.78–0.89
			Sand	1.02	0.9	1.0	172–218	0.89
Chandran and Chen [145]	Square, 0.305	0.032	Glass beads	1.58	0.1	0.4	2–97	0.71
Canada and McLaughlin [7]	Square, 0.305	0.032	Glass beads	0.65	0.1	0.2	131–538	0.71
				2.6	0.1–1.0	1.9–18.2	234–3719	0.71–0.92
Catipovic *et al.* [30]	Square, 1.0	0.051	Glass beads	1.3	0.1	0.2	78–265	0.71
				4.0	0.1	6.8	525–1408	0.71
Zabrodsky *et al.* [29]	Rectangular, 0.4 × 0.24	0.030	Fireclay	2.0	0.1	0.26	104–228	0.71
			Millet	3.0	0.1	2.01	119–441	0.71

given by

$$1 - \beta_1 = 0.45 + \frac{0.061}{(u - u_{mf}) + 0.125} \tag{68}$$

Mathur and Saxena [40] found that the percentage deviations between the predicted values based on Eqs. (67) and (68) and experimental Nusselt numbers range between about −130 and 275. However, in the majority of the cases, the scatter remained within a somewhat conservative estimate of ±100%. In certain respects, this is quite a remarkable agreement when one recalls that the entire model development is based on data taken at ambient pressures and on the physical picture of a two-phase bubbling fluidized bed. The latter is particularly controversial for beds of particles of groups IIB and III, in which turbulent heat transfer dominates.

Glicksman and Decker [27] derived an expression for the heat transfer coefficient by employing the concept of steady-state conduction and lateral mixing of the gas to evaluate the particle and gas convective contributions, respectively. The gas in excess of that required for minimum fluidization was considered to be transported through the bed as bubbles. They proposed

$$\mathrm{Nu} = (1 - \delta_B)(9.3 + 0.042\,\mathrm{Re}\,\mathrm{Pr}) \tag{69}$$

where

$$\delta_B = (\varepsilon - \varepsilon_{mf})/(1 - \varepsilon_{mf})$$

and ε may be computed from a relation given by Staub and Canada [179], namely,

$$\varepsilon = \frac{u}{1.05u + \{(1 - \varepsilon_{mf})/\varepsilon_{mf}\}u_{mf}}$$

A comparison of the values predicted by Eq. (69) and the corresponding experimental values revealed that the majority of the experimental data are overpredicted. The percentage deviations range between −30 and 105. On the basis of this comparison, Mathur and Saxena [40] assigned the most probable percentage uncertainty of Eq. (69) as −20 and 80. They [40] emphasize the most serious drawback of this correlation appears to be its tendency to overpredict the experimental data. We feel this has crept into the model development through the assumption that the gas convective contribution was directly proportional to the Reynolds number and hence to the gas density, ρ_g. This approximation is increasingly at fault as pressure increases. The experimental investigations of Xavier *et al.* [146] have shown that $h_{w\,conv}$ is proportional to $\rho_g^{1/2}$. The calculations of Mathur and Saxena [40] also revealed that the first term of Eq. (69) poorly simulated the particle convection contribution. As a result, the experimental data were overpredicted over the entire range of the Reynolds number.

Zabrodsky *et al.* [29] estimated the particle convective component by approximating the analytical equation of steady-state conduction from Zabrodsky [1]. The gas convective component was assumed to be proportional to $u^{0.2} C_{pg}\rho_g d_p$ and they recommended the following semitheoretical equation based on some of their experimental data at ambient conditions:

$$h_w = 7.2(k_g/d_p)(1 - \varepsilon)^{2/3} + 26.6u^{0.2}\rho_g C_{pg} d_p \tag{70}$$

where

$$\varepsilon = \varepsilon_{mf} + \frac{\Delta P_1}{g\rho_s H_{mf}}$$

Here, ΔP_1 is the experimental pressure drop measured across the top part of the fluidized bed above the H_{mf} level at different u values.

Mathur and Saxena [40] found that most of the data belonging to Table IV were overpredicted and the magnitude of overprediction increased with Reynolds number. At higher values of Re, the disagreement increased to as much as 670%. A careful examination revealed that the reproduction of heat transfer data at ambient pressure was much better, the percentage deviation always being between −50 and 70%. The increasing disagreement as the pressure was increased was explained along the same line as in the case of Glicksman and Decker [27]. Here, the $h_{w\,conv}$ term also has the same deficiency. The exponent of ρ_g was assumed to be one in the model development, and the numerical coefficient, 26.6, was adjusted from heat transfer data at ambient pressure. This correlation was, therefore, inferred to be satisfactory for heat transfer processes taking place at ambient pressures, and its uncertainty for such a condition is as previously stated.

Recognizing that all four correlations [28, 30, 27, and 29] are incapable of representing the available experimental data of large particles at ambient and higher pressures (Ar > 130,000), Mathur and Saxena [40] developed and proposed a correlation consistent with the present theoretical and experimental understanding of the heat transfer mechanism and capable of representing the experimental data. For large particles both at ambient and higher pressures, the heat transfer coefficient, h_w, can be expressed as

$$h_w = h_{w\,cond} + h_{w\,conv} \tag{71}$$

or, in dimensionless form, as

$$\mathrm{Nu} = \mathrm{Nu}_{cond} + \mathrm{Nu}_{conv} \tag{72}$$

The particle convective heat transfer degenerates to steady-state heat conduction for large particles because of the relatively small change in their temperatures and consequently Nu_{cond} may be approximated by a constant. However, as only a fraction of the heat transfer surface is bathed by particles, Nu_{cond} will be a function of the particle concentration, $(1 - \varepsilon)$. Following Zabrodsky *et al.* [29] and Ganzha *et al.* [147], Mathur and Saxena [40] expressed Nu_{cond} by the form

$$Nu_{cond} = C'_1(1 - \varepsilon)^{2/3} \tag{73}$$

Here C'_1 is a numerical constant, and the exponent $(\frac{2}{3})$ of $(1 - \varepsilon)$ accounts for the fact that the heat transfer is dependent on the heat transfer surface area whereas the particle concentration is a volumetric concept [1].

In evaluating the gas convective heat transfer component of the total heat transfer coefficient, they employed experimental findings, namely, $h_{w\,cond}$ is proportional to $\rho_g^{1/2}$ [146] and to $u^{0.2}$ [29]. Further, as discussed, Ar has proven very useful in characterizing the hydrodynamic and heat transfer behavior of a two-phase, gas–solid system. In view of these facts, they [40] wrote

$$Nu_{conv} = C'_2\, Ar^{0.3}\, Re^{0.2}\, Pr^{1/3} \tag{74}$$

where C'_2 is a numerical constant. Mathur and Saxena [40] evaluated the constants C'_1 and C'_2 of Eqs. (73) and (74) and proposed the following correlation on the basis of Eq. (72):

$$Nu = 5.95(1 - \varepsilon)^{2/3} + 0.055\, Ar^{0.3}\, Re^{0.2}\, Pr^{1/3} \tag{75}$$

The adequacy of the correlation of Eq. (75) was demonstrated by Mathur and Saxena [40] on the basis of a comparison of its predictions with the experimental data of Table IV. This is displayed in Figs. 10 and 11, where the data corresponding to groups IIB and III are considered. In all cases, the deviations are invariably within $\pm 35\%$, which was judged to be the probable uncertainty of the proposed correlation, Eq. (75). The scatter of the data in Fig. 10 does present an apparent concern at first sight inasmuch as a disproportionate amount of the experimental data are unpredicted. It is interesting to note that the bulk of these data are from Borodulya *et al.* [131, 144]; and almost all of their data exhibit this trend (137 data points out of the total of 143 data points). One set of these data is shown by filled circles and refers to vertical tubes, and the other set is represented by unfilled circles and refers to horizontal tubes. On the average, the former shows about twice as large a negative deviation as the latter. Whether or not this is because of the tube orientation is not clear at the present time. The existing understanding of large particle systems, where the predominant mode of heat transfer is by turbulent gas flow, suggests that tube orientation is likely to have a negligible effect. The stated accuracy of these data is judged to be $\pm 10\%$, and, within these limitations, it will be premature to attribute this systematic deviation to

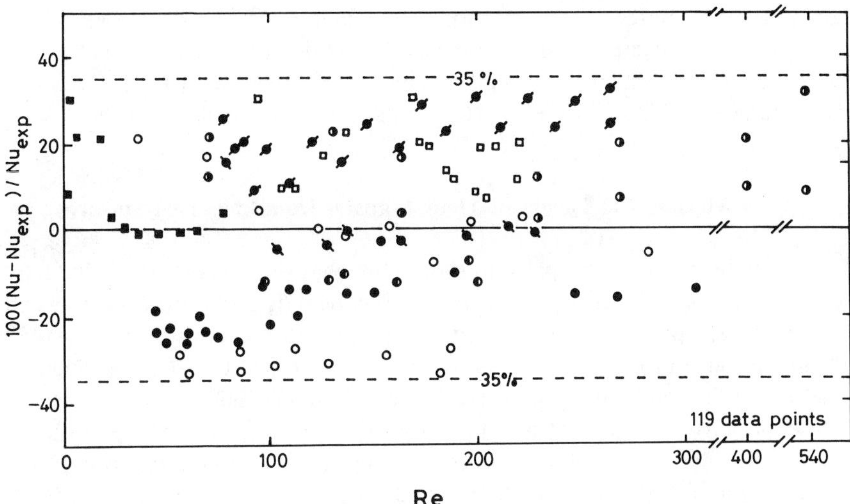

FIG. 10. Deviation of the Nusselt number predicted by Eq. (75) (Nu) from the corresponding experimental values (Nu_{exp}) for beds of group IIB particles at different values of Re (from Mathur and Saxena [40]).

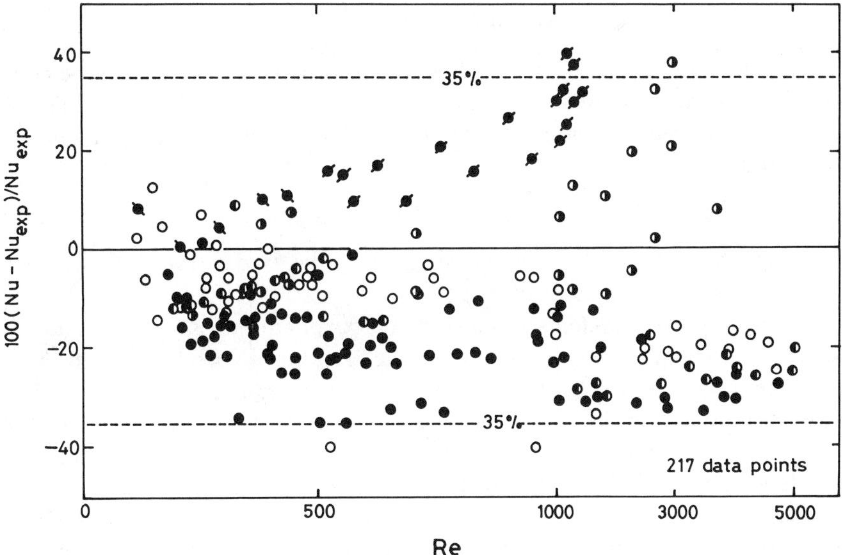

FIG. 11. Deviation of the Nusselt number predicted by Eq. (75) (Nu) from the corresponding experimental values (Nu_{exp}) for beds of group III particles at different values of Re (from Mathur and Saxena [40]).

any particular source. It was, therefore, concluded by Mathur and Saxena [40] that until more accurate data become available, the tube orientation effect for large particle systems may be neglected and the correlation of Eq. (75) may be considered reliable within $\pm 35\%$ to predict the heat transfer coefficient.

V. A Mechanistic Theory of Heat Transfer from Immersed Surfaces

It will be relevant to present now a mechanistic theory developed by Ganzha *et al.* [147, 149] to represent the heat transfer process between fluidized beds of large particles and immersed surfaces. These authors [147] assumed that in the absence of radiation the total heat transfer coefficient is the sum of the conductive and convective components. Since for large particles the conductive component is small as compared to convective component, these authors felt it sufficient to evolve a model that is simple and involves parameters that can be easily determined. They therefore assumed that there exists an orthorhombic arrangement of particles around the heat transfer surface. The formulation was further simplified by replacing the particles with equivalent cylinders with volumes the same as those of the particles and of similar unit diameter to height ratio. The diameter of these equivalent cylinders, d_c, is related to the particle diameter, d_p, by

$$d_c = (2/3)^{1/3} d_p \tag{76}$$

The orthorhombic arrangement of particles around the surface can be considered as a repetition of a unit orthorhombic cell formed by four adjacent particles with center–center interparticle separation of L. Here, L and d_p are interrelated such that

$$L = d_p[(1 - \varepsilon_{mf})/(1 - \varepsilon)]^{1/3} \tag{77}$$

For an orthorhombic arrangement of particles, the fixed bed voidage, which is the same as ε_{mf}, has a value of 0.395.

For large particle systems, it is a good approximation to assume that all the resistance to heat transfer is confined to the first row of particles near the heat transfer surface only. The heat is transferred by conduction through the gas lens between the surface and the particle. To be consistent with the equivalent volume cylinder assumption concerning the particles, it was assumed that the gas lens had a diameter equal to that of the equivalent cylinder, d_c. It was further assumed that the gas lens had a uniform thickness of δ. Thus, the heat transfer problem to be considered is that of conduction through a composite layer of gas of thickness δ and solid of thickness d_c. It was also assumed that there are no temperature-jump effects at the gas–solid interface (Saxena and Joshi [150]). The heat conduction equations for such a case, as given by

Luikov [151], were solved [147] to obtain the equivalent thickness of the uniform gas lens, δ, as

$$\delta = 0.13\, d_c \tag{78}$$

The average heat flux between the heat transfer surface and the first row of particles through the gas film and the average temperature of the composite layer of gas and particles were then evaluated, and finally $h_{\mathrm{w\,cond}}$ was determined within an estimated error of about 1.5% as follows:

$$h_{\mathrm{w\,cond}} = 1.06\frac{k_g}{\delta}\left\{1 + \frac{K(1 + K_1 K_\delta)}{(K_\varepsilon + m\sin^2\mu_1)\mu_1^2 Fo}[1 - \exp(-\mu_1^2 Fo)]\right\} \tag{79}$$

where μ_1 is the characteristic root of the equation

$$\tan\mu + K_\varepsilon \tan(K_2^{1/2} K_\delta \mu) = 0 \tag{80}$$

Here, Fo is the Fourier number for gas film and is defined as

$$Fo = \alpha_g \tau'/\delta^2 \tag{81}$$

where α_g is the thermal diffusivity of gas and is equal to $k_g/C_{pg}\rho_g$. Equation (79) can be approximated within an estimated error of less than 20% by

$$h_{\mathrm{w\,cond}} = 1.06(k_g/\delta) \tag{82}$$

and alternatively by accounting for the bed voidage as

$$h_{\mathrm{w\,cond}} = 1.02(k_g/\delta)(1 - \varepsilon)^{2/3} \tag{83}$$

or

$$h_{\mathrm{w\,cond}} = 8.95(k_g/d_p)(1 - \varepsilon)^{2/3} \tag{84}$$

A straightforward rearrangement yields

$$\mathrm{Nu}_{\mathrm{cond}} = 8.95(1 - \varepsilon)^{2/3} \tag{85}$$

The gas film thickness is larger for the case when the heat transfer surface has a curvature in comparison to a flat surface with no curvature. For a curved surface, to a good degree of approximation, Ganzha *et al.* [147] have shown that

$$\mathrm{Nu}_{\mathrm{cond}} = \frac{1.02(1 - \varepsilon)^{2/3}}{[0.114 + (h_c/d_p)]} \tag{86}$$

For beds of large particles, the Reynolds number is invariably larger than 100 and the flow around the particles is turbulent [46]. The intensity of turbulence is known to be dependent on particle size and bed voidage [152]. It would therefore follow that the gas flow around a surface immersed in a fluidized bed of large particles is turbulent. The turbulent boundary layer formed on the immersed surface is continuously disturbed by the front half of the bed particles and formed again in their wake [153, 154]. The heat transfer surface may thus be regarded as covered with a continuous arrangement of unit orthorhombic cells, which in time keep reforming as new particles arrive at the surface. The heat transfer from the surface corresponding to each of these unit cells through the turbulent boundary layer can be considered similar to that of a flat plate immersed in a turbulent gas stream. Other evidence supporting this view lies in the experimental measurements of Borodulya *et al.* [28] and Rabinovich and Sachenov [155] on heat transfer studies from immersed surfaces at high pressures in fluidized beds of large particles.

The heat transfer from a plate placed in a turbulent fluid flow is given by

$$(\mathrm{Nu}_l)_{\mathrm{conv}} = C' \,\mathrm{Re}_l^{0.8}\, \mathrm{Pr}^{0.43} \tag{87}$$

where Nu_l and Re_l are based on the characteristic length parameter [156, 157]. Ganzha *et al.* [147] have evaluated the characteristic length parameter for fluidized bed heat transfer and have finally computed the convective Nusselt number as

$$\mathrm{Nu}_{\mathrm{conv}} = C_o \,\mathrm{Re}^{0.8}\, \mathrm{Pr}^{0.43}\left[\frac{(1-\varepsilon)^{0.133}}{\varepsilon_s^{0.8}}\right] \tag{88}$$

The constant C_o was obtained by simultaneous measurements of h_w and ε_s. These authors performed such experiments with a single, 13-mm-diameter vertical tube and its staggered bundles in a restricted bed at high gas velocities and obtained a mean value of 0.12 for C_o.

As the fluidization commences, the bed voidage near the surface changes more rapidly than in the bulk of the bed in the beginning. This is because of the larger gas flow near the surface owing to the larger bed voidage than in the bulk of the bed. At higher gas flows, this rate of voidage change becomes slower because of the frictional resistance offered to the bed expansion by the immersed surface. These authors [147] proposed the variation of ε_s with G (or Re) as

$$\varepsilon_s = \varepsilon_{s,\mathrm{mf}} + 1.65A(1-\varepsilon_{\mathrm{mf}})[1-\exp(-a/A_1^2)] \tag{89}$$

Here,

$$a = 0.367 \ln[(\varepsilon_{s,\mathrm{mf}} - \varepsilon_{\mathrm{mf}})/(1-\varepsilon_{\mathrm{mf}})] \tag{90}$$

and

$$A_1 = (\mathrm{Re} - \mathrm{Re}_{\mathrm{mf}})/\mathrm{Ar}^{0.5} \tag{91}$$

Equation (50) was used to compute $\varepsilon_{s,\mathrm{mf}}$, and Eq. (89) was used to calculate ε_s for use in Eq. (88).

Combining Eqs. (86) and (88), we get the final relation for computing the

Nusselt number, namely,

$$\mathrm{Nu} = \frac{1.02(1-\varepsilon_s)^{2/3}}{0.114 + (h_c/d_p)} + \frac{0.12\,\mathrm{Re}^{0.8}\,\mathrm{Pr}^{0.43}(1-\varepsilon_s)^{0.133}}{\varepsilon_s^{0.8}} \tag{92}$$

In the above calculation of Nu, the constant C_o is obtained on the basis of experimental data for heat transfer tubes with diameters of 13 mm. A small correction must be applied to the Nusselt number obtained for tubes of different diameters according to the relation

$$(\mathrm{Nu})_{D_T} = (\mathrm{Nu})_{13\,\mathrm{mm}} \frac{[1 - \exp\{-0.1\,\mathrm{Ar}^{0.1}(13/d_p\,(\mathrm{mm}))\}]}{[1 - \exp\{-0.1\,\mathrm{Ar}^{0.1}(D_T/d_p)\}]} \tag{93}$$

This semiempirical correction is obtained on the basis of the observed variation of the maximum value of the Nusselt number with the tube diameter [158].

Ganzha *et al.* [147] tested this theory on the basis of Eq. (92), or the corresponding alternative equation for the flat surface, as follows:

$$\mathrm{Nu} = 8.95(1-\varepsilon_s)^{2/3} + 0.12\,\mathrm{Re}^{0.8}\,\mathrm{Pr}^{0.43}\left[\frac{(1-\varepsilon_s)^{0.133}}{\varepsilon_s^{0.8}}\right] \tag{94}$$

in conjunction with the experimental data of Borodulya *et al.* [8], Catipovic *et al.* [30], Chandran *et al.* [183], Zabrodsky *et al.* [29], and Golan *et al.* [184]. The calculated values, taking into account the tube curvature, were found to be in excellent agreement with the measured values and the correction for tube curvature, usually about 2–4%, could be as large as 10% [8]. The curvature correction is large whenever the tube diameter is small and particle diameter is large, that is, d_p/D_T is large. The curvature correction decreases the value of the heat transfer coefficient. This turbulent boundary-layer theory of Ganzha *et al.* [147] is further assessed on the basis of the heat transfer coefficient data of Borodulya *et al.* [131, 144] for horizontal and vertical tubes in fluidized beds of glass beads (d_p = 1.25 and 3.1 mm) and sands (0.794 and 1.225 mm) as a function of fluidizing velocity at pressures of 1.1, 2.6, 4.1, and 8.1 MPa and ambient temperature. The majority of the data points (more than 90%) could be reproduced within 15–20%. The reproducibility of the data points was judged as about ±4%, and their accuracy as ±10%. A better prediction on the basis of this theory [147] is possible only if accurate values of voidage at the heat transfer surface can be established.

VI. Heat Transfer from Immersed Rough and Finned Surfaces

Experimental work on the heat transfer coefficient of a surface with artificial roughness in a gas-fluidized bed is somewhat limited and has been reviewed by Botterill [2], Saxena *et al.* [3], and Grewal [77]. Less extensive reviews have been given by Grewal and Saxena [159], Krause and Peters [160], and Grewal *et al.* [161]. Here, only a brief reference and summary of certain salient features will be presented emphasizing both the experimental data and prediction procedures. Grewal and Saxena [159] have distinguished between the rough and finned surfaces on the basis of the mechanism of heat transfer, which is sensitively dependent on the (P_f/d_p) ratio. For $(P_f/d_p) < 1$, the heat transfer coefficient for a rough surface is always less than the smooth surface of the same size because of the increase in the heat conduction path. Such surfaces, where this mechanism is solely responsible for reducing the heat transfer rate, were designated as rough surfaces. For $(P_f/d_p) > 1$, the heat transfer coefficient value is controlled by many factors, namely, (1) the effective area of the heat transfer surface, (2) the effective area of the particle that exchanges heat with the surface, (3) the particle size range, and (4) the residence time of the particles. The relative roles of these factors may be interpreted in relation to fin geometry. For example, for V-thread tubes, the contribution of the effective area of the particle that exchanges heat with the surface predominates when $1 < (P_f/d_p) < 2$ and the net heat transfer increases. For $(P_f/d_p) > 2$, the bulk of the heat transfer coefficient is augmented by (1). The size range effect is involved and can be very involved if the proportion of fins is appreciable and their sizes are smaller than P_f. In general, whenever the heat transfer coefficient is augmented by the effective area of the heat transfer surface it seems appropriate to call it a finned surface. These remarks are only broadly helpful in classifying the rough and finned surfaces.

Many types of rough surfaces have been employed. These are transverse (serrated, continuous, or helical) and longitudinal (continuous) fins of different profiles such as rectangular, parabolic, triangular, and trapizoidal. For proper understanding of heat transfer process, it is essential to know the orientation, height, thickness, and separation of fins. The orientation of fins with respect to gas flow is also important. Because of the large number of these variable factors, intercomparison of data of different workers has been difficult, as has been the development of a correlation involving all the parameters.

Neukirchen and Blenke [162] and Gelperin *et al.* [163] have reported measurements of heat transfer rates between a rough horizontal tube and a fluidized bed. Both groups have employed three types of rough tubes, namely, longitudinal, tranverse, and a network of crosswise grooves. In the Neukirchen and Blenke [162] experiments, the groove depth was varied from 0.6 to 1.9 mm and the bed material was glass beads of 700-μm average diameter. Gelperin *et al.* [163], on the other hand, varied the groove depth from 0.5 to 0.7 mm and used quartz sand of 350-μm average diameter. The two efforts led to similar results when the groove depth was close to the particle diameter, namely, the heat transfer rate for rough tubes compared to smooth tubes of the same outside diameter either did not change at all or decreased. The decrease was particularly marked (30–40%) for tubes with crosswise grooves.

D'Albon *et al.* (164) and Vijayaraghavan and Sastri [165] have investigated the heat transfer from rough vertical tubes immersed in a fluidized bed. The former workers [164] have examined the effect of metric threads on heat transfer between a vertical cylindrical tube and a fluidized bed of sand particles with average diameters of 50 and 100 μm. For the latter case, it was found that the maximum heat transfer coefficient for a tube with threads, h_{wfb}, of height 3.9 mm was about 25% greater than a smooth tube having the same outside diameter. However, for a vertical tube with threads of height 1.7 mm, the maximum heat transfer coefficient was smaller by about the same amount as compared to a smooth tube. Vijayaraghavan and Sastri [165] studied the effect of surface roughness on heat transfer between an electrically heated vertical tube, 35 mm in diameter and 200 mm in length, and a fluidized bed of glass spheres (-250 to $+160$ μm and -630 to $+500$ μm). The tubes were roughened with transverse and longitudinal 60° V-grooves, and fine and coarse cross-knurling. The pitch for V-grooves and coarse knurling was 0.8 mm and only 0.1 mm for fine knurling. The longitudinal V-grooves produced the largest increase in the heat transfer rate for the smaller particles in comparison to the smooth tube. For coarse knurling, the increase in heat transfer rate was relatively smaller. The heat transfer rate was smaller for the transverse grooved tube as compared to the smooth tube, and it was smallest for tubes with fine cross knurling. Similar qualitative results were obtained for larger particles.

The relative efficiency of heat transfer for a finned tube as compared to a smooth tube occupying the same volume in the reactor under identical conditions of operation is defined by a factor, β, known as the heat capacity function. At a given fluidizing velocity, this is defined as the ratio of h_{wfb} to h_w. Grewal and Saxena [159] employed in the interpretation of their results the maximum values of heat transfer coefficients because in practice it would be advantageous to operate the system under conditions where the heat transfer rates are maximum. This value of β will be represented by β_{max}. Thus, β or β_{max} is directly a measure of the relative efficiency of heat transfer for a fin tube over a smooth tube occupying the same bed volume. Another way of comparing the performance of rough and finned tubes is to consider the quantity, Φ, which is defined as the ratio of the effective heat transfer coefficient for rough or finned tubes based on the total surface area of the tube, h_{wft}, to the heat transfer coefficient for a smooth tube, h_w, both referring to identical fluidized bed conditions.

A number of investigations have been reported in the literature on the performance of transverse and longitudinal finned horizontal tubes, transverse and longitudinal finned vertical tubes, spiral finned tubes, and transverse and longitudinal V-groove vertical plates. All these efforts are summarized in Table V following Grewal *et al.* [161]. The fin effectiveness factor, Φ, increases with an increase in the fluidizing gas velocity, a decrease in the fin height, an

TABLE V

SUMMARY OF HEAT TRANSFER INVESTIGATIONS OF FINNED TUBES IMMERSED IN GAS-FLUIDIZED BEDS

Reference	Tube base diameter (mm)	Tube orientation	Fin type	Fin length (mm)	Number of fins/m or Fin spacing (mm)	Fin thickness (mm)	Fin material	Air fluidizing velocity (m/s)	Bed material diameter (μm)	ϕ	β	Remarks
Petrie *et al.* [89]	19.0	Horizontal	Continuous transverse	10.0	197, 433	0.41	Aluminium	0.186–0.384	Silica sand 497, 630	0.1–0.42		Tube bundle, $P = 57$ mm
Bartel *et al.* [172]	16.0	Horizontal	Serrated transverse	10.0–25.0	315	0.6–0.8	Carbon steel	0.089–1.09	Glass beads, 114–470	0.2–0.5		Single tube
Genetti *et al.* [102]	16.0	Horizontal to vertical	Serrated transverse	19.1, 9.5	315, 236	0.64–0.76	Carbon steel and copper	0.081–1.090	Glass beads, 114–470			Single tube
Bartel and Genetti [173]	16.0	Horizontal	Serrated transverse	3.0–22.0	315	0.50	Carbon steel	0.089–1.090	Glass beads, 114–470	0.2–0.8	0.7–1.6 ($G/G_{mf} = 4$)	Tube bundle, $P = 27–122$ mm
Priebe and Genetti [174]	13.0	Horizontal	Serrated transverse, spined tube	3.0–22.0	315	0.41–0.64	Carbon steel and copper	0.163–1.120	Glass beads, 203–470		0.8–1.80 0.4–1.60 ($G/G_{mf} = 4$)	Tube bundle, $P = 27–122$ mm
Genetti and Kratovil [171]	11.5–19.0	Horizontal	Helical continuous transverse	6.0–10.5	197–709	0.4–0.64	Copper	0.136–0.949	Glass beads, 173–551		1.3–2.9 ($G/G_{mf} = 4$)	Single tube
Zabrodsky *et al.* [170]	30.0	Horizontal	Triangular	5.0–22.0	29–100		Copper	0.58–1.87	Silica sand, 260–620; Millet, 2000; Crushed fireclay, 976, 2000, 3000	0.42–1.02		Single tube; tube bundle, $P = 60, 75, 100$ mm
Natusch and Blenke [175]	25.0	Horizontal	Continuous longitudinal	10.0–40.0	2–16	2.0	Copper		Glass beads, 150–670			Single tube
Natusch and Blenke [176]	25.0	Horizontal	Continuous transverse	7.5–37.6	2–10	0.5–2.0	Copper	$G > 6G_{mf}$	Glass beads, 400–800			Single tube

Gelperin *et al.* [177]	24.0	Horizontal	Continuous transverse	4.0	0–12	4.0	Copper, aluminum		Quartz sand, 350			Single tube
Gelperin *et al.* [178]	24.0	Horizontal	Continuous transverse	Up to 10.0	6	4.0	Copper, duralumin		Quartz sand, 360			Single tube
Grewal *et al.* [161]	19.8–27.2	Horizontal	V-groove	0.68	0.79–5.08		Bronze	0.1–0.67	Silica sand 167, 504; Alumina 259	1.1–1.68	0.6–1.0	Single tube; tube bundle, $P = 1.75 - 3.50$ mm
Grewal and Saxena [159]	11.3–12.3	Horizontal	V-groove	0.21–0.68	0.24–0.79		Copper	$G/G_{mf} = 1.2$–8.5	Glass beads, 265, 367, 427		1–1.4	Single tube
Krause and Peters [160]	19.2	Horizontal	Serrated transverse	4.76–17.46	300		Carbon steel	0.110–0.530	Glass beads, 210, 430	0.23–0.65		Single tube
Staub and Canada [179]	25	Horizontal	Helical continuous transverse	12	96, 280	0.1–0.5		0.9–3.0	Glass beads, 630	0.54–0.57		Tube bundle
Chen and Withers [180]	18.8–22.6	Vertical	Helical continuous transverse	1.5–3.4	197–752		Copper	0.14–0.880	Glass beads, 130–160	0.5–1.2	1.1–2.90 ($G/G_{mf} = 3$)	Single tube
Korolev and Syromyatnikov [168]		Vertical	Longitudinal and transverse	0.4000	0.5		Copper	0.080–0.620	Corundum, 120; Chamotte, 320; polystyrene, 720;			Vertical plate
Vijayaraghavan and Sastri [165]	33.4–34.5	Vertical	Longitudinal and transverse	0.1, 0.8	1250		Copper	0.060–1.080	Glass beads, 160–300	0.56–0.83		Single tube
Genetti and Everly [181]	17.9–28.6	Vertical	Continuous groove	3.81–5.33			Copper	0.140–0.750	Glass beads, 193–417			Spiral tube
Genetti *et al.* [182]	11.5–19.1	Vertical	Helical segmental	0.32–0.87	1.0–4.7	0.4–0.79		0.170–0.780	Glass beads, 190–410			Tube bundle

increase in the fin spacing, and an increase in the thermal conductivity of fin material. A number of correlations have been recommended for computation of the heat transfer coefficient for finned tubes, and these have been summarized by Grewal [77]. Here, we will present a discussion of the V-thread tubes following our work [159, 161, 166, 167] and that of others to bring to light the mechanism of heat transfer that influences h_w in general for finned or rough tubes.

Korolev and Syromyatnikov [168] investigated the effect of surface roughness on heat transfer from an electrically heated vertical plate and fluidized beds of corundum ($d_p = 120$ μm), Chamotte ($d_p = 320$ μm), and polystyrene ($d_p = 720$ μm). The transverse and longitudinal V-grooves with 0.4-mm height and 0.5-mm pitch were machined on the surface. They also examined two other plates from which the surface roughness was obtained by treating the surfaces with a jet of corundum particles. The average height of the roughness on the two plates was 2 and 10 μm and the average pitch between projections was 0.06 and 0.10 mm, respectively. For fine particles (120 and 230 μm), when $(P_f/d_p) < 1$, an increase in surface roughness height resulted in the reduction of the heat transfer rate as compared to that for a smooth vertical plate. When the height and pitch of V-grooves were greater than the particle size, the heat transfer coefficients for rough plates were larger as compared to those for smooth plates. In the fluidized bed of larger particles ($d_p = 720$ μm), the surface roughness of these plates did not effect the heat transfer coefficient. Further, the heat transfer coefficient for the longitudinal V-groove plate was larger than that for the transverse V-groove plate.

Grewal and Saxena [159] measured the heat transfer coefficient of a horizontal copper tube with a 12.7-mm outside diameter, 60° V-threads, and pitches of 0.79, 0.40, and 0.24 mm immersed in fluidized beds of glass beads of average diameters (265, 357, and 427 μm). The heat transfer tubes were electrically heated by a calrod heater 9.4 mm in diameter and 30.5 cm long. The dependence of h_{wfb} on the fluidizing velocity, surface roughness, and particle diameter was investigated. The dependence of h_{wfb} was similar to that for smooth tubes. However, the maximum value of h_{wfb} was reached at a higher value of fluidizing velocity as compared to the smooth tube. Further, h_{wfb} values for a rough tube can be larger or smaller in comparison to a smooth tube depending upon the size of pitch in relation to the particle diameter. These authors found that β_{max} decreases from its value of unity as (P_f/d_p) is increased from zero to about 0.8 and thereafter β_{max} increases with a further increase in (P_f/d_p) and becomes greater than unity. Chen and Withers [169], on conceptual grounds, have pointed out that β will approach the value of unity again as (P_f/d_p) is continuously increased to infinity by letting P_f approach infinity as then the finned surface approaches the smooth surface. This qualitative variation of β with P_f/d_p has been explained by many workers

[159, 165, 168] by invoking such ideas as the bed porosity near the heat transfer surface, particle residence time on the surface, particle and surface contact geometry, and particle clogging.

Goel *et al.* [166] reported the heat transfer data for horizontal, 60° V-threads on tubes with 12.7- and 50.8-mm outside diameters having pitches of 5.08, 3.18, 1.58, 0.79, 0.40, and 0.24 mm immersed in fluidized beds of silica sands of average diameters (145, 167, 488, and 788 μm) at ambient temperatures and pressures as a function of fluidizing velocity. Grewal *et al.* [161, 167] have reported similar data for a 28.6-mm bronze tube for pitch values of 5.04, 3.18, 1.59, and 0.79 mm immersed in beds of silica sands (d_p = 167 and 504 μm) and alumina (259 μm). The typical design of a heat transfer tube, as used by Saxena and co-workers [159, 161, 166, 167], is shown in Fig. 12. The tube is heated by an electric calrod heater 12.7 mm in diameter and 30.5 cm long. Three iron-constantan thermocouples are bounded to the tube surface in milled grooves with technical quality copper cement. The ends of the tubes are provided with nylon supports to reduce axial heat loss to less than 1% of the total power fed to the tube.

In the calculation of the heat transfer coefficient, it is tacitly assumed that the temperature of the heat transfer surface is the same as the measured fin base temperature, and the accuracy of this assumption will depend upon the thermal conductivity of the fin material. In order to calculate the heat transfer coefficient, which is independent of the thermal properties of the fin material, it is necessary to determine the temperature distribution in the fin. Since the differential equations describing the temperature field in the fins cannot be solved by analytical methods, Grewal *et al.* [161] employed the finite-difference numerical technique to determine the temperature distribution in the fin. One-dimensional and two-dimensional heat conduction models were developed for determining the temperature distribution in a V-thread fin and

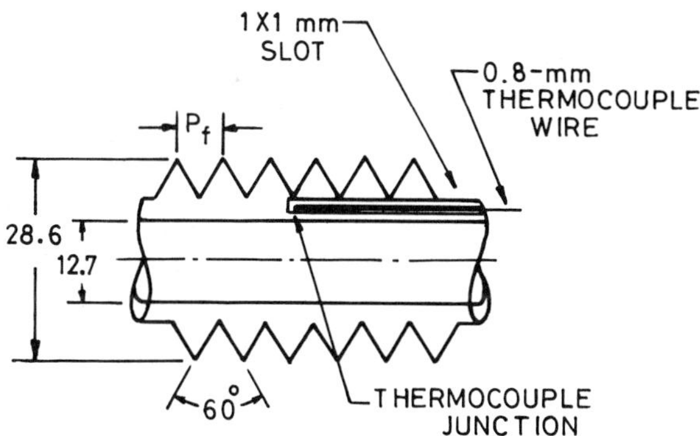

FIG. 12. Details of a typical 60° V-thread finned tube (from Grewal *et al.* [161]).

hence the corrected heat transfer coefficients. Finally, the fin efficiency was computed from the relation

$$\eta = \frac{Q}{h_{\rm wft} A_{\rm wft}(T_{\rm wb} - T_{\rm b})} \tag{95}$$

It was found that the fin efficiency decreased with an increase in the Biot number. Since the Biot number represents the ratio of conductive to convective resistances, a large value Biot number means that the conductive resistance is higher as compared to the convective resistance. As a result, the temperature drop in the fin is significant with fins of low thermal conductivity or of large fin height. The percentage error in fin efficiency based on the one-dimensional model is less than 1% for values of the Biot number less than unity. The values of the Biot number for the experiments of Saxena and co-workers were between 0.002 and 0.028, much below unity. Therefore, it was concluded that inaccuracy in using the one-dimensional model for the analysis of the V-thread fin was negligible. The calculated values of the corrected heat transfer coefficients, $h_{\rm wft}$ and $h_{\rm wfb}$, by Grewal *et al.* [161], were obtained by using the one-dimensional model.

For a given gas–solid fluidized bed, the amount of heat removed from the heater surface depends on the average bed porosity in the vicinity of the heater or gas conduction paths between solid particles and the heater, the effective area of the heat transfer surface, the effective area of the particles that exchange heat with the heater, and the particle residence time at the surface. The variation of β with an increase in $(P_f/d_{\rm p})$ can be explained by Fig. 13. The surface area of a 60° V-thread tube with a 0.24-mm pitch is about 1.98 times the surface area of a smooth tube with the same outside diameter, and the heat transfer rates for the former are smaller for all the three sizes of glass beads [159]. This was explained by referring to Fig. 13a and b. As the value of

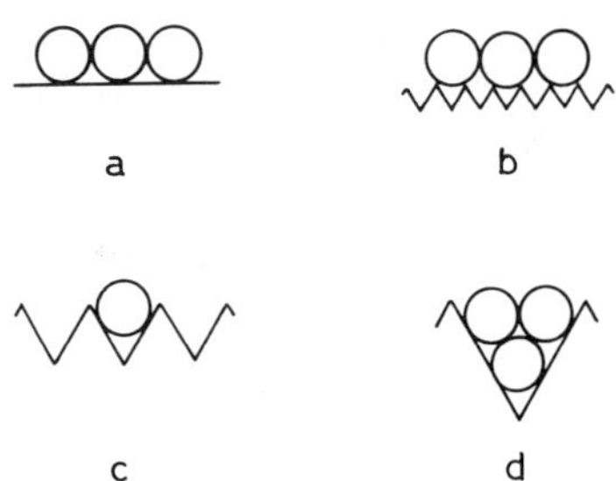

FIG. 13. Particle and heat transfer surface contact for (a) a smooth tube; (b) a tube with fine roughness, $P_{\rm f}/d_{\rm p} = 0.5$; (c) a tube with matching roughness $P_{\rm f}/d_{\rm p} = 1.2$; and (d) a tube with coarse roughness, $P_{\rm f}/d_{\rm p} = 2.5$ (from Grewal and Saxena [159]).

(P_f/d_p) is increased from zero, the value of the average porosity or conduction path between the heat transfer surface and the first row of solid particles increases, as is evident from a comparison of Fig. 13a and b. The resistance to heat flow is greater for the case shown in Fig. 13b than for the case corresponding to Fig. 13a because of a longer heat conduction path or a larger value of porosity in the vicinity of the heat transfer surface. Further, because of the irregularities of the surface, the particle residence time close to the heat transfer surface will increase. Thus, as (P_f/d_p) increases from 0 to about 0.8, the heat conduction path increases while the particle residence time increases. As a result, $h_{wfb\,max}$ or β_{max} for rough tubes decreases.

Fig. 13c represents a typical case when (P_f/d_p) is greater than 1 but less than 2. In this case, most of the particles have multiple contact with the heat transfer surface and therefore more of the surface area of the glass sphere becomes available for heat transfer from the heated tube surface in comparison to the case of a smooth tube, Fig. 13a. This will enhance the heat transfer rate. The increase in $h_{wfb\,max}$ for a V-thread tube $(P_f/d_p = 1.8)$ is about 34% over a smooth tube. The actual increase in the value of $h_{wfb\,max}$ will depend upon whether or not any of the V-threads have been clogged by bed particles. When the particle size is much smaller than the pitch, that is, when $(P_f/d_p) > 2$, there will be only a limited number of particles having multiple contact with the heat transfer surface and their number will further decrease with an increase in (P_f/d_p). Thus, the rate of increase of β would decrease with the increase in (P_f/d_p). The maximum heat transfer coefficient for rough tubes with $(P_f/d_p) = 3$ was found by Grewal and Saxena [159] to be about 40% greater than for smooth tubes. The increase in the heat transfer rate for larger values of (P_f/d_p) is mainly because of the increase in the effective surface area of the heater and the decrease in the particle residence time [161]. Goel *et al.* [166] inferred on the basis of their results that the size range of the particles in addition to the mean particle diameter must also be considered.

The β_{max} values of Grewal *et al.* [159, 161] for a transverse-thread horizontal tube, Vijayaraghavan and Sastri [165] for a longitudinal V-groove vertical tube, and Korolev and Syromyatnikov [168] for a vertical plate are in good agreement with each other. This is because of the fact that in all four cases not only the fin shapes but also their orientations with respect to air flow are identical. The values of β_{max} of Chen and Withers [169], Genetti and Kratovil [171], Vijayaraghavan and Sastri [165], and Zabrodsky *et al.* [170] are not, in general, in good agreement with the values of Grewal *et al.* [159, 161]. This is mainly because of the difference in the area of heat transfer surface provided by finned tubes for the same (P_f/d_p) ratio and the orientation of the fins with respect to the flow.

From a practical stand point, the knowledge of the heat transfer capacity function, β, is very significant for design calculations. A simple functional dependence of β on (P_f/d_p) within an uncertainty of $\pm 20\%$, which ignores its dependence on such factors as D_T, particle size range, and fluidizing velocity, is proposed by Goel *et al.* [166] on the basis of their data in the range $0.5 < (P_f/d_p) < 10.5$ as

$$\beta = 0.977 + (P_f/d_p)^{0.21} \tag{96}$$

Another factor that has been considered to represent the performance of finned tubes is the fin effectiveness factor, ϕ. Grewal *et al.* [161] proposed the correlation

$$\phi = 0.55 + 1.077[1.0 - 0.77(d_p/D_T)^{-0.068}(P_f/d_p)^{-0.171}] \tag{97}$$

on the basis of their data, which could be correlated within $\pm 10\%$. The experimental data of Vijayaraghavan and Sastri [165] for longitudinal V-groove vertical tubes is in excellent agreement with predictions based on Eq. (97). The data of Zabrodsky *et al.* [170] and Vijayaraghavan and Sastri [165] for a transverse V-groove vertical tube could be reproduced within a maximum error of $\pm 20\%$. However, the values of ϕ reported by Chen and Withers [169] were generally larger as compared to the calculated values based on the proposed correlation. This is probably because of the different fin geometry and orientation employed by these authors in relation to that of Grewal *et al.* [159, 161].

It may be pointed out that β and ϕ are related to each other for a given geometry of the finned surface. For V-shaped fins, Goel *et al.* [166] have given that

$$\phi = \frac{2\beta}{\pi D_T}\sqrt{P_f^2 + \pi^2(D_T - 0.866P_f)^2} \tag{98}$$

From the definition of the fin effectiveness factor, it follows that

$$\mathrm{Nu}_{\mathrm{ftT}} = \mathrm{Nu}_{\mathrm{T}}\,\phi \tag{99}$$

Grewal *et al.* [161] and Goel *et al.* [166] have correlated their data within an uncertainty of $\pm 25\%$ on the basis of Eq. (99) and relations for Nu_{T} given by Eqs. (36) and (38), respectively.

VII. Heat Transfer from Tube Bundles

The knowledge of the heat transfer coefficient for a tube bundle configuration appropriate for optimum gas and solids movement in the bed is of paramount importance to the designers of fluidized-bed combustors. Many works that summarize the state of the art are due to Saxena *et al.* [3], Botterill [2], Gelperin and Einshtein [5], and Grewal [77]. Therefore, only a brief account will be presented here, with emphasis on the most recent investigations. In general, the tubes constituting the bundle can be arranged in an in-line arrangement, Fig. 14, or in a staggered tube arrangement, Fig. 15. In each case, the size of the tube bank is controlled by choosing the appropriate number of tube rows and columns. The congestion of the tube bank is governed by the distance between the centers of two consecutive tubes, which is referred to as the pitch. In the in-line arrangement, the separation between consecutive horizontal and vertical tubes is represented by S_H and S_V, respectively, and the tube bank by the $S_H \times S_V$ bundle. The in-line arrangement usually has a square configuration in a horizontal plane so that the pitch in the X and Y directions is the same as shown in Fig. 14. If $S_H = S_V$, the tubes will be located on the edges of a cube of size S_H. In a staggered tube arrangement, the horizontal tubes in a bundle are located at the vertices of equilateral triangles, as shown in Fig. 15. Other configurations corresponding to isosceles triangles are also frequently used. The horizontal pitch significantly affects the heat transfer from a tube bundle, while the influence of the vertical pitch is relatively insignificant. This result can be qualitatively explained by the fact that a closer arrangement of tubes in a horizontal plane (perpendicular to the direction of the gas flow) hampers solids mixing more appreciably than does the decrease of tube spacing in a vertical plane. This observation is confirmed by experiments in which the pitches were varied in a limited range, and caution must be exercised in extrapolating this conclusion.

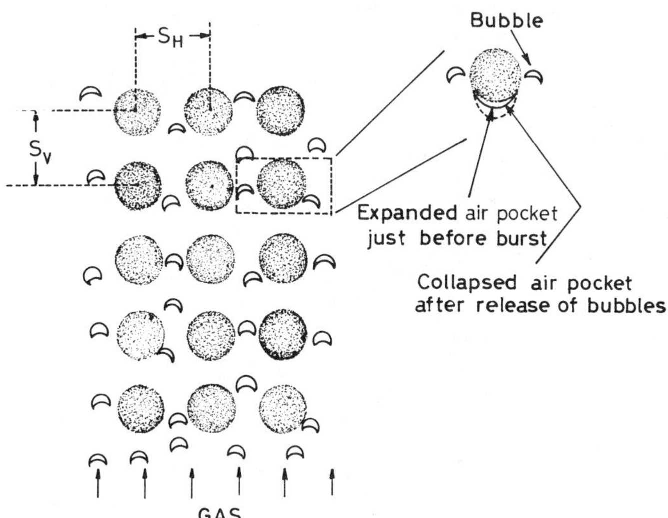

FIG. 14. Flow pattern for bubbles traveling up through an in-line tube bank (from Doherty and Saxena [190]).

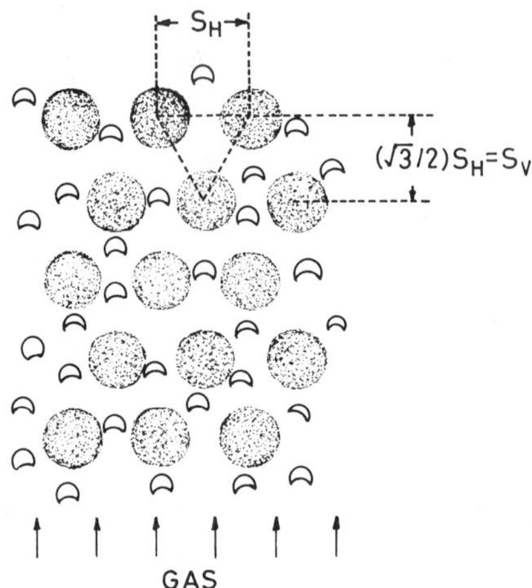

FIG. 15. Flow pattern for bubbles traveling up through a staggered tube bank (from Doherty and Saxena [190]).

Saxena [185] reported measurements of heat transfer coefficients of horizontal copper tubes and tube banks of 12.7-mm outside diameters in fluidized beds of silica sands with average diameters of 167 and 504 μm. The bed cross section was 30.5 cm by 30.5 cm and was equipped with a distributor consisting of two perforated steel plates with a coarse cloth sandwiched between them [186]. His results indicated that the heat transfer coefficient from a single tube to the fluidized bed was not influenced by adding two or four more rows as long as the horizontal pitch $\geq 4D_T$ and vertical pitch $\geq 2D_T$ for staggered tube bundles, and the pitches were $\geq 4D_T$ for in-line tube bundles.

Borodulya *et al.* [8] conducted experiments in a rectangular fluidized bed, 60 × 60 cm, of spherical millet particles with 2-mm average diameters and measured the heat transfer coefficient for a 14-mm copper tube and tube bundles of in-line and staggered (equilateral triangular) arrangements of different pitches as a function of fluidizing velocity. Their results revealed that h_w values are independent of S_H and S_V values as long as these are greater or equal to $2D_T$ for in-line tube bundle arrangements. A similar conclusion is valid for staggered tube bundles for $S_H \geq 2D_T$ and $S_V \geq \sqrt{3}\,D_T$. It was also noticed that h_w values are more sensitive to S_H than to S_V. Further work by these authors [187] with sands of average particle diameters (250 and 660 μm) in fluidized beds of 60 × 60 cm and 60 × 30 cm in cross section and tube bundles of in-line (square) and staggered (equilateral triangular) configurations revealed that $h_{w\,max}$ values were almost constant and equal to that for a single tube as long as P was varied in the range $2D_T < P < 6D_T$.

Grewal and Saxena [188, 189] reported heat transfer results with horizontal single tubes of copper ($D_T = 12.7$ mm) and bronze ($D_T = 28.6$ mm) and of their staggered tube bundles with tubes located at the vertices of equilateral triangles with pitch varying between 1.75 D_T and 9.0 D_T and immersed in square fluidized beds of silica sands ($d_p = 167$ and 504 μm) and alumina ($d_p = 259$ μm) as a function of air fluidizing velocity. These authors found that the heat transfer coefficients remain unaltered for 12.7-mm tube bundles as long as the relative pitch (P/D_T) was between 4.5 and 9.0, and for 28.6-mm tube bundles for a relative pitch of 3.5 and greater. The pitch effect was found in the former case for $(P/D_T) = 2.25$ and in the latter case for $(P/D_T) = 1.75$. The decrease in h_w values for these cases varied between 5 and 8.5% with the reduction in relative pitch and was attributed to the obstruction of solid particle movement when the tubes were spaced closely in a congested tube bundle configuration.

Zabrodsky *et al.* [29] reported heat transfer coefficients for in-line and staggered horizontal tube bundles and air fluidized beds of spherical millet ($d_p = 2$ mm) and nonspherical clay ($d_p = 3$ mm) as a function of fluidizing velocity, 0.6–2.8 m/sec. Horizontal duraluminum tubes, 30 mm in outer diameter were used and 30-mm-diameter semicylinders were fixed to the side walls of the column against the horizontal rows of tubes in order to simulate the extension of the rows of tubes. In the staggered tube bundle, the tubes were spaced with equal horizontal, S_H, and vertical, S_V, pitches: $S_H \times S_V = 100 \times 100$, 60×60, and 45×45 mm. The in-line bundles were $S_H \times S_V = 100 \times 100$, 100×50, 80×80, 60×60, and 50×50 mm. In all cases, the influence of pitch was negligible for $S_V = S_H > 2D_T$. It was also noted that the change in S_H from 100 to 50 mm, with S_V being constant and equal to 100 mm, significantly reduced h_w. The influence of the vertical pitch, on the other hand, was negligible as the h_w values were identical for 100×100 and 100×50 bundles. This result was explained qualitatively by the fact that closer arrangement of tubes in a horizontal plane (perpendicular to the direction of the gas flow) hampered solids mixing more appreciably than did the decrease of tube spacing in a vertical plane. These authors have also proposed the idea of particle bridging between adjacent tubes below which there are gas voids almost free of solids. Each of these reduces the heat transfer rate, and the closer the tubes are, the lower is the gas velocity at which bridge formation takes place. With an increase in gas velocity, the mobility of these bridges increases and more frequent bridge breakdown occurs and this enhances the rate of heat transfer. This phenomenon has been conjectured and employed to explain the observed dependence of h_w on u for an in-line 50×100 bundle in a fire clay bed.

Doherty and Saxena [87, 190] reported their heat transfer results for in-line and staggered tube bundles immersed in fluidized beds of silica sand

($d_p = 488$ μm) and red silica sand ($d_p = 777$ μm) at ambient temperature as a function of fluidizing velocity. Brass tubes, 28.6-mm in diameter, and aluminium tubes, 50.8 mm in diameter, were used. For the former tubes, the relative pitches were 2.66 and 5.32 and 2.66 for the in-line and staggered tube arrangements, respectively. For the latter tubes, the relative pitches were 1.50 and 3.00 and 1.50 for the in-line and staggered tube arrangements, respectively. These authors have discussed the dependence of h_w on gas velocity in two ranges, lower and higher. This discussion is revealing for bubble dynamics and hence will be reproduced here [87].

In the lower gas velocity range, for a single tube there are only a few bubbles present in the bed and solids mixing is poor. However, the tube banks capture some of the rising bubbles, forming gas pockets underneath the tubes. The expanded air pocket then collapses, forming small bubbles that bring about relatively better mixing in the bed than do the larger bubbles (Fig. 14). This results in higher h_w values for the tube bundles than for the single tube. It is important to note that h_w values are greatest for the staggered tube arrangement where the additional bubbling has the greatest effect (Fig. 15). On the other hand, in the higher velocity range, the tube banks have lower h_w values compared to the single tube. The in-line tube bundles for the same relative pitch have values which are even smaller than for the staggered tube bundles. A possible explanation for this may be that the trapped air pocket expands downward until it reaches a maximum size and then collapses, generating bubbles. For in-line arrangements, these bubbles travel up the bed in the channels formed in between the adjacent columns of the tubes. For the staggered tube bundles, these bubbles encounter the tubes of the next upper row in the arrangement and this results in good mixing and larger h_w values than for the in-line tube bundles (Figs. 14 and 15).

Borodulya *et al.* [131] reported h_w values for vertical tube bundles immersed in fluidized beds of glass beads (d_p = 1.25 and 3.1 mm) and sand (d_p = 0.794 and 1.225 mm) for pressures ranging from 1.1 to 8.1 MPa. The bundles are made from 13-mm-diameter and 76-mm-long wooden cylinders arranged in an equilateral triangular configuration with pitches equal to 19.5, 29.3, and 39.0 mm. The central tube in each case served the purpose of the heat transfer probe, and it was made by winding a 70-μm-diameter copper tube. The wire turns were held in position by glue and were machined to a depth of half of the wire diameter to obtain a smooth surface finish. For particles of a given size, the h_w values depended upon pressure and increased monotonically with it. For a given particle size and pressure, the h_w values also depended upon the tube pitch. This dependence was sensitive to pressure and at higher pressures the difference in h_w values was larger for the same difference in pitch than that at a lower pressure. The h_w values for widely spaced tubes were greater than for closely packed tubes in a bundle under otherwise identical conditions. In general, h_w values exhibited the conventional dependence on G, that is the values increased with increasing G in the beginning, attained a maximum value, and then decreased with further a increase in G. However, the sharpness of the maximum seemed to depend upon the system pressure, and it decreased as the pressure was increased. These data also indicated that for spherical glass beads as d_p was increased from 1.25 to 3.1 mm, the h_w increased, though the difference was relatively pronounced at the highest pressure and was practically negligible at the lowest pressure.

Borodulya *et al.* [144] also reported h_w values for horizontal square in-line tube bundles ($D_T = 13$ mm) of three different pitches (19.5, 29.3, and 39.0 mm) immersed in fluidized beds of glass beads ($d_p = 1.25$ and 3.1 mm) and sands ($d_p = 0.794$ and 1.225 mm) as a function of fluidizing velocity at pressures of 1.1, 2.6, 4.1, and 8.1 MPa. It was clear from these data that h_w followed in all cases the same qualitative variation with respect to fluidizing velocity, system pressure, change in particle diameter, and pitch. In all cases, for a given particle and tube bundle at a fixed pressure, the h_w values increased with an increase in mass fluidizing velocity. The increase was rapid at low values and became relatively slower at higher values for G. The h_w versus G plots exhibited a flat maximum for these large particles at these high operating pressures. It was also noted that h_w was consistently larger at higher pressures over the entire velocity range. For the conditions of their experiments, no systematic effect of tube pitch was found on h_w. For otherwise identical conditions, the h_w values were greater for larger particles than for smaller particles, and this trend became more pronounced with increasing pressure. The reason for this is that, for small particles, particle convection plays a dominant role, whereas for large particles, the gas convection is more important and its contribution to h_w enhances as the pressure increases.

It is somewhat surprising that the effect of tube pitch was not found even when its value was as low as 19.5 mm or 1.5 D_T. This is not in agreement with the observations made at ambient pressures by Zabrodsky *et al.* [29], Borodulya *et al.* [8, 28], Grewal and Saxena [188, 189], and Saxena [185]. This work [144] suggested that at high pressures, because of the increased value of gas density, the formation of particle bridging between the adjacent tubes, "Lee" stacks on the downstream side of the tube, or in general stagnant solids in the region around a horizontal tube is probably prevented. As a result, smooth fluidization occurred even for a relatively tight bundle.

Four correlations for estimating $h_{w\,max}$ or Nu_{max} for horizontal tube bundles are available in the literature and have been evaluated by Borodulya *et al.* [187] and Grewal [77] and are given here.

Gelperin *et al.* [191] for in-line tube bundles,

$$Nu_{max} = 0.79\,Ar^{0.22}\left(1 - \frac{D_T}{S_H}\right)^{0.25} \tag{100}$$

for $215 \leq Ar \leq 2200$ and $1.5D_T \leq S_H \leq 10D_T$.

Gelperin *et al.* [192] for staggered tube bundles,

$$\mathrm{Nu}_{\max} = 0.74\,\mathrm{Ar}^{0.22}\left[1 - \frac{D_T}{S_H}\left(1 + \frac{D_T}{S_V + D_T}\right)\right]^{0.25} \tag{101}$$

for $200 \leq \mathrm{Ar} \leq 2400$, $2D_T \leq S_H \leq 9D_T$, and $D_T \leq S_V \leq 10D_T$.

Chekansky *et al.* [193] for staggered tube bundles,

$$\mathrm{Nu}_{\max} = 28.2\rho_s^{0.2}k_g^{-0.4}d_p^{0.64}\left(\frac{P - D_T}{d_p}\right)^{0.04}\left(\frac{D_T}{D_{20}}\right)^{-0.12} \tag{102}$$

for $16 \leq (P - D_T)/d_p \leq 63$.

Grewal *and Saxena* [189] for staggered tube bundles,

$$\mathrm{Nu}_{\max} = 0.9\left(\mathrm{Ar}\frac{D_{12.7}}{D_T}\right)^{0.21}\left(\frac{C_{ps}}{C_{pg}}\right)^{0.2}\left[1 - 0.21\left(\frac{P}{D_T}\right)^{-1.75}\right] \tag{103}$$

for $75 \leq \mathrm{Ar} \leq 20{,}000$ and $1.75D_T \leq P \leq 9D_T$.

These correlations have been compared by Borodulya *et al.* [187] with their own experimental data on in-line and staggered tube bundles and by Grewal and Saxena [77, 189] with the data of Gelperin *et al.* [192], Howe and Aulisio [194], Borodulya *et al.* [187], Bartel and Genetti [173], Chandran *et al.* [183], Staub and Canada [179], Canada and McLaughlin [7], Bansal *et al.* [195], Xavier and Davidson [196], Grewal and Hajicek [197], Grewal and Saxena [189], and the Aerojet Energy Conversion Company [198]. These studies [77, 187–189] have revealed that the Grewal and Saxena [189] correlation of Eq.(103) is generally the most appropriate, and its uncertainty of prediction is about $\pm 20\%$. This correlation, originally developed for staggered tube bundles, is also found adequate for in-line tube bundles. This correlation is recommended only for small particles of group I of the Saxena and Ganzha [14] powder classification scheme.

Grewal [200] proposed to employ the correlation of h_w for a single horizontal tube of Grewal and Saxena [79] and the correction factor (CF) of the following form to account for the effect of the relative pitch of the tube bundles:

$$CF = 1 - 0.21(P/D_T)^{-1.75}$$

for $1.75 \leq (P/D_T) \leq 9$ and $75 \leq \mathrm{Ar} \leq 20{,}000$.

His final expression is

$$\mathrm{Nu}_T = 47(1 - \varepsilon)\left[\left(\frac{GD_T\rho_s}{\rho_g\mu_g}\right)\left(\frac{\mu_g^2}{d_p^3\rho_s^2 g}\right)\right]^{0.325} \times \left(\frac{\rho_s C_{ps} D_T^{3/2} g^{1/2}}{k_g}\right)^{0.23} \mathrm{Pr}^{0.3}[1 - 0.21(P/D_T)^{-1.75}] \tag{104}$$

Grewal [200] found that the relation of Eq. (104) could generally reproduce the available data within a scatter of $\pm 25\%$. It is recommended for estimation of Nu_T for horizontal tube bundles and fluidized beds of group I particles of the Saxena and Ganzha [14] powder classification scheme. It was also shown that a somewhat similar correlation of Bansal *et al.* [201] was relatively less satisfactory, a conclusion that was also substantiated for bundles of horizontal finned tubes [161].

Staub [26] developed a model for horizontal tube bundles in which he accounted for the restrictions on the motions of large particles in the bed caused by the tubes. Staub contended that the gas flow in a fluidized bed of large particles is in the turbulent regime and proposed the following normalized relation for the computation of the heat transfer coefficient:

$$\mathrm{Nu} = \left[1 + \left(\frac{150}{d_p}\right)^{0.73} \left(\frac{\rho_s u_s}{\rho_g u}\right)\right]^{0.45} \mathrm{Nu}_g \tag{105}$$

Here d_p is in μm. The relation of Eq. (105) is for $20\ \mu\text{m} < d_p < 1000\ \mu\text{m}$, and d_p is taken as $1000\ \mu$m for $1000\ \mu\text{m} < d_p < 3000\ \mu\text{m}$. The solids circulation velocity, u_s, is expressed in the following form by employing the concept of mixing length:

$$u_s = 0.42(1 - \varepsilon)S_H^{0.4} \tag{106}$$

Here, Nu_g is calculated from the Colburn equation for heat transfer [199] for the gas flow across the tube bundle in the absence of solid particles. This model is tested against experimental data by Borodulya *et al.* [8, 144] and Zabrodsky *et al.* [29] with unsatisfactory success. The theoretical model predictions are appreciably smaller than the experimental values.

Grewal *et al.* [161] found on the basis of their experimental data for smooth horizontal and 60° V-groove horizontal tube bundles that the effect of tube pitch on the heat transfer coefficient is similar for the two cases. On this basis, they proposed to use Eqs. (99), (97), and (104) to compute the heat transfer coefficient for horizontal finned tube bundles and a gas–solid fluidized bed of group I particles of Saxena and Ganzha [14] with an estimated uncertainty of $\pm 25\%$.

VIII. Heat Transfer to Immersed Surfaces at High Temperatures

A. HEAT TRANSFER FLUXES AND COEFFICIENTS

The long and controversial history of the heat transfer between a fluidized bed and either its containing vessel or an immersed surface has been discussed at length by various authors [1–4, 77, 202], and here we will present only a concise state of the art in relation to experimental measurement techniques and theoretical models to predict the total heat transfer coefficient, h_w, and the radiative heat transfer coefficient, h_{wr}. The heat transfer process at high temperatures in gas-fluidized beds consists of particle convection, gas convection, and radiation. The total heat flux from the bed to an immersed surface at a relatively lower temperature, q, is given by

$$q = (q_{pce} + q_{gce} + q_{re})(1 - f_B) + (q_{gcB} + q_{rB})f_B \tag{107}$$

The fluxes due to particle convection, gas convection, and radiation are

$$q_{pc} = q_{pce}(1 - f_B) \tag{108}$$

$$q_{gc} = g_{gce}(1 - f_B) + q_{gcB}f_B \tag{109}$$

and

$$q_r = q_{re}(1 - f_B) + q_{rB}f_B \tag{110}$$

Combining Eqs. (107) through (110), we get

$$q = q_{pc} + q_{gc} + q_r \tag{111}$$

In general, q_{pce} and q_{gce} are different since the mean boundary-layer thickness on the surface is different for the emulsion and the bubble phases. The mean boundary-layer length is of the order of a particle diameter in the emulsion phase [147] and of the order of a bubble diameter in the bubble phase. The radiant fluxes in the emulsion, q_{re}, and the bubble, q_{rB}, are also different. In the former case, the surface exchanges radiation with the particles adjacent to it (the first layer of particles), which are at a temperature less than that of the bed core; whereas in the latter case, the immersed surface "sees" and exchanges radiation with the bed core directly. In beds of small particles (groups I and IIA of Saxena and Ganzha [14]), the heat transfer due to gas convection is relatively small, and consequently, for such beds, Eq. (107) simplifies to

$$q = (q_{pce} + q_{re})(1 - f_B) + q_{rB}f_B \tag{112}$$

For group III particle beds, the absence of bubbling leads to the following simplification of Eq. (107):

$$q = q_{pce} + q_{gce} + q_{re} \tag{113}$$

The general form of the heat flux equation, Eq. (107), must therefore be used for beds of group IIB particles.

The simultaneous radiative and convective heat transfer processes in fluidized beds must also be examined from the perspective of the definition of heat transfer coefficients since the temperature differences defining the two processes are different. The convective heat transfer process (both gas and particle) is governed by the bed to surface temperature difference, $(T_b - T_s)$, whereas the radiative flux is governed by the difference in temperature of the first layer of particles and the heat transfer surface, $(T_1 - T_s)$. Thus, the convective and radiative heat transfer coefficients may be appropriately defined as

$$h_{w\,conv} = (q_{pc} + q_{gc})/(T_b - T_s) \tag{114}$$

and

$$h_{wr} = q_r/(T_1 - T_s) \tag{115}$$

The total heat transfer coefficient, h_w, is

$$h_w = h_{w\,conv} + h_{wr} \tag{116}$$

For group III particles, the first layer of particles are almost at the bed temperature owing to the large heat capacity of the particles, and Eq. (115) may be written as

$$h_{wr} = q_r/(T_b - T_s) \tag{117}$$

Consequently, for group III particles, the total heat transfer coefficient, h_w, may be defined on the basis of Eqs. (114), (115), and (110) as

$$h_w = q/(T_b - T_s) \tag{118}$$

In practice, however, Eq. (118) is used for beds of particles of all sizes as an approximation since T_1 is difficult to measure. Equations (117) and (118) are used rather routinely to determine h_{wr} and h_w, respectively.

B. EXPERIMENTAL TECHNIQUES OF MEASUREMENT OF h_w AND h_{wr}

The total and radiative heat transfer coefficients between a fluidized bed and an immersed surface at high temperatures have been measured by a limited number of investigators, and these are listed in a historical sequence in Table VI. This table also identifies the experimental techniques used by each investigator, as well as the ranges of average particle sizes and temperatures. The techniques used for measuring h_w and h_{wr} may be divided into three different categories in both cases for the sake of convenience of understanding. A brief description of each category follows, as detailed by Mathur and Saxena [221].

A schematic representation of the three different techniques used for measuring h_w is shown in Fig. 16. Figure 16A illustrates the principle of the methods that employ the change in temperature with time of a heated metallic probe dropped in the bed to obtain the rate of heat transfer. This technique is designated as category 1 in Table VI. The category II methods use a fluxmeter

TABLE VI

HIGH-TEMPERATURE HEAT TRANSFER INVESTIGATIONS IN GAS-FLUIDIZED BEDS

Reference	h_w measurement method category	h_{wr} measurement method category	d_p (μm)	T_b (K)
Jolley [203]	I		800	1048–1253
Il'chenko *et al.* [204]	II	1	570–1750	700–1700
Szekely and Fisher [205]		3	200–300	<600
Wright *et al.* [206]	III		1800–2900	1073–1173
Baskakov *et al.* [22]	I, II		350–1250	1123
Yoshida *et al.* [207]	III	2	180	823–1273
Thring [208]	III		582, 928	923–1273
Basu [209]	II	1	325–500	973–1273
Panov *et al.* [210]	I	2	500–6000	800–1200
Vadivel and Vedamurthy [211]	II	1	4000, 6000	1023
Flamant [212]		3	250	600–1400
Botterill *et al.* [213]	I, II	2	370–3000	573–1273
George and Welty [214]	II		2140, 3230	810–1005
Ozkaynak *et al.* [215]	II	1	1030	<760
Zhang *et al.* [216]	II		1340–1580	1073–1273
Alavizadeh *et al.* [217]	II		520–3230	812–1050
Goshayeshi *et al.* [218]	II		2140, 3230	810–1005
Tuzla *et al.* [219]	II		465–1400	773
Zhang and Xie [220]	II	1	1000, 4000	1073–1313

and measure the temperature difference across a slab of a material of known thermal conductivity, as shown in Fig. 16B. The category III methods, shown in Fig. 16C, employ a water-cooled probe to measure the incident flux, which together with the temperature difference between the probe surface and the bed is used to measure h_w. Both category II and III probes have water flowing through them, the distinction lies in the fact that the cooling water is not used to measure the incident flux in category II methods as it is in category III methods, instead it is used only to remove the heat conducted through the fluxmeter.

The various methods for measuring h_{wr} are also classified into three categories. Category 1 methods employ a quartz or silica window to block the convective fluxes and allow only the radiative flux to reach a fluxmeter. The category 2 methods involve the measurement of heat transfer by two probes: one with a high-emissivity surface, and the other with a low-emissivity surface. The measurement from the high-emissivity surface probe is used to calculate h_w, and from the low-emissivity probe to calculate h_{wc}. The difference in the heat transfer rates measured by the two probes is ascribed to radiation. The category 3 measurement methods involve passing of the radiative flux only in the bed and measuring the resulting temperature change.

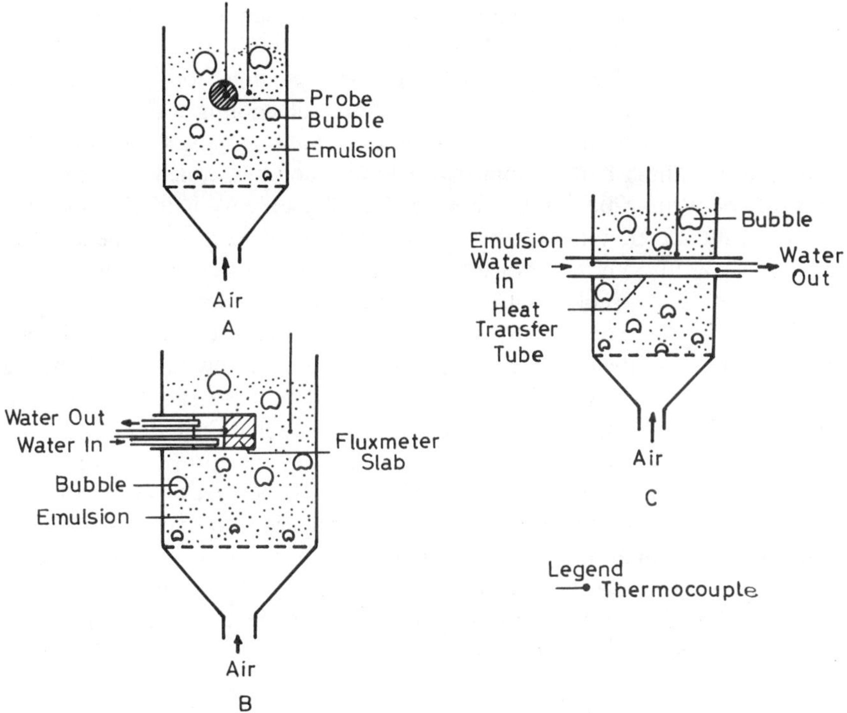

FIG. 16. Schematic representation of the three categories of high-temperature h_w measurement techniques (from Mathur and Saxena [221]).

These different methods for the measurement of h_w and h_{wr} may now be assessed to identify the techniques best suited for high temperatures in terms of the ease of fabrication of the probe and of its operation, unambiguous interpretation of the measured quantities, and accuracy of the measured coefficients. The last two criteria eliminate the category I methods, as they have been found to yield values of h_w greater than those obtained from the methods of categories II and III [22, 213]. The reason for this discrepancy is easy to understand in view of the different conditions under which heat transfer takes place in the two cases. In the category II and III probes, the particles adjacent to the probe surface are at a temperature that is smaller than that of the bed, and this temperature remains constant with time. In the case of the category I probes, however, the particles adjacent to the probe surface are initially at the bed temperature, and their temperature steadily decreases with time. As a result, q and h_w measured by category I probes are greater than those measured by probes belonging to categories II and III. Category III probes for the measurement of h_w are also not very attractive. The relatively large uncertainties associated with temperature measurement of the coolant outlet stream lead to correspondingly large uncertainties in the measured values of h_w [111]. Additionally, category III probes are not suited for use at high bed temperatures at which there is a possibility of the cooling water vaporizing into steam. This latter problem can be circumvented by using a coolant with a comparatively high boiling point, for example, sodium, which has been employed by Thring [208]. In this perspective, the category II methods, which employ a fluxmeter, appear to be most attractive for the measurement of h_w.

It hardly needs to be emphasized that the simultaneous measurement of h_w and h_{wr} is most desirable as only then an unambiguous interpretation of the results for the same operating conditions is possible. This eliminates the h_{wr} measurement methods of category 3 as these do not allow the simultaneous measurement of the two coefficients. The methods of category 2, based on utilizing two probes with high and low surface emissivities, are likely to yield erroneous results owing to the change in surface emissive properties with time because of the continuous abrasion of the surface by particles. Further, a technique that measures q_r directly is preferable over methods of this category, which obtain q_r in an indirect manner. The use of a quartz glass window to isolate the convective flux, characteristic of category I methods, has been commonly used. The general drawback of this technique is that the finite conductivity of quartz glass causes some convective flux to be conducted to the fluxmeter. However, the convective flux can be either reduced by experimental arrangement or accurately estimated by a detailed calculation. In the perspective of this discussion, a probe for the measurement of the total heat transfer and its radiative component at high temperatures has been designed and developed by Mathur and Saxena [221] and will be described in the following.

C. A Typical Design of the High-Temperature Heat Transfer Probe

The probe design of Mathur and Saxena [221] will be discussed as a typical illustration of the probes of this category and general comments will be presented for the methodology and interpretation developed by workers for such probes. The probe is shown in Fig. 17. The probe has a nominal diameter of 66.3 mm and consists of a water inlet section, A, two fluxmeters, B and C, and a water outlet section, D. The middle section of the probe accommodates two fluxmeters, each housed in a 304-stainless steel, 50.8-mm tube. These tubes are transversely inserted into the main probe, a 66.3-mm outer diameter, 3.9-mm thick, 304-stainless steel pipe. The lengths of the tubes housing the fluxmeters B and C are 20.6 and 25.4 mm, respectively. The thickness of these tubes are 2.4 and 4.8 mm, respectively.

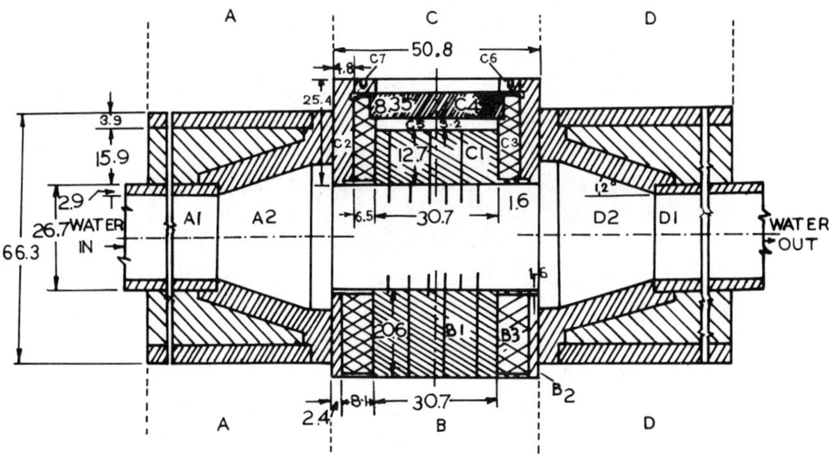

FIG. 17. Design diagram of a high-temperature heat transfer probe. All dimensions are in millimeters. See text for details (from Mathur and Saxena [221]).

The water inlet section, A, consists of the water intake pipe, A1, of stainless steel 304, having an outer diameter of 26.7 mm, and a thickness of 2.9 mm, which is insulated from the probe pipe by a layer of Fiberfrax T-30R insulation, 15.9 mm in thickness. The purpose of the diverging section, A2, is primarily to provide mechanical support to the water intake pipe, A1, as well as to smooth out the water flow past the interior faces of the fluxmeters. The space between the divergent section and the probe pipe is also filled with insulating Fiberfrax LoCon felt. The water outlet section, D, is similar to the water inlet section, A, both in design and purpose. The water flows out of the reducer section, D2, and the water outlet pipe, D1.

The fluxmeter, B, designed to measure the total heat flux and the total heat transfer coefficient, is a 30.7-mm-diameter, 420-stainless steel plug, B1, inserted into the main housing pipe, B2, and insulated from it by a 8.1-mm-thick annular ring of Zircar thermal insulation, B3. The thermal conductivity of the plug material and its thickness are precisely known, and the temperature difference across it is measured by a set of six thermocouples, three on each of the two faces.

The thermal fluxmeter, C, designed to isolate and measure the radiation component of the total heat flux and the radiative heat transfer coefficient, is also a plug of 420-stainless steel, 30.7 mm in diameter, and 12.7 mm thick. A 6.35-mm-thick polished quartz glass window, C4, is provided to isolate the thermal radiation from the total incident thermal flux, and only the former is transmitted to the plug, C1. The plug, C1, is welded to the main housing pipe, C2, with a 5.5-mm-thick Zircar insulation lining, C3, in between. The quartz window, C4, is held tightly on a shoulder of the Zircar insulation via a threaded ring, C6, mating with threads in the pipe housing, C2. The ring rotation is accomplished through two pinholes, C7. The air gap, C5, between the quartz window, C4, and the exterior face of the plug, C1, is 3.2 mm thick. The heat conduction or natural convection occurring in this air film augmenting the heat transfer process is corrected by a theoretical calculation [221]. All of the thermal radiation incident on the quartz window is not transmitted to the plug, C1, for measurement, and a part is scattered and absorbed. These losses have been accounted through the knowledge of the transmissivity of the quartz glass. The temperatures of the two faces of the plug, C1, are established by six thermocouples, as in the case of B1.

The thermocouples used in the heat transfer probe are chromel–alumel (K-type) and are encased in grounded inconel sheaths filled with magnesium oxide as an electrical insulator. The thermocouple leads are brought out through the water outlet pipe and are connected to a Leeds and Northrup 934 Numatron with a 16-point switch for digital display. The water flowing through the probe removes the heat added through the two fluxmeters and enables the setting up of thermal equilibrium in both of them. The fluxes incident on the probe are given by

$$q = k_p(T_{Bo} - T_{Bi})/X_B \tag{119}$$

and

$$q_c = k_p(T_{Co} - T_{Ci})/X_C \tag{120}$$

respectively. Here, T_{Bo} and T_{Co} and T_{Bi} and T_{Ci} are the temperatures of the outer and inner surfaces of fluxmeters B and C, respectively. The heat transfer coefficients are calculated from these relations:

$$h_w = q/(T_b - T_{Bo}) \tag{121}$$

and

$$h_{wr} = q_r/(T_b - T_{Bo}) \tag{122}$$

where q_r is the radiative heat flux incident on the fluxmeter C, and is obtained from q_c after applying a correction for the convective flux leakage. This leakage has been accounted for by other investigators who have used category 1 methods for measuring h_{wr} in different ways. Il'Chenko *et al.* [204], Vadivel and Vedamurthy [211], Alavizadeh *et al.* [222], and Zhang and Xie [220] calibrated the temperature difference across the fluxmeter against a known radiative flux emanating from a blackbody. Basu [209] accounted for the conductive flux by a theoretical calculation, details of which are not given by him. Ozkaynak *et al.* [215] minimized the leaked convective flux reaching the fluxmeter by enclosing the fluxmeter in an inner cavity, also covered by glass.

Air was blown in the space in between the outer glass window (contacting the bed) and the inner window (enclosing the fluxmeter) so as to maintain the inner window at constant and uniform temperature irrespective of the bed temperature. Mathur and Saxena [221] developed a correction procedure based on the thermal analysis of the fluxmeter and have reported details of the same, which are not being reproduced here for the sake of brevity.

D. HEAT TRANSFER COEFFICIENTS AT HIGH TEMPERATURES

Values of h_w and h_{wr} have been reported by a number of workers as a function of temperature and other parameters, and these will be discussed in this section in reference to the measured values of Mathur and Saxena [221] for beds of 559-μm and 751-μm sand particles. These results are shown in Figs. 18–21, in each case for various temperatures and fluidizing velocities. Here, h_w is seen to increase with an increase in T_b. This increase is attributed to two sources: the increase in gas thermal conductivity at higher temperatures, which leads to increases in q_{pce}, q_{gce}, and q_{gcB}, and higher radiative fluxes, q_{re} and q_{rB}, because of the larger value of $(T_1^4 - T_s^4)$.

The variation of h_w with d_p is more complex. At comparatively low bed temperatures (773–823 K), Panov *et al.* [210], Botterill *et al.* [213], George and Welty [214], and Tuzla *et al.* [219] found h_w to decrease as d_p increased from about 465 to about 6000 μm. The decrease is quite rapid for beds with average particle sizes less than about 1500 μm, but thereafter the decrease became quite gradual for larger values of d_p. For higher bed temperatures

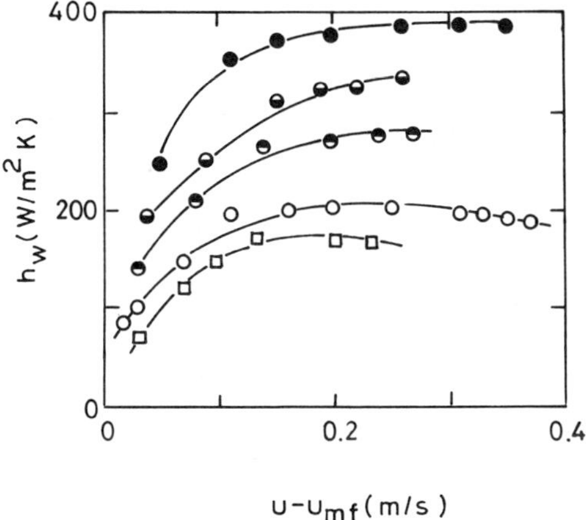

FIG. 18. Variation of h_w with $(u - u_{mf})$ in a bed of 559-μm sand particles at five different bed temperatures: 985 K (●), 820 K (◒), 675 K (◓), 495 K (○), and 385 K (□) (from Mathur and Saxena [221]).

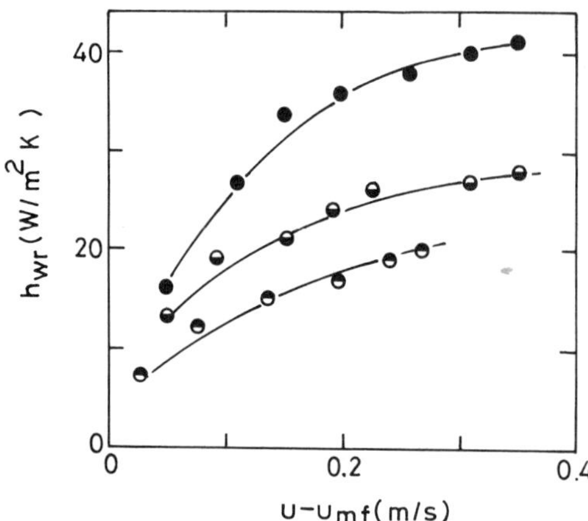

FIG. 19. Variation of h_{wr} with $(u - u_{mf})$ in a bed of 559-μm sand particles at various bed temperatures: 985 K (●), 820 K (◒), and 675 K (◓) (from Mathur and Saxena [221]).

($>$1000 K) and for values of d_p up to about 1150 μm, an initial decrease in h_w with an increase in d_p has been observed [208, 210, 213]. For d_p greater than 1150 μm, h_w increases (or remains constant) with an increase in d_p [206, 211, 214, 216]. Consequently, the difference between h_w values at low and high temperatures is found to increase with an increase in particle size. Basu [209] attributed this trend to the smaller change in T_1 in beds of small particles because of a change in T_b as compared to that in large particle beds. This

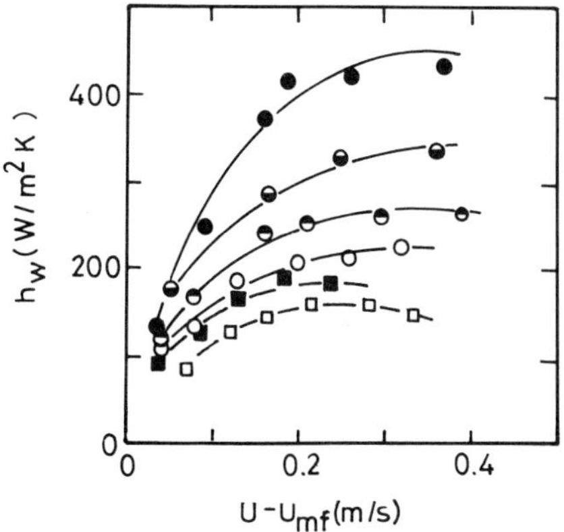

FIG. 20. Variation of h_w with $(u - u_{mf})$ in a bed of 751-μm sand particles at six different bed temperatures: 1175 K (●), 915 K (◒), 785 K (◓), 675 K (○), 510 K (■), and 395 K (□) (from Mathur and Saxena [221]).

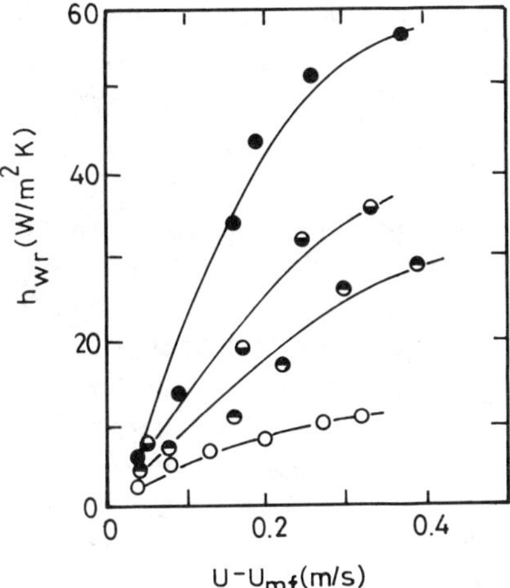

FIG. 21. Variation of h_{wr} with $(u - u_{mf})$ in a bed of 751-μm sand particles at various bed temperatures; 1175 K (●), 915 K (◒), 785 K (◓), 675 K (○) (from Mathur and Saxena [221]).

causes radiative flux to be comparatively small in small particle beds as compared to that in large particle beds at the same bulk bed temperature. Consequently, the difference in h_w values at low and high temperatures goes on increasing as d_p increases.

The dependence of h_w on u is not quite unambiguous. Wright *et al.* [206], Basu [209], Vadivel and Vedamurthy [211], Alavizadeh *et al.* [217], and Tuzla *et al.* [219] have found h_w to initially increase with an increase in u and then become constant. The d_p in these investigations ranged from 400 to 5000 μm, and bed temperature from 773 to 1223 K. Botterill *et al.* [213] found a similar trend for a bed of 1150-μm alumina at 1153 K, but found h_w to continuously increase with an increase in u for a bed of alumina at 773 K up to a u/u_{mf} value of about 10. On the other hand, Zhang *et al.* [216] found h_w to decrease as u increased in the range $3.5 < u/u_{mf} < 4.5$ for a bed of 1340-μm particles. It is not possible to explain this discrepancy at this time on the basis of the available details.

The effect of u on q_r is also the subject of some controversy. Il'Chenko *et al.* [204], Baskakov *et al.* [22], Basu [209], and Ozkaynak *et al.* [215] did not observe any variation in h_{wr} with changes in u. However, Szekely and Fisher [205], Flamant [212], Botterill *et al.* [213], Alavizadeh *et al.* [217], and Zhang and Xie [220] found h_{wr} (or q_r) to increase with an increase in u.

The variation of h_w and h_{wr} with u clearly shows the effect of increased bed voidage on radiation. Initially, h_{wr} increases rapidly as u increases beyond u_{mf}, but then the increase becomes much more gradual as u increases further. This indicates that once the bed is "well fluidized," further increases in u have only a marginal effect on h_{wr}. It is also interesting to note that the presence or absence of radiation is also clearly indicated by the trend of the $h_w - u$ curves. The data at lower temperatures ($T_b < 675$ K), where radiation is found to be negligible, show that h_w initially increases with an increase in u, becomes approximately constant, and then decreases slowly as u is increased further. By contrast, at higher bed temperatures (>675 K), h_w is never observed to decrease with an increase in u. This is because of the contribution of h_{wr}, which increases continuously with increases in u. Mathur and Saxena [221] found that the calculated h_{wc} ($=h_w - h_{wr}$) values are well predicted with the correlation of Grewal and Saxena [79] over the entire ranges of u and T_b. The correlation overpredicts the experimental data for the smallest values of ($u - u_{mf}$), but for all the data obtained for $(u/u_{mf}) > 1.2$, the agreement is within 25%. It is therefore concluded by Mathur and Saxena [221] that the correlation of Grewal and Saxena [79] is capable of predicting the convective component of the heat transfer coefficient with adequate precision even at high temperatures.

Mathur and Saxena [221] presented a comparison of their heat transfer coefficients data with those available in the literature under comparable conditions with a view to sustain some confidence in the design and operation of heat transfer probes. Thring [208] measured h_w for beds of 582- and 928-μm particles at 1100 K. The measured h_w values ranged from 550 to 635 W/m^2 K for the 582-μm bed and from 470 to 625 W/m^2 K for the 928-μm bed. Mathur and Saxena [221] found $h_{w\,max}$ to be about 400 W/m^2 K for the 559-μm bed at 985 K. The trend in the increase of $h_{w\,max}$ with T_b indicates that $h_{w\,max}$ would be about 470 W/m^2 K at 1100 K for this bed. This is lower than the range of values obtained by Thring [208]. This difference is probably due to the difference in the placement of the heat transfer probe in the two investigations. The probe used by Thring was lowered vertically into the bed, whereas in the present study, the probe has been placed horizontally in the bed. The presence of Thring's vertical probe in the 0.152-m i.d. bed probably caused turbulent mixing in the narrow annular bed leading to higher values of h_w. Tuzla *et al.* [219] found $h_{w\,max}$ to be about 400 W/m^2 K for a bed of 465-μm sand at 773 K. By comparison, the $h_{w\,max}$ value for the 559-μm particle bed is interpolated as about 300 W/m^2 K at 773 K. The lower value is expected as it corresponds to a larger average particle size. The measurements of Alavizadeh *et al.* [217] indicate an $h_{w\,max}$ of 349 W/m^2 K for a 520-μm lone grain bed at 812 K. A corresponding value of 355 W/m^2 K of Mathur and Saxena [221] is for the 559-μm bed at 820 K. For this condition, Alavizadeh *et al.* [217] found h_{wr} as 18 W/m^2 K. The comparable value found by Mathur and Saxena [221] is 28 W/m^2 K. These approximate comparisons show that the probe of Mathur and Saxena [221] leads to values that are generally in the range of reported literature data.

E. HIGH-TEMPERATURE HEAT TRANSFER MODELS

Various models have been developed to describe the high-temperature heat transfer process in gas-fluidized beds. The models can be classified into three categories depending on how the emulsion phase is modeled. The first category models treat the emulsion phase as discrete particles, the second as a continuum with averaged properties, and the third as a stack of plane parallel plates. Szekely and Fisher [205] developed a first category model in which heat transfer is assumed to occur between the surface and the first row of particles. This is equivalent to the large particle approximation, and the change in the temperature of the particles during their residence time is assumed to be very small. However, a simultaneous approximation leads to the assumption of a uniform spatial temperature distribution within the particle. This latter assumption is valid for small particles only and is at odds with the first assumption. The two assumptions together imply that the particles do not lose, or gain any heat during their stay at the heat transfer surface. The model predicts radiation to be insignificant at temperatures less than 1273 K.

The second category models are from Yoshida *et al.* [207], Chen and Chen [223], and Glicksman and Decker [224]. The models of Yoshida *et al.* [207] and Glicksman and Decker [224] are similar inasmuch as both consider the emulsion phase heat transfer to be described by the packet model of Mickley and Fairbanks [143]. The additional heat transfer in the packet due to radiation is accounted for in both the models by the use of an enhanced, effective bed thermal conductivity that accounts for particle convective and radiative components of heat transfer. The model of Yoshida *et al.* [207] calculates the radiative flux in the bubble phase by modeling it as radiative transfer between a hemisherical surface and its equatorial plane, both at constant though different temperatures. The most significant difference between these two models is that the model of Glicksman and Decker [224] introduces an additional thermal resistance in the form of a gas film between the immersed surface and the packet (Gelperin and Einshtein [5]). The heat transfer across this film is assumed to occur by steady-state conduction and by radiation (as between two plane parallel plates). Chen and Chen [223] modeled the heat transfer process by a nonlinear differential formulation of the simultaneous radiative–conductive flux in the emulsion packet. The radiative flux in the emulsion is described by the two-flux model, which assumes a forward flux and a backward flux to traverse through the emulsion undergoing absorption and scattering. The general limitation of the second category models is that they require knowledge of the averaged bed properties, such as effective bed thermal conductivity, packet residence time, and bubble fraction. In addition, the more rigorous model of Chen and Chen [223] requires the bed absorption and scattering cross sections, which are difficult to evaluate. These parameters are either not known reliably at all or only at conditions other than those at which the model is being tested.

The third category models evaluate heat transfer between arbitrarily defined bed regions, each separated from the other by a plane parallel plate. Thus, Vedamurthy and Sastri [225] assume radiation to occur only between adjacent plates, along with conduction through the emulsion phase in between the plates. The emulsion phase is assumed transparent to radiation. Bhattacharya and Harrison [226] modified the model of Vedmurthy and Sastri [225] by assuming the emulsion to participate in the radiative transfer, and each plate to radiatively interact with 25 plates on either side. Thring [208] maintained the assumption of the radiative transparency of the emulsion, but decreased the thickness of the gas film between the immersed surface and the bed to 0.08 d_p from the value of 0.5 d_p assumed by both Vedmurthy and Sastri [225] and Bhattacharya and Harrison [226]. In another variation of the model of Vedamurthy and Sastri [225], Zhang *et al.* [216] assumed the gas film thickness to be 0.154 d_p and the radiative transfer between plates to be more than one plate deep. A significantly different approach has been adopted by Kolar *et al.* [202, 227], who model the emulsion as a series of alternate slabs of gas and solids. The heat transfer through the solid slabs is by conduction alone, while that across the gas slabs occurs by both conduction and radiation. Borodulya and Kovensky [228] model the parallel plates as two-dimensional planes with particles arranged on them in a regular grid. The radiation incident on each plate is assumed to undergo reflection, radiation, and scattering, and the absorptivity, reflectivity, and transmissivity of an individual plate are calculated on the basis of the geometric configuration of the particles. The limiting values of these parameters, as the number of plates in the stack approaches infinity, are considered to be the effective bed values.

The solution of the third category methods requires numerical computation of the finite-difference equations approximating the differential equations describing the heat transfer process in the bed. Each model has been compared with limited experimental data, and no detailed comparison or assessment of these models has been made with data of different groups of workers over a wide range of operating variables. This is an intricate and time-consuming task and has therefore not yet been attempted. The models that are easier to use, and for which the parameters required in computation can be evaluated with a relatively high level of precision, have also not been subjected to extensive detailed testing. The prediction of h_w at high temperatures from the models of Yoshida *et al.* [207] and Glicksman and Decker [224] is next compared with the experimental data obtained by Mathur and Saxena [221].

The total heat transfer coefficient, h_w, predicted by the model of Yoshida *et al.* [207] is given by the expression

$$h_w = (1 - \delta_B)(k_{eff}\rho_e C_{ps}/\tau)^{1/2} + \delta_B \sigma_s \varepsilon_t (T_b^2 + T_s^2)(T_b + T_s) \qquad (123)$$

The effective packet density is assumed to be

$$\rho_e = \rho_s(1 - \varepsilon_{mf}) \tag{124}$$

The packet residence time and the bubble fraction were calculated from the correlations by Thring [208], and the effective thermal conductivity was calculated from the model of Kunii and Smith [229].

Figures 22 and 23 show the comparison of the h_w values predicted by Eq. (123) with the experimental values of Mathur and Saxena [221] for the two particles. The predicted values are greater than the experimental values except at very low fluidizing velocities. The overprediction is greater for the 559-μm sand beds than for the 751-μm beds when compared at the same temperature of 675 K. This trend clearly points out that the lack of a gas film resistance in the model of Yoshida *et al.* [207] causes it to predict h_w values higher than those obtained experimentally. In this particle size range, and probably for all I and IIA particle beds, it is estimated that the model of Yoshida *et al.* [207] overpredicts the experimental data by about 50%.

Figures 24 and 25 show similar comparisons of the experimental h_w values with those predicted by the model of Glicksman and Decker [224]. The emulsion phase flux, q_e, is obtained from

$$q_e = \frac{(T_b - T_s)}{R_w + R_e} \tag{125}$$

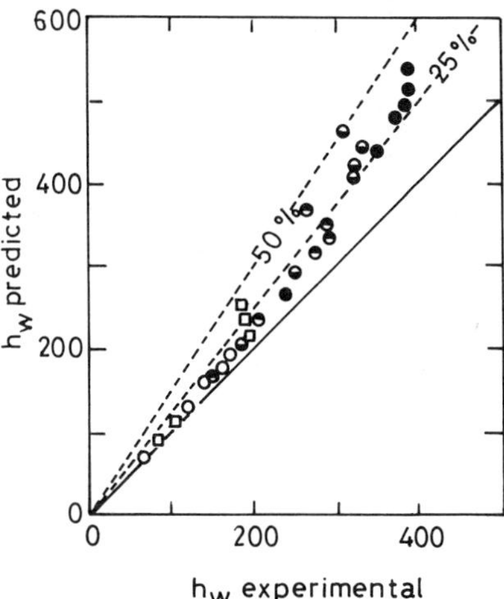

FIG. 22. Comparison of the experimental h_w values for the bed of 559-μm sand particles with the corresponding values predicted by the model of Yoshida *et al.* [207] at bed temperatures of 985 K (●), 820 K (◒), 675 K (◓), 495 K (□), and 385 K (○) (from Mathur and Saxena [221]).

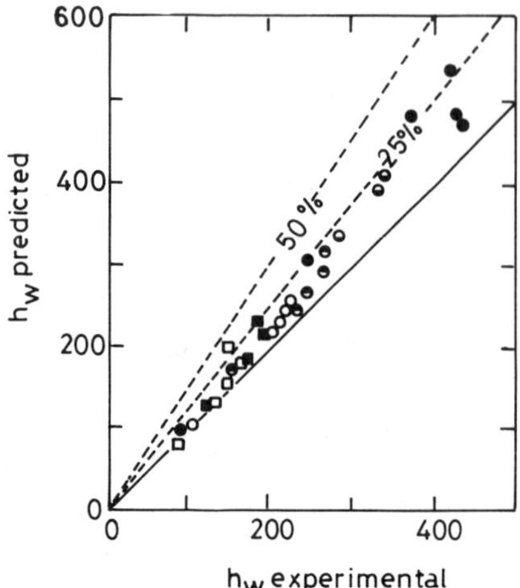

FIG. 23. Comparison of the experimental h_w values for the bed of 751-μm sand particles with the corresponding values predicted by the model of Yoshida *et al.* [207] at bed temperatures of 1175 (●), 915 K (◒), 785 K (◓), 675 K (○), 510 K (□), and 395 K (■) (from Mathur and Saxena [221]).

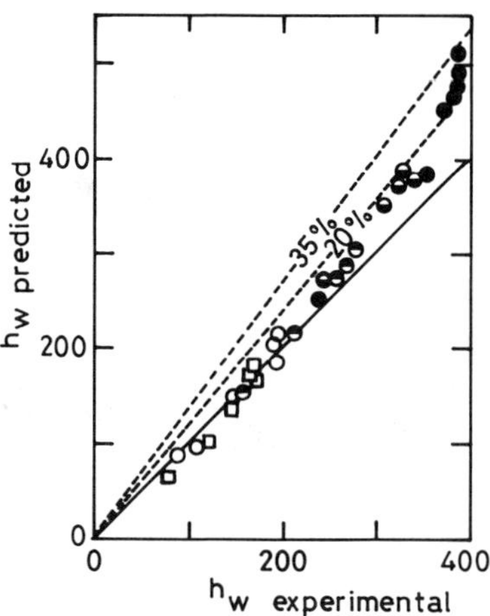

FIG. 24. Comparison of the experimental h_w values for the bed of 559-μm sand particles with the corresponding values predicted by the model of Glicksman and Decker [224] at bed temperatures of 985 K (●), 820 K (◒), 675 K (◓), 495 K (○), and 385 K (□) (from Mathur and Saxena [221]).

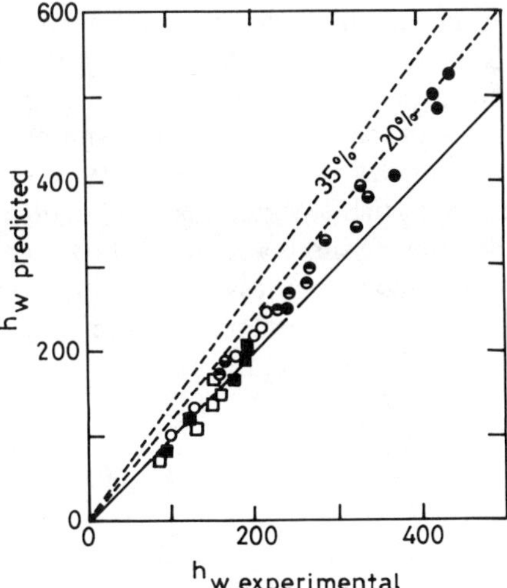

FIG. 25. Comparison of the experimental h_w values for the bed of 751-μm sand particles with the corresponding values predicted by the model of Glicksman and Decker [224] at bed temperatures of 1175 K (●), 915 K (◒), 785 K (◓), 675 K (○), 510 K (■), and 395 K (□) (from Mathur and Saxena [221]).

where

$$\frac{1}{R_w} = \frac{6k_g}{d_p} + \frac{\sigma_s(T_1^2 - T_s^2)(T_1 + T_s)}{(1/\varepsilon_t) + (1/\varepsilon_b) - 1} \tag{126}$$

and

$$\frac{1}{R_e} = (4k_{eff}\rho_e C_{ps}/\pi\tau)^{1/2} \tag{127}$$

ρ_e, τ, and f_B are calculated as described earlier, and k_{eff} is calculated from

$$k_{eff} = k_{eo} + k_r \tag{128}$$

where k_{eo} is the thermal conductivity at ambient temperature from the model of Kunii and Smith [229]. The radiation component, k_r, has been calculated, following the recommendation of Glicksman and Decker [224], by the following expression from Hill and Wilheim [230]:

$$k_r = (8/9)\, d_p \sigma_s T_b^3 \tag{129}$$

q_e is calculated from Eq. (125) but with R_w set equal to zero. Equation (112) is then used to determine q. Finally, h_w is computed using the relationship of Eq. (118).

The predictions from this model are in good agreement with the experimental data, mostly being within $\pm 20\%$. The model overpredicts the h_w values at high temperatures, and this overprediction increases with an increase in temperature. This is probably due to the plane parallel approximation of the tube-interface geometry. Also, the reliability of Eq. (126) has not been established, and the use of other correlations for k_e at high temperatures could result in better predictions. It should be noted, however, that the maximum overprediction by this model is less than 35%, which is remarkable considering its simplicity. It is, therefore, recommended that this model be used for estimating h_w at high temperatures.

IX. Concluding Remarks

From this discussion, it is clear that the heat transfer rate from an immersed surface in a gas-fluidized bed depends upon a number of factors, such as the size, size range, shape, and properties of the bed particles; operating conditions, namely, temperature, pressure, fluidizing velocity, and properties of the gas; shape, size, surface finish, and orientation of the heat transfer surface; the relative size of the heat transfer surface and the fluidized bed; and on the design of the gas distributor plate. Some brief remarks about the last feature are in order, as the rest have already been discussed in earlier sections of this chapter.

A large variety of gas distributor plates have been employed in fluidized beds of varying sizes and for different applications varying from laboratory, pilot-plant, demonstration, and commercial-size units. In a recent review article, Saxena *et al.* [231] have discussed the design principles and operating characteristics of a wide variety of gas distributor plates that have been used so far with varying degrees of success. Grewal *et al.* [232] investigated the influence of two different perforated plate distributors on h_w from a smooth, horizontal, 12.7-mm-diameter, heated copper tube immersed in square fluidized beds of glass beads of three different particle sizes (265, 357, and 427 μm). The first distributor (PCI) consisted of two perforated steel plates with a coarse cloth sandwiched between them, and a silk screen was used to cover the downstream side of the perforated plates. Holes, 4 mm in diameter, were drilled in both the plates in a square array at a pitch of 12.7 mm, the free area of the perforated plates was 7.7%. The second distributor (PCII) also consisted of two perforated steel plates with the same coarse cloth (as in PCI) sandwiched between them. Holes, 10.2 mm in diameter, were drilled in both the plates at a triangular pitch of 14 mm. The free area of the perforated plates was 37.5%. Their measurements of the pressure drop as a function of fluidizing velocity revealed that the pressure drop across the cloth predominated in both cases. The h_w values for the two distributors differed from each other under otherwise identical conditions. They attributed these differences to bubble size and bubble frequency and the resulting solids mixing patterns, which controlled particle residence time and packing density on the heat transfer surface and hence the heat transfer rate. Bubble dynamics is sensitively dependent on the distributor design [186, 231, 233], and hence h_w will, in general, depend upon the distributor type, particularly for small particles and relatively smaller fluidizing velocities, where the heat transfer rates are sensitively dependent on particle dynamics.

In passing, mention may be made of the heat transfer between the distributor plate and the bed [234]. The heat transfer coefficient was defined by making an energy balance between the various energy transfer processes taking place in the system, namely, thermal energies brought in and removed by the fluidizing gas and those exchanged between the distributor plate and the bed particles. This heat transfer coefficient was found to increase with the increase in fluidizing gas velocity, acquiring a maximum value at fluidization because of increased solids movement and slowly decreasing thereafter with an increase in fluidizing velocity because of the decrease in solids concentration in the region close to and above the distributor plate. Such a heat transfer rate was found to be independent of bed height and will also be independent of bed temperature as long as radiation is not appreciable. The heat transfer coefficient was found to increase as the particle diameter was decreased. From the knowledge of heat transfer rates for five different kinds of solids, an empirical correlation was proposed [234], namely,

$$\frac{h_D d_p}{k_g} = 0.09\left(\frac{\rho_s C_{ps}}{\rho_g C_{pg}}\right)^{0.4}\left(\frac{d_p \rho_g u}{\mu_g}\right)^{0.15} \tag{130}$$

for

$$0.3 < \left(\frac{d_p \rho_g u}{\mu_g}\right) < 40$$

Ho *et al.* [235] investigated the heat transfer between an immersed horizontal tube and shallow fluidized beds in the grid region, as compared to that in the bubbling zone, and examined its dependence on particle size, slumped bed height, superficial gas velocity, distributor open area, distributor hole size, distributor hole number, and location of the heating tube. They found a different dependence of h_w on operating parameters in the grid region as compared to those in the bed region in which bubbling occurs. For example, h_w monotonically increases in the grid region with an increase in the fluidizing velocity, while, as discussed earlier, h_w exhibits a maximum in the bubbling bed region with an increasing gas flow rate. The distributor open area ratio, distributor hole pattern, pitch, and hole numbers were found to have minimal or little effect on h_w in the bubbling bed zone but have a systematic, pronounced effect in the grid zone. These results have highlighted the importance of air jetting velocities in the grid region. The effect of slumped bed height was found to be negligible on h_w in the grid region. Similarly, the influence of particle size on h_w in the grid zone was found to be similar to that in the bubbling zone, namely, h_w increases with a decrease in d_p.

It would thus appear that more investigations of heat transfer from the distributor plate to the bed, and in the grid region from the particles to the immersed surfaces, are in order to establish the hydrodynamics and mechanistic details of the heat transfer process.

It has been shown that the particle classification scheme of Saxena and Ganzha [14], which is based on a physical understanding of the fluid-flow picture around the particles and considers the resulting interparticle force field [41], is capable of explaining the hydrodynamic behavior of fluidized beds [17, 32, 42–44] and also provides a firm basis for understanding the heat transfer process between immersed surfaces and gas-fluidized beds of small and large particles at ambient and high temperatures and pressures. All the available data and correlations are found consistent with this powder classification scheme. In light of this powder classification scheme and on the basis of its thorough assessment in conjunction with the available experimental data, it has been possible to make definite recommendations for computing heat transfer coefficients. These will be summarized in the following.

For particles belonging to groups I and IIA, probably the most reliable predictions of Nu_{max} and Nu_T can be made on the basis of Eq. (31) of Grewal and Saxena [80, 98] and Eq. (36) of Grewal and Saxena [79] for horizontal smooth cylindrical tubes. On the other hand, for vertical tubes, reliable estimates of Nu_{max} are possible from Eq. (56) of Mathur *et al.* [110] and of Nu_T from Eq. (64) of Wender and Cooper [141]. For large particles belonging to groups IIB and III, the tube orientation is not important and Eq. (75) of Mathur and Saxena [40] provides a good prediction formula for Nu based on particle diameter. The mechanistic theory of Ganzha *et al.* [147] is found adequate to describe the heat transfer rates in general for small and large particles at ambient as well as at high temperatures and pressures. However, the greatest limitation of this theory is in our ability to predict the heat transfer surface voidage values, ε_s and $\varepsilon_{s,mf}$. It is gratifying to note that experimental techniques are being developed by Saxena *et al.* [83–85] to measure these.

We are relatively more limited in our ability to predict h_w for rough and finned surfaces. However, our understanding for 60° V-shaped fins is much more elaborate, and Eq. (99) in conjunction with Eq. (97) is recommended for numerical estimates of $Nu_{ftT} \cdot Nu_T$ may be estimated as explained in the previous paragraph.

For bundles of horizontal tubes and small particles (Group I), we seem to have a reliable base for predicting heat transfer rates. The correlations of Eq. (103) of Grewal and Saxena [189] for Nu_{max} and Eq. (104) of Grewal [200] for Nu_T are recommended. Similarly, the computation based on Eq. (99) in conjunction with Eqs. (104) and (97) is found adequate by Grewal *et al.* [161] for 60° V-grooved horizontal tube banks. Experimental work with particles of groups II and III with bundles of rough and finned tubes needs to be conducted to propose reliable estimation procedures. Such experimental work needs to be undertaken as a priority in view of its relevance to the technology of fluidized-bed combustion of coal.

The status of high-temperature heat transfer in fluidized beds is discussed with special reference to radiative heat transfer. Experimental measurement techniques and theoretical prediction models are reviewed in three proposed classified categories in both the cases. On the basis of limited experimental data and analyses, it is tentatively recommended that the model of Glicksman and Decker [224] be employed to estimate the magnitude of total and radiative heat transfer coefficients. Refinements and upgrading of this model are recommended, but the need of additional reliable data to provide more detailed clues about the shortcomings of this model, as well as of some of the others, is very much in order. High-temperature heat transfer probes need to be developed for reliable and accurate measurements of h_w and h_{wr}. A typical design developed by Mathur and Saxena [221] is described at length as an illustration. Further experimental work is urgently needed. This will provide the basis for an adequate understanding of the mechanistic details of the high-temperature heat transfer process.

NOMENCLATURE

a	defined by Eq. (90)
A	empirical constant in Eq. (49)
A_1	defined by Eq. (91)
Ar	Archimedes number defined by Eq. (6)
Ar_b	defined by Eq. (22)
A_{wft}	total surface area of a finned tube
C	an adjustable parameter in Eq. (46)
C'	constant in Eq. (87)
C_0	constant in Eq. (88)
C_1	defined by Eq. (19)
C_1'	numerical constant in Eq. (73)
C_2	defined by Eq. (20)
C_2'	numerical constant in Eq. (74)
C_D	drag coefficient of a particle in a fluid, Eq. (6)
C_R	correlation factor, Eq. (64)
C_{pg}	specific heat of fluidizing gas at constant pressure
C_{ps}	specific heat of solid particles at constant pressure
d_c	diameter of the particle equivalent cylinder of unit diameter to height ratio
d_o	outer diameter of the cylindrical heat transfer tube
d_p	average particle diameter, Eq. (55)
d_{pi}	arithmetic average diameter of the successive screens
D_b	fluidized bed diameter
D_T	outside diameter of the heat transfer tube
$D_{12.7}$	heat transfer tube, 12.7 mm in diameter

D_{20}	heat transfer tube, 20 mm in diameter
f_B	time fraction during which a probe is exposed to bubbles
f_0	fraction of time the heat transfer tube is covered by bubbles
Fo	Fourier number for gas film $= \alpha_g \tau'/\delta^2$, Eq. (81)
g	acceleration due to gravity
G	superficial mass fluidizing velocity
G_{mf}	mass velocity at minimum fluidizing condition
h_c	thickness of tube segment
h_D	heat transfer coefficient between bed and distributor plate
h_w	total average heat transfer coefficient for a smooth tube
h_{wr}	radiative heat transfer coefficient between the bed and an immersed surface
$h_{w\,cond}$	conductive heat transfer coefficient between the bed and an immersed surface
$h_{w\,conv}$	convective heat transfer coefficient between the bed and an immersed surface
$h_{w\,max}$	maximum heat transfer coefficient of a smooth tube
h_{wfb}	total heat transfer coefficient for finned tube based on the surface area of a smooth tube with outside diameter equal to finned tube tip diameter
h_{wft}	total heat transfer coefficient for finned tube based on actual surface area of the finned tube
$h_{wfb\,max}$	maximum heat transfer coefficient for finned tube based on the surface area of a smooth tube with outside diameter equal to finned tube tip diameter
H	bed height
H_s	static or slumped bed height
H_{mf}	bed height at minimum fluidizing condition
k_g	gas thermal conductivity
k_p	thermal conductivity of the probe material
k_r	component of k_{eff} due to radiation
k_s	solid particle thermal conductivity
k_{eff}	effective bed thermal conductivity
k_{eo}	effective bed thermal conductivity at ambient temperature
K_1	dimensionless thermal conductivity ratio, k_g/k_s
K_2	dimensionless thermal diffusivity ratio, α_g/α_s
K_δ	dimensionless thickness ratio, d_c/δ
K	dimensionless ratio, $k_g C_{pg} P_g / k_s C_{ps} P_s$
Kn	Knudsen number, $2l_1/d_p$, Eq. (48)
l	Characteristic length parameter
l_1	modified mean-free-path of gas molecules, Eq. (48)
L	center-to-center distance between adjacent equivalent cylinders
m	dimensionless function, $K_2^{1/2} K_\delta$
M	molar mass
N	nondimensional contact time or number of transfer units, Eq. (46)
Nu	Nusselt number based on particle diameter, $h_w d_p/k_g$
Nu_{cond}	conductive component of particle Nusselt number, $h_{w\,cond} d_p/k_g$
Nu_{conv}	convective component of particle Nusselt number, $h_{w\,conv} d_p/k_g$
Nu_{max}	maximum value of Nusselt number, $h_{w\,max} d_p/k_g$
Nu_T	Nusselt number based on tube diameter, $h_w D_T/k_g$
$(Nu)_{13\,mm}$	value of the particle Nusselt number obtained on the basis of data for h_w of a 13-mm diameter tube
Nu_{D_b}	Nusselt number based on bed diameter, $h_w D_b/k_g$
$(Nu)_{D_T}$	value of the particle Nusselt number obtained on the basis of data for h_w of D_T diameter tube
Nu_g	Nusselt number for gas flow without solids, $h_w D_T/k_g$
$(Nu_l)_{conv}$	convective Nusselt number based on characteristic length parameter, $h_{w\,conv} l/k_g$
Nu_{ftT}	Nusselt number for a rough or finned tube based on tube diameter, $h_{wft} D_T/k_g$
P	tube pitch, i.e., center-to-center distance of adjacent tubes; gas pressure

P_f pitch for V-thread tubes, i.e., the distance between two identical points of consecutive threads

Pr Prandtl number, $\mu_g C_{pg}/k_g$

q heat flux at the probe surface

q_e heat flux in the emulsion phase

q_r radiative heat flux at the probe surface

q_{gc} gas convective heat flux at the probe surface

q_{pc} particle convective flux at the probe surface

q_{rB} bubble phase radiative flux at the probe surface

q_{re} emulsion phase radiative flux at the probe surface

q_{gcB} bubble phase gas convective flux at the probe surface

q_{gce} emulsion phase gas convective flux at the probe surface

q_{pce} emulsion phase particle convective flux at the probe surface

Q electrical power supplied to the heater

R universal gas constant

R_e emulsion phase heat transfer resistance

R_p packet contact resistance, Eq. (52)

R_t thermal contact resistance, Eq. (54)

R_w gas film heat transfer resistance at the probe wall

Re particle Reynolds number, $G d_p/\mu_g$ or $u\rho_g d_p/\mu_g$

Re_{mf} particle Reynolds number at minimum fluidization, $G_{mf} d_p/\mu_g$ or $u_{mf}\rho_g d_p/\mu_g$, Eq. (5)

Re_{D_b} Reynolds number based on bed diameter, $u\rho_g D_b/\mu_g$

Re_l Reynolds number based on characteristic length parameter, Gl/μ_g

S_H horizontal tube pitch, i.e., center to center horizontal tube spacing

S_V vertical tube pitch, i.e., center to center vertical tube spacing

T temperature

T_1 temperature of the particles adjacent to the heat transfer surface

T_b bed temperature

T_s surface temperature of the heat transfer probe

T_{Bi} temperature of the inner face of fluxmeter B

T_{Bo} temperature of the outerface of fluxmeter B

T_{ci} temperature of the inner face of fluxmeter C

T_{co} temperature of the outer face of fluxmeter C

T_{wb} fin-base temperature

u superficial gas fluidizing velocity

u_s average solids velocity

u_{mf} superficial gas velocity at minimum fluidization

x_i weight fraction of particles retained in the ith and $(i + 1)$th sieves

X_B thickness of the fluxmeter B

X_c thickness of the fluxmeter C

Z a group as defined by Eq. (45)

Greek Symbols

α_g thermal diffusivity of gas, $k_g/C_{pg}\rho_g$

α_s thermal diffusivity of solid particles, $k_s/C_{ps}\rho_s$

β heat transfer capacity function defined as the ratio of the effective heat transfer coefficient for a finned tube, h_{wfb}, to the heat transfer coefficient for a smooth tube, having the same outside diameter as the tip diameter of the finned tube, h_w, at the same fluidizing velocity

β_{max} ratio of the maximum heat transfer coefficient for a rough tube to its value for a smooth tube, having the same outside diameter and at the same fluidizing velocity

β_1 time fraction that a tube is covered by bubbles

ΔP_1 experimental pressure drop measured across the top part of the fluidized bed above the H_{mf} level at a given fluidizing velocity

ΔP_{mf}	pressure drop across the bed at the minimum fluidization condition	μ_g	gas viscosity
δ	gas film thickness for a flat surface	μ	a quantity defined by Eq. (80)
δ_B	bubble fraction in the bed	μ_1	a characteristic root of Eq. (80)
ε	porosity or voidage in general	ρ_b	bed density
ε_b	bulk bed voidage	ρ_e	emulsion phase density
ε_s	bed voidage near heat transfer surface	ρ_g	gas density
$\bar{\varepsilon}_s$	mean bed voidage near heat transfer surface	ρ_s	density of solid particles
ε_t	emissivity of heat transfer probe surface	σ	thermal accommodation coefficient for gas–solid interface
ε_{mf}	bulk bed voidage at minimum fluidization	σ_s	Stefan–Boltzmann constant
$\varepsilon_{s,mf}$	bed voidage near heat transfer surface at minimum fluidization	τ	packet residence time at the heat transfer surface
η	fin efficiency, Eq. (95)	τ'	time
		ϕ	fin effectiveness factor, h_{wft}/h_w, Eq. (97)
		ϕ_s	sphericity of solid particles

ACKNOWLEDGMENTS

This work is supported in part through a grant from the National Science Foundation in a joint India–United States research program. Partial support of this work from the Center of Research for Sulfur in Coal is also acknowledged. The cooperation and hospitality from the Banaras Hindu University and in particular from the staff of the Department of Chemical Engineering and Technology are gratefully acknowledged.

REFERENCES

1. S. S. Zabrodsky, "Hydrodynamics and Heat Transfer in Fluidized Beds." MIT Press, Cambridge, Massachusetts, 1966.
2. J. S. M. Botterill, "Fluid Bed Heat Transfer." Academic Press, New York, 1975.
3. S. C. Saxena, N. S. Grewal, J. D. Gabor, S. S. Zabrodsky, and D. M. Galershtein, *Adv. Heat Transfer* **14,** 149 (1978).
4. S. C. Saxena and J. D. Gabor, *Prog. Energy Combust. Sci.* **7,** 73 (1981).
5. N. I. Gelperin and V. G. Einshtein, *In* "Fluidization" (J. F. Davidson and D. Harrison, ed.), p. 471. Academic Press, New York, 1971.
6. N. M. Catipovic, G. N. Jovanovic, and T. J. Fitzgerald, *Am. Inst. Chem. Eng. J.* **24,** 543 (1978).
7. G. S. Canada and M. H. McLaughlin, *Am. Inst. Chem. Eng. Symp. Ser.* **74,** 27 (1978).
8. V. A. Borodulya, V. L. Ganzha, S. N. Upadhyay, and S. C. Saxena, *Int. J. Heat Mass Transfer* **23,** 1602 (1980).
9. D. Geldart, *Powder Technol.* **7,** 285 (1973).
10. D. Kunii and O. Levenspiel, "Fluidization Engineering." Krieger, Huntington, New York, 1977.
11. F. A. Zenz and D. F. Othmer, "Fluidization and Fluid-Particle Systems." Van Nostrand-Reinhold, Princeton, New Jersey, 1960.

12. J. R. Grace, *In* "Handbook of Multiphase Systems" (G. Hetsroni, ed.), Hemisphere, New York, 1982.
13. N. Seki, S. Fukusako, and K. Torikoshi, *Trans. Am. Soc. Mech. Eng., J. Heat Transfer* **101,** 386 (1979).
14. S. C. Saxena and V. L. Ganzha, *Powder Technol.* **39,** 199 (1984).
15. J. S. M. Botterill, Y. Teoman, and K. R. Yuregir, *Am. Inst. Chem. Eng. Symp. Ser.* **77,** 330 (1981).
16. G. Geldart and R. R. Cranfield, *Chem. Eng. J.* **3,** 211 (1972).
17. J. S. M. Botterill, Y. Teoman, and K. R. Yuregir, *Powder Technol.* **31,** 101 (1982).
18. S. C. Saxena and G. J. Vogel, *Trans. Inst. Chem. Eng.* **55,** 184 (1977).
19. R. R. Pattipati and C. Y. Wen, *Ind. Eng. Chem. Process Des. Dev.* **20,** 705 (1981).
20. S. Ergun, *Chem. Eng. Prog.* **48,** 89 (1952).
21. V. K. Maskaev and A. P. Baskakov, *J. Eng. Phys.* **24,** 411 (1973).
22. A. P. Baskakov, B. V. Berg, O. K. Vitt, N. F. Filippovsky, V. A. Kirakosyan, J. M. Goldobin, and V. K. Maskaev, *Powder Technol.* **8,** 273 (1973).
23. A. P. Baskakov and N. M. Suprun, *Int. Chem. Eng.* **12,** 324 (1972).
24. A. O. O. Denloye and J. S. M. Botterill, *Powder Technol.* **19,** 197 (1978).
25. J. S. M. Botterill and A. O. O. Denloye, *Chem. Eng. Sci.* **33,** 509 (1978).
26. F. W. Staub, *Trans. Am. Soc. Mech. Eng., J. Heat Transfer* **101,** 391 (1979).
27. L. R. Glicksman and N. Decker, *Proc. Int. Conf. Fluid. Bed Combust., 6th* **3,** 1152 (1980).
28. V. A. Borodulya, V. L. Ganzha, and A. I. Podberezsky, *In* "Fluidization" (J. R. Grace and J. M. Matsen, eds.), p. 201. Plenum, New York, 1980.
29. S. S. Zabrodsky, Yu G. Epanov, D. M. Galershtein, S. C. Saxena, and A. K. Kolar, *Int. J. Heat Mass Transfer* **24,** 571 (1981).
30. N. M. Catipovic, G. N. Jovanovic, T. J. Fitzgerald, and O. Levenspiel, *In* "Fluidization" (J. R. Grace and J. M. Matsen, eds.), p. 225. Plenum, New York, 1980.
31. J. S. M. Botterill and Y. Teomen, "Fluidization" (J. R. Grace and J. M. Matsen, eds.), p. 93. Plenum, New York, 1980.
32. J. S. M. Botterill, Y. Teoman, and K. R. Yuregir, *Chem. Eng. Commun.* **15,** 227 (1982).
33. V. D. Goroshko, R. B. Rozenbaum, and D. M. Todes, *Izv. Vyssh. Uchbn. Zaved. Neft. Gaz.* **1,** 125 (1958).
34. S. S. Zabrodsky, High temperature fluidized bed installations. *Energia* (*Moscow*) 38 (1971).
35. A. O. O. Denloye and J. S. M. Botterill, *Chem. Eng. Sci.* **32,** 461 (1977).
36. S. S. Zabrodsky, N. V. Antonishin, and A. L. Parnas, *Can. J. Chem. Eng.* **54,** 52 (1976).
37. S. C. Saxena and V. L. Ganzha, *Powder Technol.* **44,** 115 (1985).
38. A. Mathur and S. C. Saxena, "Trends in Electric Utility Research" (C. W. Bullard and P. J. Womeldorff, eds.), p. 305. Pergamon, New York, 1984.
39. S. C. Saxena and A. Mathur, *Annu. Pittsburgh Coal Conf., 1st* 411 (1984).
40. A. Mathur and S. C. Saxena, *Energy* **11,** 843 (1986).
41. A. Mathur and S. C. Saxena, *Powder Technol.* **45,** 287 (1986).
42. A. Lucas, J. Arnaldos, J. Casal, and L. Puigjaner, *Chem. Engl. Commun.* **41,** 121 (1986).
43. S. C. Saxena, A. Mathur, and Z. F. Zhang, *Am. Inst. Chem. Eng. J.* **33,** 500 (1987).
44. A. Mathur, S. C. Saxena, and Z. F. Zhang, *Powder Technol.* **47,** 247 (1986).
45. M. E. Aerov and O. M. Todes, "Hydraulic and Thermal Fundamentals of the Operation of Steady-State Fluidized Granular Bed Apparatus." Khimia, Leningrad, 1968.
46. M. Leva, "Fluidization." McGraw-Hill, New York, 1959.
47. R. Clift, J. R. Grace, and M. E. Weber, "Bubbles, Drops and Particles." Academic Press, New York, 1978.
48. D. C. Chitester, R. M. Kornosky, L. S. Fan, and J. P. Danko, *Chem. Eng. Sci.* **39,** 253 (1984).
49. S. C. Saxena and A. Mathur, *Ind. Eng. Chem. Process Des. Dev.* **26,** 859 (1987).

50. R. R. Pattipati and C. Y. Wen, *Ind. Eng. Chem. Process Des. Dev.* **21,** 785 (1982).
51. T. Mii, K. Yoshida, and D. Kunii, *J. Chem. Eng. Jpn.* **6,** 100 (1972).
52. B. Singh, G. R. Rigby, and T. G. Callcott, *Trans. Inst. Chem. Eng.* **51,** 93 (1973).
53. J. Broughton, *Trans. Inst. Chem. Eng.* **52,** 105 (1974).
54. A. Desai, H. Kikukawa, and A. H. Pulsifer, *Powder Technol.* **16,** 143 (1977).
55. M. E. Crowther and J. C. Whitehead, *In* "Fluidization" (J. F. Davidson and D. L. Keairns, eds.), p. 55. Cambridge Univ. Press, London and New York, 1978.
56. D. F. King and D. Harrison, *In* "Fluidization" (J. R. Grace and J. M. Matsen, eds.), p. 101. Plenum, New York, 1980.
57. S. P. Babu, B. Shah, and A. Talwalkar, *Am. Inst. Chem. Eng. Symp. Ser.* **74,** 176 (1978).
58. N. S. Grewal and S. C. Saxena, *Powder Technol.* **26,** 229 (1980).
59. V. Thonglimp, N. Hiquily, and C. Lagurie, *Powder Technol.* **38,** 233 (1984).
60. C. Y. Wen and Y. H. Yu, *Chem. Eng. Prog. Symp. Ser.* **62,** 100 (1966).
61. P. Bourgeois and P. Grenier, *Can. J. Chem. Eng.* **46,** 325 (1968).
62. J. F. Richardson and M. A. Da St. Jeronimo, *Chem. Eng. Sci.* **34,** 1419 (1979).
63. Z. X. Zheng, R. Yamazaki, and G. Jimbo, *Kagaku Kogaku Ronbunshu* **11,** 115 (1985).
64. S. Shrivastava, A. Mathur, and S. C. Saxena, *Am. Inst. Chem. Eng. J.* **32,** 1227 (1986).
65. J. S. M. Botterill, Y. Teoman, and K. R. Yuregir, *Powder Technol.* **30,** 95 (1981).
66. S. C. Saxena and N. S. Grewal, *Powder Technol.* **30,** 96 (1981).
67. V. A. Borodulya, V. L. Ganzha, and V. I. Kovensky, "Hydrodynamic and Heat Transfer in a Pressurized Fluidized Bed." Nauka I Technika, Minsk, 1982.
68. J. S. M. Botterill and M. Desai, Powder Technol. **6,** 231 (1972).
69. T. M. Knowlton, *Proc. Annu. Meet. Am. Inst. Chem. Eng., 67th, Washington, D. C., December* Pap. No. 9B (1974).
70. B. V. Chesnokov, M. G. Slinko, and V. Sh. Kernerman, *Chim. Prom. (Moscow)* **11,** 767 (1961).
71. G. P. Sechenov and V. S. Altshuler, *Gornaya Prom.* **11,** 12 (1958).
72. I. P. Mikhlenov, D. G. Traber, and V. B. Sarkiez, *Zh. Prikhadnoy Chim. Leningrad* **33,** 2206 (1960).
73. T. Varadi and J. R. Grace, "Fluidization"(J. F. Davidson and D. L. Keairns, eds.), p. 55. Cambridge Univ. Press, London and New York, 1978.
74. V. S. Altshuler and G. P. Sechenov, Protsessy v. Pseuvdoozhyzhennom Sloye Pod Davleniem, Moscow Academy of Sciences, 1963.
75. N. M. Boguslavskiy and T. Kh. Melik-Akhnazarov, Psevdoozhyzhenie v Khimicheskoy Technologii, Moscow GOSINTI, 1960.
76. N. S. Grewal, E. S. Sorenson, and G. Goblirsch, *Chem. Eng. Commun.* **39,** 43 (1985).
77. N. S. Grewal, *In* "Handbook of Heat and Mass Transfer, Vol. I: Heat Transfer Operations" (N. P. Cheremisinoff, ed.), p. 609. Gulf Publ., Houston, 1986.
78. L. R. Glicksman and D. Decker, *Am. Inst. Chem. Eng. Symp. Ser.* **77,** 341 (1981).
79. N. S. Grewal and S. C. Saxena, *Int. J. Heat Mass Transfer* **23,** 1505 (1980).
80. N. S. Grewal and S. C. Saxena, *Ind. Eng. Chem. Process Des. Dev.* **20,** 108 (1981).
81. R. S. Verma and S. C. Saxena, *Energy* **8,** 909 (1983).
82. D. H. Glass and D. Harrison, *Chem. Eng. Sci.* **19,** 1001 (1964).
83. S. C. Saxena, D. C. Patel, and D. Kathuria, *Am. Inst. Chem. Eng. J.* **33,** 672 (1987).
84. S. C. Saxena and D. C. Patel, *Int. J. Particulate Powder Technol.,* in press (1988).
85. S. C. Saxena and D. C. Patel, *Am. Inst. Chem. Eng. Symp. Ser., Fluidization Engineering: Fundamentals and Applications,* in press (1988).
86. P. J. Shah, S. N. Upadhyay, and S. C. Saxena, *Natl. Heat Mass Transfer Conf., 6th, Indian Inst. Technol., Madras, India* HMT-21-81 (1981).
87. J. A. Doherty and S. C. Saxena, *Proc. Int. Conf. Fluidized Bed Combust., 8th* **III,** 1389 (1985).
88. H. A. Vreedenberg, *Chem. Eng. Sci.* **9,** 52 (1958).
89. J. C. Petrie, W. A. Freeby, and J. A. Buckham, *Chem. Eng. Prog. Symp. Ser.* **64,** 45 (1968).

90. N. I. Gelperin, V. G. Einstein, and N. A. Romanova, *Khim. Mashinostr.* No. 5, 13 (1963).
91. I. Goel and S. C. Saxena, *Proc. Int. Fluidized Bed Conf., 7th, Philadelphia, October 1982; DOE/METC*/83-48, **2,** 804 (1983).
92. D. C. Cherrington, L. P. Golan, and F. G. Hammitt, *Proc. Int. Conf. Fluidized Bed Combust., 5th, Washington, D. C.* 184 (1977).
93. L. P. Golan, R. D. Cherrington, C. E. Sraporough, and S. C. Weiner, *Chem. Eng. Prog.* **75,** 63 (1979).
94. D. G. Traber, V. B. Sarkicz, and I. P. Muchlenov, *Zhu. Prikladnoy Chim. (Leningrad)* **33,** 2197 (1960).
95. L. B. Rabinovich and G. O. Sechenov, *J. Eng. Phys. (Minsk)* **22,** 789 (1972).
96. A. M. Xavier, D. F. King, J. F. Davidson, and D. Harrison, *In* "Fluidization" (J. R. Grace and J. M. Matsen, eds.), p. 209. Plenum, New York, 1980.
97. N. N. Varygin and I. G. Martyushin, *Khim. Mashinostr.* **5,** 6 (1959).
98. N. S. Grewal, *Lett. Heat Mass Transfer* **9,** 377 (1982).
99. B. R. Andeen and L. R. Glicksman, *ASME-AIChE Heat Transfer Conf., St. Louis, August 9-11* 76-HT-67 (1976).
100. V. G. Ainshtein, *In* "Hydrodynamics and Heat Transfer in Fluidized Beds" (S. S. Zabrodsky, ed.), p. 270. MIT Press, Cambridge, Massachusetts, 1966.
101. N. I. Gelperin, V. Ya. Kruglikov, and V. G. Ainshtein, *In* "Heat Transfer Between a Fluidized Bed and a Surface" (V. G. Ainshtein and N. I. Gelperin, eds.). *Int. Chem. Eng.* **6,** 67 (1966).
102. W. E. Genetti, R. A. Schmall, and E. S. Grimmett, *Am. Inst. Chem. Eng. Prog. Symp. Ser.* **67,** 90 (1971).
103. A. N. Ternovskaya and Yu. G. Korenberg, "Pyrite Kilning in a Fluidized Bed." *Izd. Khimiya (Moscow)* 1971.
104. N. S. Grewal, Experimental and theoretical investigations of heat transfer between a gas-solid fluidized bed and immersed tubes. Ph. D. thesis, University of Illinois at Chicago, 1979.
105. H. Martin, *Chem. Eng. Process* **18,** 157 (1984).
106. H. Martin, *Chem. Eng. Process* **18,** 199 (1984).
107. J. A. Doherty, R. S. Verma, S. Shrivastava, and S. C. Saxena, *Energy* **11,** 773 (1986).
108. H. Martin, *Chem. Eng. Commun.* **13,** 1 (1981).
109. H. Martin, *Int. Chem. Eng.* **22,** 30 (1982).
110. A. Mathur, S. C. Saxena, and A. Chao, *Ind. Eng. Chem. Process Des. Dev.* **25,** 156 (1986).
111. T. R. White, A. Mathur, and S. C. Saxena, *Chem. Eng. J.* **32,** 1 (1986).
112. J. C. Chen and J. G. Withers, *Am. Inst. Chem. Eng. Symp. Ser.* **74,** 327 (1974).
113. N. V. Antonishin, *In* "Hydrodynamics and Heat Transfer in Fluidized Beds" (S. S. Zabrodsky, ed.), p. 291. MIT Press, Cambridge, Massachusetts, 1966.
114. C. A. Baerg, J. Klassen, and P. E. Gishler, *Can. J. Res.* **28F,** 287 (1950).
115. H. S. Mickley and C. A. Trilling, *Ind. Eng. Chem.* **41,** 1135 (1949).
116. H. S. Mickley, D. F. Fairbanks, and R. D. Hawthorne, *Chem. Eng. Prog. Symp. Ser.* **57,** 51 (1961).
117. A. B. Whitehead and A. D. Young, *Proc. Int. Symp. Fluidization, Eindhoven* p. 284 (1967).
118. A. B. Whitehead, D. C. Dent, and A. D. Young, *Powder Technol.* **1,** 149 (1967).
119. A. B. Whitehead, G. Gartside, and D. C. Dent, *Chem. Eng. J.* **1,** 175 (1970).
120. A. B. Whitehead and A. D. Young, *Proc. Int. Symp. Fluidization, Eindhoven* p. 294 (1967).
121. W. H. Park, W. K. Kang, D. E. Capes, and G. L. Osberg, *Chem. Eng. Sci.* **24,** 851 (1969).
122. J. Werther and O. Molerus, *Int. J. Multiphase Flow* **1,** 123 (1973).
123. J. Werther, *Am. Inst. Chem. Eng. Symp. Ser.* **70,** 53 (1974).
124. J. Werther, *Fluidization Technol.* **1,** 215 (1975).
125. M. M. Chen, B. T. Chao, and J. Liijegren, *In* "Fluidization" (D. Kunii and R. Toei, eds.), p. 203. Engineering Foundation, New York, 1983.
126. J. S. Lin, M. M. Chen, and B. T. Chao, *Am. Inst. Chem. Eng. J.* **31,** 465 (1985).

127. P. N. Rowe and D. J. Everett, *Trans. Inst. Chem. Eng.* **50,** 42 (1972).
128. P. N. Rowe and H. Masson, *Chem. Eng. Sci.* **35,** 1443 (1980).
129. P. N. Rowe and H. Masson, *Trans. Inst. Chem. Eng.* **59,** 177 (1981).
130. W. Volk, C. A. Johnson, and H. H. Stotler, *Chem. Eng. Prog.* **58,** 44 (1963).
131. V. A. Borodulya, V. L. Ganzha, A. I. Podberezsky, S. N. Upadhyay, and S. C. Saxena, *Int. J. Heat Mass Transfer* **26,** 1577 (1983).
132. A. O. O. Denloye, Ph. D. thesis, University of Birmingham, 1970.
133. M. Kimura, N. Kono, and T. Kameda, *Chem. Eng. J.* **19,** 397 (1955).
134. T. F. Ozkaynak and J. C. Chen, *Am. Inst. Chem. Eng. J.* **26,** 544 (1980).
135. R. D. Toomey and H. F. Johnstone, *Chem. Eng. Prog. Symp. Ser.* **5,** 51 (1953).
136. R. Wunder, Dissertation, Tech. Univ., Muchen, 1980; tabulated in H. Martin, *Chem. Eng. Process* **18,** 199 (1984).
137. N. Ziegler, L. B. Koppel, and W. T. Brazelton, *Ind. Eng. Chem. Fundam.* **3,** 325 (1964).
138. C. O. Miller and A. K. Logwinuk, *Ind. Eng. Chem.* **43,** 1220 (1951).
139. H. A. Vreedenberg, J. Appl. Chem. Suppl. S26 (1952).
140. H. A. Vreedenberg, *Chem. Eng. Sci.* **11,** 274 (1960).
141. L. Wender and G. T. Cooper, *Am. Inst. Chem. Eng. J.* **4,** 15 (1958).
142. H. L. Olin and O. C. Dean, *Pet. Eng.* **25,** C-23 (1953).
143. H. S. Mickley and D. F. Fairbanks, *Am. Inst. Chem. Eng. J.* **1,** 374 (1955).
144. V. A. Borodulya, V. L. Ganzha, A. I. Podberezsky, S. N. Upadhyay, and S. C. Saxena, *Int. J. Heat Mass Transfer* **27,** 1219 (1984).
145. R. Chandran and J. C. Chen, *Am. Inst. Chem. Eng. J.* **28,** 907 (1982).
146. A. M. Xavier, D. F. King, J. F. Davidson, and D. Harrison, *In* "Fluidization" (J. R. Grace and J. M. Matsen, eds.), p. 209. Plenum, New York, 1980.
147. V. L. Ganzha, S. N. Upadhyay, and S. C. Saxena, *Int. J. Heat Mass Transfer* **25,** 1531 (1982).
148. J. Kubie, *Int. J. Heat Mass Transfer* **27,** 153 (1984).
149. V. L. Ganzha and S. C. Saxena, *Int. J. Heat Mass Transfer* **27,** 153 (1984).
150. S. C. Saxena and R. K. Joshi, "Thermal Accommodation Coefficient and Adsorption of Gases." McGraw Hill, New York, 1979.
151. A. V. Luikov, "Theory of Thermal Conductivity" (in Russian), p. 288. Gostechizdat, Moscow, 1952; English translation, "Analytical Heat Diffusion Theory" (J. P. Hartnett, ed.), p. 411. Academic Press, New York, 1968.
152. T. R. Galloway and B. H. Sage, *Chem. Eng. Sci.* **30,** 495 (1970).
153. O. Levenspiel and J. S. Walton, *Chem. Eng. Prog. Symp. Ser.* **50,** 7 (1954).
154. D. T. Wasan and K. S. Ahluwalia, *Chem. Eng. Sci.* **24,** 1535 (1969).
155. L. B. Rabinovich and G. P. Sechenov, *Eng. Phys. Zh.* (in Russian) **22,** 789 (1972).
156. F. P. Incorpera and D. P. DeWitt, "Fundamentals of Heat Transfer," p. 328. Wiley, New York, 1981.
157. V. P. Isachenko, O. A. Osipova, and A. S. Sukomel, "Heat Transfer," p. 209. Mir, Moscow, 1977.
158. V. A. Borodulya, V. L. Ganzha, and A. I. Podberezsky, "Problemy Teple-i-Massoobmena v Processach Goreniya, Ispolzuyemych v Energetice," p. 141. Minsk, 1980.
159. N. S. Grewal and S. C. Saxena, *Trans. Am. Soc. Mech. Eng., J. Heat Transfer* **101,** 397 (1979).
160. W. B. Krause and A. R. Peters, *ASME-AIChE Heat Transfer Conf. Orlando July 27–30* HT-48 (1980).
161. N. S. Grewal, T. K. Cheung, and S. C. Saxena, *Ind. Eng. Chem. Process Des. Dev.* **24,** 458 (1985).
162. B. Neukirchen and H. Blenke, *Chem. Ing. Tech.* **45,** 307 (1973).
163. N. I. Gelperin, V. G. Einshtein, I. N. Toskubaev, and S. K. Vasiliev, *Trudy MITKhT, Moscow, Vyp.* **11,** 116 (1974).

164. G. D'Albon, D. Peretz, V. Cernescu, I. Bendescu, V. Clotan, and Gh. Lozonschi, *Bull. Inst. Politechnic din Jasi*, **XVII,** N3-4, 101 (1971).
165. M. R. Vijayaraghavan and V. M. K. Sastri, *Conf. Future Energy Prod. Int. Center Heat Mass Transfer, Dubrovnic* 571 (1975).
166. I. Goel, S. C. Saxena, and A. F. Dolidovich, *Trans. Am. Soc. Mech. Eng., J. Heat. Transfer* **106,** 91 (1984).
167. N. S. Grewal, T. K. Chueng, and S. C. Saxena, *Am. Soc. Mech. Eng.* 84-HT-110 (1984).
168. V. N. Korolev and N. I. Syromyatnikov, *J. Eng. Phys.* **28,** 698 (1975).
169. J. C. Chen and J. G. Withers, *Am. Inst. Chem. Eng. Natl. Heat Transfer Conf., 15th, San Francisco, August* Pap. No. 34 (1975).
170. S. S. Zabrodsky, A. I. Tamarin, A. F. Dolidovich, G. I. Palchonok, and Yu. G. Epanov, *In* "Fluidization" (J. R. Grace and J. M. Matsen, eds.), p. 195. Plenum, New York, 1980.
171. W. E. Genetti and M. T. Kratovil, *Am. Inst. Chem. Eng. Natl. Am. Inst. Chem. Eng. Meet., 83rd, Houston, March* Pap. No. 46 (1977).
172. W. J. Bartel, W. E. Genetti, and E. S. Grimmet, *Chem. Eng. Prog. Symp. Ser.* **116,** 85 (1971).
173. W. J. Bartel and W. E. Genetti, *Chem. Eng. Prog. Symp. Ser.* **69,** 85 (1973).
174. S. J. Priebe and W. E. Genetti, *Am. Inst. Chem. Eng. Symp. Ser.* **73,** 38 (1977).
175. H. J. Natusch and H. Blenke, *Verfahrenstechnik* **8,** 286 (1974).
176. H. J. Natusch and H. Blenke, *Chem. Ing. Tech.* **7,** 293 (1973).
177. N. I. Gelperin, V. G. Einshtein, and I. N. Toskubaev, *Khim. Tekhnol. Topliv. Masel* **2,** 42 (1972).
178. N. I. Gelperin, V. G. Einshtein, I. N. Toskubaev, S. K. Vasiliev, and N. Ryspaev, *Trudy TIT Ch T, Vyp* **1,** 165 (1974).
179. F. W. Staub and G. S. Canada, *In* "Fluidization" (J. F. Davidson and D. L. Keairns, eds.), p. 339. Cambridge Univ. Press, London and New York, 1978.
180. J. C. Chen and J. G. Withers, *Am. Inst. Chem. Eng. Symp. Ser.* **74,** 327 (1978).
181. W. E. Genetti and D. Everly, *Annu. Meet. Am. Inst. Chem. Eng., 71st, Miami Beach, November* (1978).
182. W. E. Genetti, S. P. Yurich, and D. W. Vanderhoof, *ASME-AIChE Natl. Heat Transfer Conf.* 80-HT-118 (1980).
183. R. Chandran, J. C. Chen, and F. W. Staub, *Trans. Am. Soc. Mech. Eng. Ser. C, J. Heat Transfer* **102,** 152 (1980).
184. L. P. Golan, G. V. LaLonde, and S. C. Weiner, *Proc. Int. Conf. Fluidized Bed Combust., 6th* 1173 (1980).
185. S. C. Saxena, *Lett. Heat Mass Transfer* **6,** 225 (1979).
186. S. C. Saxena, A. Chatterjee, and R. C. Patel, *Powder Technol.* **22,** 191 (1979).
187. V. A. Borodulya, V. L. Ganzha, A. I. Zheltov, S. N. Upadhyay, and S. C. Saxena, *Lett. Heat Mass Transfer* **7,** 83 (1980).
188. N. S. Grewal and S. C. Saxena, *Am. Soc. Mech. Eng., New York* 80-HT-119 (1980).
189. N. S. Grewal and S. C. Saxena, *Ind. Eng. Chem. Process Des. Dev.* **22,** 367 (1983).
190. J. A. Doherty and S. C. Saxena, *Proc. Annu. Pittsburgh Coal Conf., 2nd* p. 861 (1985).
191. N. I. Gelperin, V. G. Einshtein, and A. V. Zaikov-ski, *Khim. Mashinoster (Moscow)* No. 3, 17 (1968).
192. N. I. Gelperin, V. G. Einshtein, and L. A. Korotyanskaya, *Int. Chem. Eng.* **9,** 137 (1969).
193. V. V. Chekansky, B. S. Sheindlin, D. M. Galershtein, and K. S. Antonyuk, *Tr. Spets. Konstr. Byurs. Avtomat. Neftepererab. Neftekhim.* **3,** 143 (1970).
194. W. C. Howe and C. Aulisio, *Chem. Eng. Prog.* **73,** 69 (1977).
195. R. K. Bansal, P. V. Kadaba, and P. V. Desai, *ASME/AIChE Natl. Heat Transfer Conf., Orlando* 80-HT-115 (1980).
196. A. M. Xavier and J. F. Davidson, *In* "Fluidization" (J. F. Davidson and D. L. Keairns, eds.), p. 333. Cambridge Univ. Press, London and New York, 1978.

197. N. S. Grewal and D. R. Hajicek, *Int. Heat Transfer Conf., 7th, Munich* HX23 (1982).
198. Aerojet Energy Conversion Company, Final Technical Report submitted to the Department of Energy under Contract No. AC03-78ET 11343, entitled, "Evaluation of Fluid Bed Heat Exchanger Optimization Parameters," Sacramento, California, 1980.
199. W. H. McAdams, "Heat Transmission," 3rd ed., p. 271. McGraw Hill, New York, 1954.
200. N. S. Grewal, *Powder Technol.* **30,** 145 (1981).
201. R. K. Bansal, P. V. Kadaba, and P. V. Desai, *ASME-AIChE Natl. Heat Transfer Conf., Orlando, July* 80-HT-115 (1980).
202. A. K. Kolar, N. S. Grewal, and S. C. Saxena, *Int. J. Heat Mass Transfer* **22,** 1695 (1979).
203. L. J. Jolley, *Fuel* **28,** 114 (1949).
204. A. I. Il'Chenko, V. S. Pishakov, and K. E. Makhorin, *J. Eng. Phys.* **14,** 321 (1968).
205. J. Szekely and R. J. Fisher, *Chem. Eng. Sci.* **24,** 833 (1969).
206. S. J. Wright, R. Hickman, and H. C. Ketley, *Br. Chem. Eng.* **15,** 1551 (1970).
207. K. Yoshida, T. Ueno, and D. Kunii, *Chem. Eng. Sci.* **29,** 77 (1974).
208. R. H. Thring, *Int. J. Heat Mass Transfer* **20,** 911 (1977).
209. P. Basu, *Am. Inst. Chem. Eng. Symp. Ser.* **74,** 187 (1978).
210. O. M. Panov, A. P. Baskakov, Yu. M. Goldobin, N. F. Fillipovskii, and Yu. S. Mazur, *J. Eng. Phys.* **36,** 275 (1979).
211. R. Vadivel and V. N. Vedamurthy, *Proc. Int. Conf. Fluidized Bed Combust., 6th* **III,** 1159 (1980).
212. G. Flamant, *Am. Inst. Chem. Eng. J.* **28,** 529 (1982).
213. J. S. M. Botterill, Y. Teoman, and K. R. Yuregir, *Powder Technol.* **39,** 177 (1984).
214. A. H. George and J. R. Welty, *Am. Inst. Chem. Eng. J.* **30,** 482 (1984).
215. T. F. Ozkaynak, J. C. Chen, and T. R. Frankenfield, *In* "Fluidization" (D. Kunii and R. Toei, eds.). Engineering Foundation, New York, 1984.
216. H. Zhang, K. Cen, and G. Huang, *Int. Chem. Eng.* **24,** 158 (1984).
217. N. Alavizadeh, Z. Fu, R. L. Adams, J. R. Welty, and A. Goshayeshi, *Int. Symp. Heat Transfer, Beijing* (1985).
218. A. Goshayeshi, J. R. Welty, R. L. Adams, and N. Alavizadeh, *Am. Inst. Chem. Eng. Symp. Ser.* **81,** 34 (1985).
219. K. Tuzla, S. Biyikli, and J. C. Chen, *Proc. Int. Conf. Fluidized Bed Combust., 8th* 159 (1985).
220. H. Zhang and C. Xie, *Proc. Int. Conf. Fluidized Bed Combust., 8th, Houston,* 142 (1985).
221. A. Mathur and S. C. Saxena, *Am. Inst. Chem. Eng. J.* **33,** 1124 (1987).
222. N. Alavizadeh, R. L. Adams, J. R. Welty, and A. Goshayeshi, *AIChE/ASME Natl. Heat Transfer Conf., 22nd, Niagara Falls* (1984).
223. J. C. Chen and K. L . Chen, *Chem. Eng. Commun.* **9,** 225 (1981).
224. L. Glicksman and N. A. Decker, *Proc. Int. Conf. Fluidized Bed Combust., 8th, Houston* 45 (1985).
225. V. N. Vedamurthy and V. M. K. Sastri, *Int. J. Heat Mass Transfer* **17,** 1 (1974).
226. S. C. Bhattacharya and D. Harrison, *Eur. Congr. Particle Technol., Nuremberg* p. 23, Session K2 (1977).
227. A. K. Kolar, N. S. Grewal, S. C. Saxena, and J. D. Gabor, *Num. Heat Transfer* **1,** 425 (1978).
228. V. A. Borodulya and V. I. Kovensky, *Int. J. Heat Mass Transfer* **26,** 277 (1983).
229. D. Kunii and J. M. Smith, *Am. Inst. Chem. Eng. J.* **1,** 374 (1955).
230. F. B. Hill and R. H. Wilhelm, *Am. Inst. Chem. Eng. J.* **5,** 486 (1959).
231. S. C. Saxena, D. Sathiyamoorthy, and C. V. Sundaram, *Adv. Transp. Processes* **7,** 241 (1987).
232. N. S. Grewal, S. C. Saxena, A. F. Dolidovich, and Z. Z. Zabrodsky, *Chem. Eng. J.* **18,** 197 (1979).
233. S. N. Upadhyay, S. C. Saxena, and F. T. Ravetto, *Powder Technol.* **30,** 155 (1981).
234. G. T. Zhang and F. Ouyang, *Ind. Eng. Chem. Process Des. Dev.* **24,** 430 (1985).
235. T. C. Ho, R. C. Wang, and J. R. Hopper, *Am. Inst. Chem. Eng. J.*, in press (1988).
236. A. R. Noe and J. G. Knudsen, *Am. Inst. Chem. Eng. Symp. Ser.* **64,** 202 (1968).

Variational Solutions of Complex Heat and Mass Transfer Problems

N. M. TSIRELMAN

Ufa Aviation Institute, 450000 UFA, USSR

I. Introduction

The design of modern technology necessarily involves calculation of unsteady-state temperature fields and the relevant fields of temperature gradients that largely determine the strength of the assemblies and parts of constructions. Similarly, it is impossible to create energy-saving technological processes and the corresponding equipment without being able to determine the fields of heat and mass transfer potentials, and the mean-volumetric values of these potentials at different time instants from the start of heating or cooling (or of drying).

The solution of the problem concerning the history of the fields of heat and mass transfer potentials in solid bodies requires a preliminary determination of the boundary conditions, that is, of the heat and mass transfer situation on the bounding surfaces. Very often, this situation corresponds to convective heat and mass transfer conditions, which are formed in liquid or gas flows through a channel. Note that mathematically, convective heat and mass transfer in an internal problem coincides with the problems of transfer in solid bodies.

Analytical (exact and approximate) descriptions of heat and mass transfer phenomena have been most successful for one-dimensional bodies of materials with constant thermophysical characteristics and fixed boundaries and with the intensity of convective heat and mass transfer being constant over time. Here, mention should be made of the outstanding contributions of the Soviet scientists Luikov, Mikhailov, Petukhov, Kartashov, Goldfarb,

Tyomkin, and Tsoi [1–7] *et al.*, and of the non-Soviet scientists Carlslaw, Jaeger, Schneider, Mikhailov [8–10], and others.

A review is given of the available methods of approximate analytical solution of complex heat and mass transfer problems, including variational methods. Construction of a variational description of nonstationary heat conduction for a multidimensional region with variable thermophysical properties of a medium and with variable intensity of convective heat transfer is shown. Based on the Kantorovich method, a corresponding system of Euler–Lagrange equations is written out and examples are given for obtaining analytical approximations in one- and multidimensional cases.

II. Complex Problems of Heat and Mass Transfer

The development of quantitative methods for complex phenomena of heat and mass transfer requires accounting for all important aspects that correspond to the real progress of the processes. This imposes certain difficulties, such as the dependence of the thermophysical properties of the body material or the moving medium (heat and mass conductivity, heat and mass capacity) on the problem arguments, that is, on the coordinates of the body point and on time. Examples are furnished by variations in the properties of a heat-protecting shield in the process of burnoff, changes in the properties of plastics and resins in the process of solidification and in operation, the heterogeneity of the properties of composite materials, variation in the properties of bodies being dried, the difference in the coefficients of transfer in a turbulent flow, etc. The problem is made much more involved by the time dependence of the convective heat and mass transfer intensity on the bounding surfaces of bodies. The presence of moving boundaries adds to the complexity of heat and mass conduction calculations even in the case where the boundary motion is specified in time (deformation of bodies in the processes of pressure shaping and cutting; evaporation of liquid droplets in a gas flow, etc.).

All these factors, which complicate the quantitative analysis, although remaining within the framework of the class of linear heat and mass transfer theory problems, require the development of special methods of solution in order to allow for their specific behavior.

The development of quantitative methods becomes more involved when the thermophysical characteristics depend on heat and mass transfer potentials (temperature, concentration, partial pressure, etc). Extreme complexity is added to the construction of heat and mass transfer laws in the case in which a phase change occurs in a body (for example, solidification and melting), so that it is necessary to calculate the motion of the solid–liquid interface from an additional condition (the Stefan condition). Moreover, on the body boundary, heat can be supplied or removed according to the law of the fourth orders of temperature (the Stefan–Boltzmann law), by the free convection law, and so on. These cases already involve nonlinear problems of heat and mass transfer that mathematically differ in principle from linear problems by the increased level of complexity.

A general difficulty in the development of quantitative methods is the necessity to allow for a complex shape of a solid body or of a channel cross section (the problem of multidimensionality).

Exact analytical solutions of the linear heat conduction problem with the thermophysical characteristics of a body being dependent only on the coordinates are constructed only in the case of a semi-infinite body with a fixed boundary for a specific case of a power or exponential dependence of heat capacity and (or) of the heat conduction coefficient on the coordinate [1, 11, 12]. Exact analytical solutions, when the intensity of external heat transfer depends on time, have been obtained by Gordov [13] only for a solid cylinder (one-dimensional case) with constant thermophysical characteristics of material and external heat transfer intensity and the surrounding medium temperature simultaneously changing in time according to linear, exponential, or harmonic laws.

In the case of the specified motion of the boundaries of a body with constant thermophysical properties, the greatest successes have been achieved by Grinberg [14] and Kartashov [15]. Grinberg obtained a special functional transformation that converted the boundary-value problem in the region with moving boundaries into a mobile system of coordinates, in which the transformed heat conduction equation allowed an exact solution of the first boundary-value problem by the classical method of separation of variables. Note that application of this method to problems with boundary conditions of the second or third kind leads to boundary conditions of the third kind with a time-variable heat transfer coefficient in the transformed problem and its exact solution becomes difficult. E. M. Kartashov developed a method for constructing Green's functions in noncylindrical regions and illustrated its effectiveness for a region with uniform motion of one of the boundaries and with a boundary condition of the first kind.

The algorithm for obtaining exact solutions of a nonlinear problem with temperature-dependent thermophysical characteristics for a half-space with a fixed boundary, on which boundary conditions of the first kind are assigned for partial (specifically arranged) relationships between the heat conduction coefficient and temperature, was suggested by Luikov [16] and Kozdoba [17]. An exact analytical solution of the problem with phase transition of substance (the Stefan problem) has also been obtained for a half-space with constant thermophysical characteristics in the old and the new phases [8].

It should be noted that exact analytical solutions of the previously mentioned complex linear and nonlinear problems, though they pertain to specific cases, are very valuable for the theory of heat and mass transfer because they serve as verifying standards for developing approximate analytical and numerical methods of solving more complex problems, including those concerned with practical application.

However, an exact analytical solution of complex heat and mass transfer problems in a linear case (where complexity is due to the space and time dependence of the thermophysical characteristics, the time dependence of the convective heat transfer coefficient, the presence of moving boundaries with a prescribed law of motion, and the complex shape of a body) and in a nonlinear case (where complexity is attributed to the dependence of material thermophysical properties on potentials of transfer, the presence of phase transition of a substance, and the complex shape of a body) is impossible in a general case in the presence of the complicating factors acting either jointly or separately. Therefore, to analyze these complex phenomena, it is necessary to develop effective quantitative methods that would possess the main property of analytical solutions, that is, the property of parametrization, which admits the case of representation, necessary for the analysis and simplicity of calculations.

III. Methods of Solution of Complex Heat and Mass Transfer Problems

Consideration of the previous development of quantitative methods for the description of heat and mass transfer processes shows that only three of these methods are capable of studying complex phenomena: (1) physical and analog modeling, (2) numerical methods (the methods of grids and of finite elements), and (3) approximate analytical methods.

Here, the capabilities of the method of physical and analog modeling will not be discussed; it will only be noted that thanks to the work of Soviet scientists M. V. Kirpichev, A. A. Gukhman, L. I. Sedov, L. A. Kozdoba, Yu. M. Matsevity, and others' great successes have been attained in this area.

As to the approximate analytical method for constructing solutions of complex heat and mass transfer problems, three categories should probably be distinguished (1) the methods possessing a small or limited degree of generality, (2) the methods of direct solution of boundary-value problems with a high degree of generality, and (3) variational methods with a high degree of generality.

The first category includes the methods of approximate solution of heat and mass transfer problems that are applicable [due to successful substitutions, linearization, the use of the method of small parameter (perturbations), averaging of functional corrections, etc.] only in the case of special variations of the thermophysical properties of material with the coordinates, time, or temperature (potential of mass conduction), suitable for a simple geometric region and more often with fixed boundaries. The available advances in this category of methods have been analyzed in detail in the works of Luikov and Kozdoba [12, 16, 17]. This category also includes particular methods of accounting for the time-variable character of the convective heat transfer intensity [18]. The same situation is observed when body boundaries moving according to a given law are accounted for. A detailed consideration of the methods, which do not possess generality, is given elsewhere [19]. A great number of specific methods for solving problems of heat conduction in the presence of phase transition in a body is surveyed in detail by Kozdoba [20]. The number of works of this kind increases with due attention being paid to accounting for complex shapes of bodies; for example, Novikov [21], by introducing a special system of coordinates, transformed the curvilinear trapezium-type region to a canonic form, and, by making a variety of assumptions, he obtained an approximate solution for the nonstationary heat conduction problem with constant thermophysical characteristics and fixed body boundaries.

A high degree of generality is exhibited by methods involving direct solution of boundary-value problems. These include the Galerkin method (I. G. Bubnov–B. G. Galerkin), which is not associated with any variational formulation of the problem, and thus can be successfully applied to any differential equation (and it is not restricted to differential equations). The number of studies in which the Bubnov–Galerkin method was employed for the practical solution of a wide variety of applied problems (including nonlinear ones) is tremendous; some of them can be found in articles by Kozdoba [17] and Mikhlin [22]. Proofs of the convergence of this method are also provided in these articles.

Closely allied to the Bubnov–Galerkin method is the method of the heat balance integral of Goodman [17, 23, 24], which allowed one to obtain a number of approximate analytical solutions with thermophysical characteristics of material depending on coordinates, time, or temperature in a one-dimensional case and also with phase transition of substance. Note that the implications used when applying the heat balance integral virtually coincide with those stated in the work of Veinik [25].

Contiguous with the Bubnov–Galerkin method are the methods of collocations and moments, which comprise the method of weighted residues [17, 26]. Emphasis should be placed on the application of the Bubnov–Galerkin method to the solution of boundary-value problems. which are preliminarily subjected to the Laplace integral transformation with respect to the time variable and are reduced, with respect to the transform, to the solution of the boundary-value problem in the remaining space variables [7]. Also, the combined application of the Bubnov–Galerkin method and the method of characteristics should be mentioned [27].

It is generally known that before the appearance of electronic computers the most complicated problems of mechanics were solved by variational methods. It will be assumed here that if there exists such an integral J (functional J), for which the condition of the vanishing of the first variation ($\delta J = 0$) leads to the Euler–Lagrange equations that coincide with the initial boundary-value problem, it means that the variational principle is valid and that the variational formulation of the problem is found. If there is no such functional, then it will be assumed that there is no variational description of the problem either.

In the special case in which the solutions of the boundary-value problem also extremize the functional J (as indicated by the constancy of the sign of the second variation $-\delta^2 J > 0$ or $\delta^2 J < 0$), it will be regarded that the classical variational formulation of the problem is found.

A variational solution of an initial boundary-value problem consists of finding functions that provide the indicated functional with a stationary point ($\delta J = 0$) or an extremum value ($\delta J = 0$ and $\delta^2 J > 0$ or $\delta^2 J < 0$). Most commonly, these solutions are approximate. However, this does not belittle the importance of variational methods, because in the overwhelming majority of cases, complex problems are solved only approximately. A sufficiently high accuracy of the variational solutions makes them suitable for use in engineering practice. The sequence of functions that minimize the functional J and whose linear combination gives an approximate solution of the boundary-value problem is constructed first by the Ritz method (Rayleigh–Ritz), in which the form of the solution is chosen *a priori* and only then "the best" values of constants involved are selected (S. G. Mikhlin suggests that this procedure be called the Ritz process rather than the Ritz method [22]). In applying the Ritz method, the questions arise as to in which case will the minimizing sequence converge to a solution of a variational problem, what is the order of approximation, and what is the error in the nth approximation. Though the substantiation of these questions has been the concern of numerous works of outstanding Soviet and non-Soviet mathematicians (N. M. Krylov, N. N. Bogolyubov, S. L. Sobolev, L. V. Kantorovich, S. G. Mikhlin, R. Courant, K. Freidrichs, and many others), it can hardly be solved in a general case. It should be noted that the justification of the Ritz method has been made only for steady-state processes.

The most advantageous method for obtaining the solution of variational problems, which possess parameterization properties, is, to the author's present opinion, the Kantorovich method of reduction to ordinary differential equations. This method fits in between an exact solution of the problem and an approximate one obtained by the Ritz method. In this case, the solution is sought in such a form that it encompasses the undetermined functions of one variable. Besides its high accuracy, a major advantage of the Kantorovich method is that only a part of the expression, which gives the solution, is selected a priori. When constructing the sequence, which minimizes the functional J, use can undoubtedly be made of other methods ("processes", according to S. G. Mikhlin), such as the methods of Bubnov–Galerkin and of Treftz, and the least square method. Note that for those problems, associated with variational ones, the Bubnov–Galerkin method is related to the Ritz method and sometimes is equivalent to the latter in the sense that it leads to the same solution. Special mention should be made of the high potentialities of the least square method, which uses, when constructing the minimizing sequences, a functional such as the sum of residual squares. For the parabolic-type equation, this method was successfully used by Anderssen [29, 30].

IV. Variational Methods for Solving Problems of Unsteady-State Heat and Mass Transfer

It is evident that variational descriptions of the previously mentioned problems can conveniently be classified into three groups according to the way they are constructed (1) variational descriptions based on ideas of irreversible thermodynamics, (2) variational descriptions constructed by analogy with methods of classical mechanics, and (3) variational descriptions based on the principle of mathematical equivalence to the initial boundary-value problem (without resorting to different physical considerations or assumptions when varying the functionals J).

The first group includes the so-called Biot variational principle, the essential idea of which was taken up in many works of M. Biot and others [16, 17, 31–35], and therefore it will not be given here.

Note that the variational equation used by M. Biot is valid, but it is not a variation of some functional. Virtually no variational principle is obtained, and the Biot method is related to variational ones only provisionally, following up on the tradition. Not all of the quantities introduced by M. Biot are physically meaningful, for example, the heat flux field vector H is defined as

$$-c\rho T = \operatorname{div} H \tag{1}$$

and this makes it difficult to obtain a good approximation. In the same way, it is actually impossible for two- and three-dimensional problems to select the generalized coordinates q_i in the equation

$$H = H(q_i, x, y, z, \tau) \tag{2}$$

that characterize the time-varying configuration of the field H. When formulating the problems with the second kind of boundary conditions (i.e., the density of heat flux into a wall is prescribed) and with the third kind of boundary conditions (i.e., the conditions of convective heat transfer are given), the use of the Biot principle, even in a one-dimensional case, becomes possible only after the total energy balance equation has been imposed as an additional constraint on the change in time of $q_i(\tau)$ [33]. Almost all of the result obtained by the Biot methods relate to a half-space and an infinite plate [33–35].

Attention should evidently be drawn to the sharp and justified criticism of the Biot method by Finlayson and Scriven [36, 37]. Although the number of publications using the Biot method decreased drastically, efforts are nevertheless continuing to extend the method to interrelated heat and mass transfer processes having a finite rate of propagation of perturbations [38–42].

The other group of works of this type employs the fact that, when applied to stationary states, the three familiar postulates of nonequilibrium thermodynamics lead to the principle (functional) of minimum entropy, which is valid in the case of constant thermal conductivity. This approach gives, with corresponding variation, the Euler–Lagrange equation for steady-state heat conduction processes.

The principle of minimum entropy is applicable to the description of a limited class of phenomena. A more general criterion, which describes the steady state of certain continuous systems, is the functional called the generalized origination of entropy or the local potential. It is interesting to note that for the use of this functional-local potential in solving practically important problems of heat and mass transfer it is necessary to know the sought-after fields of temperature or concentration, that is, it is necessary to undertake a number of attempts to "guess" the actual distributions. The details of the overcoming this difficulty are discussed in the monograph by R. Schechter [43].

Rosen and Chambers [44–46] suggested variational methods for nonstationary processes having a changing structure of the local potential. These methods are suitable only for stationary states. In their approach, they extended the concept of the local potential and abandoned the variation of some functions of the integrand; however, they allowed other functions, called the local thermodynamic variables to vary.

All constructions [44–46] in connection with the use of the extended concept of a local potential are not very convincing despite attempts to rely on the physical sense of the operations. Moreover, it should also be noted that up to now it has been possible to construct solutions by the method of the local potential only for the simplest problems.

Methods based on the analogy of problems of nonstationary heat conduction with methods of classical mechanics are being constructed by Vujanovic and co-workers. They describe the transfer phenomena by Lagrange equations for the Lagrangian function and the generalized coordinates [47–49]. They introduce into the consideration the Lagrangian function of the form

$$L = \left[\frac{c\rho\tau_r}{2}\left(\frac{\partial T}{\partial \tau}\right)^2 - \frac{\lambda}{2}\sum_{i=l}^{3}\left(\frac{\partial T}{\partial x_i}\right)^2\right]\exp(\tau/\tau_r) \tag{3}$$

and show that the Euler-Lagrange equation for L

$$\frac{\partial L}{\partial T} - \frac{\partial}{\partial \tau}\frac{\partial L}{\partial(\partial T/\partial \tau)} - \sum_{i=l}^{3}\frac{\partial}{\partial x_i}\frac{\partial L}{\partial(\partial T/\partial x_i)} = 0 \tag{4}$$

is equivalent to the well-known hyperbolic equation

$$c\rho\tau_r\frac{\partial^2 T}{\partial \tau^2} + c\rho\frac{\partial T}{\partial \tau} = \lambda\nabla^2 T. \tag{5}$$

This result obtains from the variation of the functional the integral of action.

$$J = \int_{\tau_1}^{\tau_2}\int_{\Omega} L\, d\Omega\, d\tau \tag{6}$$

It is not difficult to see that functional [Eq. (6)] does not include the boundary-value conditions of the heat conduction problem, and, therefore, when it is used, additional initial and boundary conditions must be included into the variational description. This reduces the capabilities of the above-mentioned variational principle. It does give the possibility of obtaining results for the simplest case of heat propagation in a half-space and in a plate with constant thermophysical characteristics and with the volumetric heat capacity being linearly dependent on temperature at the given temperature on the bounding surfaces.

The third group of variational principles includes the works of Lebon and Lambermont, who constructed a variational description of the boundary-value problem of unsteady-state heat conduction supposedly equivalent to it in the mathematical sense [50, 51]. Identical with their functional is the functional of Krajewski [52]—a fact that was commented on by Lebon and Lambermont [53] and Krajewski [54].

It will now be shown that the articles by Lebon and Lambermont [50, 53], Lebon and Mathieu [51], and Krajewski [52, 54] are incorrect in the sense that the authors did not find a variational description of the following problem for the process of unsteady-state heat conduction in the geometrical region Ω (Fig. 1):

$$c\rho(T)\frac{\partial T}{\partial \tau} = \operatorname{div}[\lambda(T)\operatorname{grad} T] + q_v(T) \qquad \tau > 0, \qquad (x, y, z) \in \Omega \tag{7}$$

$$T(0, M) = T_0(M), \qquad M \in \Omega \tag{8}$$

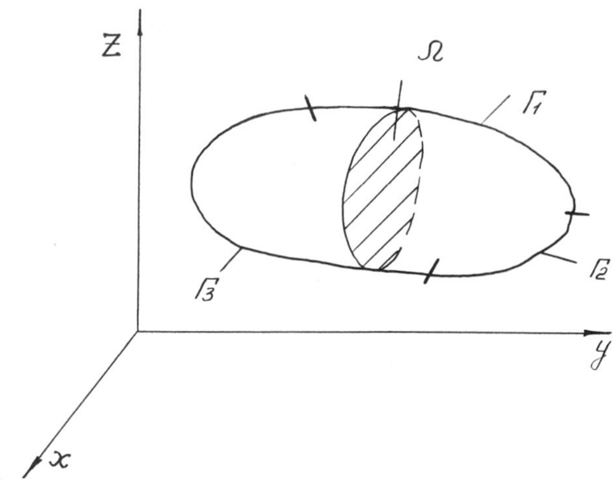

FIG. 1. Region of unsteady-state heat conduction.

$$T\big|_{\Gamma_1} = T_w(M, \tau), \qquad \tau > 0 \tag{9}$$

$$-\lambda(T)\frac{\partial T}{\partial n}\vec{n}\bigg|_{\Gamma_2} = q(T), \qquad \tau > 0 \tag{10}$$

$$\left[-\lambda(T)\frac{\partial T}{\partial n}\vec{n} = \alpha(T^m - T_c^m)\right]\bigg|_{\Gamma_3}, \qquad \tau > 0 \tag{11}$$

Here, the boundary conditions of Eqs. (9), (10), and (11) of the first, second, and third kinds, respectively, are prescribed on the "segments" of the surface Γ_1, Γ_2, and Γ_3, respectively, and Eq. (8) is the initial condition. The authors of Refs. [50–54] state that the problem in Eqs. (7)–(11) is reduced to the following variational problem:

$$\begin{aligned} J(T) = &\int_0^t \left\{ \int_\Omega \left[\frac{\lambda^2(T)}{2}(\nabla T)^2 - \int_{T^*}^T q_v(\Theta)\lambda(\Theta)\,d\Theta \right.\right. \\ &\left.\left. + F(x, y, z, t) \int_{T^*}^T c\rho(\Theta)\lambda(\Theta)\,d\Theta \right] d\Omega \right\} d\tau \\ &+ \int_0^t \left\{ \int_{\Gamma_2} \left[\int_{T^*}^T q(\Theta)\lambda(\Theta)\,d\Theta \right] ds \right\} d\tau \\ &+ \alpha \int_0^t \left\{ \int_{\Gamma_3} \left[\int_{T^*}^T (\Theta^m - T_c^m)\lambda(\Theta)\,d\Theta \right] ds \right\} d\tau \\ &+ \frac{1}{2}\int_\Omega (T - T_0)^2\,d\Omega = \text{minimum} \end{aligned} \tag{12}$$

where

$$F(x, y, z, t) = \left.\frac{\partial T(x, y, z, \tau)}{\partial \tau}\right|_{\tau = t}$$

is being "frozen" for subsequent variations, and T^* is the characteristic temperature. Calculate the first variation of functional (12).

$$\delta J(T) = \int_0^t \left\{ \int_\Omega [\lambda'(T)\,\delta T(\nabla T)^2 \lambda(T) + \lambda^2(T)\,\nabla T \nabla\,\delta T - q_v(T)\lambda(T)\,dT + F(x, y, z, t)c\rho(T)\lambda(T)\,\delta T]\,d\Omega \right\} d\tau + \int_0^t \left\{ \int_{\Gamma_2} [q(T)\lambda(T)\,\delta T]\,ds \right\} d\tau + \alpha \int_0^t \left\{ \int_{\Gamma_3} (T^m - T_c^m)\lambda(T)\,\delta T \times ds \right\} d\tau + \int_\Omega [T - T_0]\,\delta T\,d\Omega \tag{13}$$

Making use of the Ostrogradsky–Gauss equation, obtain

$$\delta J(T) = \int_0^t \left\{ \int_\Omega [F(x, y, z, t)c\rho(T) - \operatorname{div}(\lambda(T)\nabla T) - q_v(T)]\lambda(T)\,\delta T\,d\Omega \right\} d\tau + \int_0^t \left\{ \int_{\Gamma_1 + \Gamma_2 + \Gamma_3} \lambda^2(T)\vec{n}\,\nabla T\,\delta T\,ds \right\} d\tau + \int_0^t \left\{ \int_{\Gamma_2} [q(T)\lambda(T)\,\delta T]\,ds \right\} d\tau + \alpha \int_0^t \left\{ \int_{\Gamma_3} [(T^m - T_c^m)\lambda(T) \times \delta T]\,ds \right\} d\tau + \int_\Omega [(T - T_0)]\,\delta T\,d\Omega = \int_0^t \left\{ \int_\Omega [F(x, y, z, t)c\rho(T) - \operatorname{div}(\lambda(T)\nabla T) - q_v(T)]\lambda(T)\,\delta T\,d\Omega \right\} d\tau + \int_0^t \left\{ \int_{\Gamma_2} [\lambda(T)\vec{n}\,\nabla T + q(T)]\lambda(T)\,\delta T\,ds \right\} d\tau + \int_0^t \left\{ \int_{\Gamma_3} [\lambda(T)\vec{n}\,\nabla T + \alpha(T^m - T_c^m)]\lambda(T)\,\delta T\,ds \right\} d\tau + \int_\Omega [(T - T_0)]\,\delta T\,d\Omega + \int_0^t \left\{ \int_{\Gamma_1} \lambda^2(T)\vec{n}\,\nabla T\,\delta TT\,ds \right\} d\tau = 0 \tag{14}$$

According to the main lemma of the variational calculus [55], the assumption that $\delta T|_{\Gamma_1} \equiv 0$ yields $\delta J(T) = 0$ if and only if

$$c\rho(T)\frac{\partial T(x, y, z, \tau)}{\partial \tau}\bigg|_{\tau=t} = \text{div}(\lambda(T)\nabla T) + q_v(T) \tag{7'}$$

$$(T - T_0)|_{\tau=0} = 0 \tag{8'}$$

$$-\lambda(T)\vec{n}\nabla T|_{\Gamma_2} = q(T), \qquad \tau > 0 \tag{10'}$$

$$[-\lambda(T)\vec{n}\nabla T = \alpha(T^m - T_c^m)]|_{\Gamma_3}, \qquad \tau > 0 \tag{11'}$$

As is seen from the comparison of the problem in Eqs. (7)–(11) with the system of Euler equations, Eqs. (7′), (8′), (10′), and (11′), Eq. (7) is not obtained, since

$$\frac{\partial T(x, y, z, \tau)}{\partial \tau}\bigg|_{\tau=t} \neq \frac{\partial T(x, y, z, \tau)}{\partial \tau}$$

whereas the boundary condition, Eq. (9), is totally absent.

To summarize: the problem in Eqs. (7)–(11) was not reduced by Lebon and Lambermont [50, 53], Lebon and Mathieu [51], and Krajewski [52, 54] to the variational problem for functional (12).

This conclusion is fully consistent with the results of Atherton and Homsy [56] about the impossibility, in principle, of constructing a variational description for nonlinear problems.

Consideration of the proposed variational descriptions shows that they are equivalent to the formulated linear boundary-value problem of heat and mass transfer only with the use of the convolution-type functional. Apparently, such a functional was first constructed by Tao [57] to determine the temperature field in round and elliptic tubes with plug and laminar modes of flow; uniform temperature distribution in the inlet section was assumed, and the simplest boundary conditions of the first kind were considered with the thermophysical characteristics of the hydrodynamical stabilized flow being constant. Almost simultaneously the same functional was suggested by Ainola [58] and somewhat later by Tokarenko and Kravchenko [59] for the problems of unsteady-state heat conduction with constant thermophysical material characteristics.

Convolution-type functionals, whose structure is associated in the most simple way with boundary-value problems of unsteady-state heat and mass conduction in solid bodies of complex shapes (including those with a moving boundary) and with the internal problem of unsteady- and steady-state convective heat transfer with variable thermophysical characteristics, are given by Tsirelman [60, 61, 62, 63, 64] and Tsirelman and Bronstein [65, 66].

V. Variational Description of Heat Transfer Processes with Variable Thermophysical Characteristics including Convective Heat Transfer

The construction of a variational description will now be shown in which use is made of the convolution-type functional for a multidimensional problem corresponding to the process of unsteady-state heat conduction in an arbitrary region Ω bounded by the segments Γ_1, Γ_2, and Γ_3 of the boundary surface Γ (Fig. 1). Γ_1, Γ_2, and Γ_3 can also denote the closed surfaces of inner enclosures in a body or a portion of the body in the case of a multiply connected region. Let the thermophysical properties of the body material be dependent on the coordinates of the points of the region $\bar{\Omega}$ ($M \in \bar{\Omega} = \Omega \cup \Gamma$) and on time, $\tau - \lambda = \lambda(M, \tau)$, $c\rho = c\rho(M, \tau)$. There are sources of volumetric heat generation in the body whose power depends on the coordinates and time, $-q_v = q_v(M, \tau)$. The initial temperature distribution, $T_0(M)$, is known, and on the surfaces Γ_1, Γ_2, and Γ_3 boundary conditions of the first, second, and third kinds are prescribed. Thus, the temperature $T_w(M, \tau)$, heat flux density $q(M, \tau)$ or the surrounding medium temperature $T_c(M, \tau)$ respectively, depend arbitrarily on the coordinates of the points on the surfaces Γ_1, Γ_2, and Γ_3 and on time, with the coefficient of convective heat transfer $\alpha(M, \tau)$ being dependent on the coordinates of Γ_3 and on time.

A mathematical formulation of such a problem has the form

$$c\rho(M,\tau)\frac{\partial T(M,\tau)}{\partial \tau} = \frac{\partial}{\partial x}\left[\lambda(M,\tau)\frac{\partial T}{\partial x}\right] + \frac{\partial}{\partial y} \times \left[\lambda(M,\tau)\frac{\partial T}{\partial y}\right] + \frac{\partial}{\partial z}\left[\lambda(M,\tau)\frac{\partial T}{\partial z}\right] + q_v(M,\tau), \qquad M \in \Omega, \qquad \tau > 0 \tag{15}$$

$$T(M,0) = T_0(M), \qquad \tau = 0, \qquad M \in \Omega \tag{16}$$

$$T(M,\tau) = T_w(M,\tau) \qquad \tau > 0, \qquad M \in \Gamma_1 \tag{17}$$

$$-\lambda(M,\tau)\nabla T\vec{n} = q(M,\tau), \qquad \tau > 0, \qquad M \in \Gamma_2 \tag{18}$$

$$-\lambda(M,\tau)\nabla T\vec{n} = \alpha(M,\tau)[T(M,\tau) - T_c(\tau)], \qquad \tau > 0, \qquad M \in \Gamma_3 \tag{19}$$

The solution of Eqs. (15)–(19) will be sought in the interval $0 < \tau < t$.

For this purpose, for continuous functions $f(M, \tau)$ determined in the region $\Omega \times [0, t]$, the symmetrization transformation $f(M, \tau) \rightarrow \bar{f}(M, \tau)$ will be introduced such that

$$\bar{f}(M,\tau) = \begin{cases} f(M,\tau), & 0 \le \tau \le t \\ f(M, 2t - \tau), & t \le \tau \le 2t \\ 0, & \tau > 2t \end{cases} \tag{20}$$

The functions $\bar{f}(M, \tau)$ thus constructed do have the symmetry property

$$\bar{f}(M, 2t - \tau) = \bar{f}(M, \tau), \qquad 0 < \tau < 2t$$

as shown graphically in Fig. 2.

Note that the functions $f(M, \tau)$ can be nondifferentiable at $\tau = t$. Then, within the interval $t - \varepsilon < \tau < t + \varepsilon$, where $\varepsilon > 0$ is a sufficiently small quantity, the functions $\bar{f}(M, \tau)$ can be smoothed so that they are differentiable for all τ within $0 < \tau < 2t$ (dashed line in Fig. 2). It is also evident that in practical applications the nondifferentiability of $\bar{f}(M, \sigma)$ in the vicinity of $\tau = t$ is inessential, since solutions of the given problem for $0 < \tau < t$ are of interest.

In what follows, the functions $f(M, \tau)$ are understood to represent the functions $c\rho(M, \tau)$, $\lambda(M, \tau)$, and $\alpha(M, \tau)$ from Eqs. (15)–(19).

As already mentioned, solution of the boundary-value problem, Eqs. (15)–(19), meets with insuperable difficulties even for a one-dimensional region because of the dependence of the thermophysical characteristics and (or) of the convective heat transfer coefficient on the coordinates of the region Ω and (or) on time τ. It is expedient to integrate Eqs. (15)–(19) on the basis of the variational description of the formulated problem with the use of the following convolution-type functional:

$$\begin{aligned} J(T) = {} & \int_0^{2t} \int_\Omega \left\{ \frac{\partial}{\partial x}\left[\bar{\lambda}(M, \tau)\frac{\partial T(M, \tau)}{\partial x}\right] + \frac{\partial}{\partial y}\left[\bar{\lambda}(M, \tau)\frac{\partial T(M, \tau)}{\partial y}\right] \right. \\ & + \frac{\partial}{\partial z}\left[\bar{\lambda}(M, \tau)\frac{\partial T(M, \tau)}{\partial z}\right] - \overline{c\rho}(M, \tau) \\ & \left. \times \frac{\partial T(M, \tau)}{\partial \tau} + \frac{1}{2}\overline{\left(\frac{\partial c\rho(M, \tau)}{\partial \tau}\right)} T(M, \tau) + 2q_v(M, \tau) \right\} \\ & \times T(M, 2t - \tau)\, d\Omega\, d\tau - \int_\Omega \overline{c\rho}(M, 0)[T(M, 0) - 2T_0(M)] \\ & \times T(M, 2t)\, d\Omega - \int_0^{2t} \int_{\Gamma_1} [2\bar{\lambda}(M, \tau) T_w(M, \tau) \nabla T(M, 2t - \tau) \\ & \times \vec{n} - \bar{\lambda}(M, \tau) T(M, \tau) \nabla T(M, 2t - \tau)\vec{n}]\, ds\, d\tau \\ & - \int_0^{2t} \int_{\Gamma_2} [2q(M, \tau)T(M, 2t - \tau) + \bar{\lambda}(M, \tau) T(M, 2t - \tau) \nabla T(M, \tau)\vec{n}]\, ds\, d\tau \\ & - \int_0^{2t} \int_{\Gamma_3} \{\bar{\lambda}(M, \tau) T(M, 2t - \tau) \nabla T(M, \tau)\vec{n} \\ & + \bar{\alpha}(M, \tau)[T(M, \tau) - 2T_c(\tau)] T(M, 2t - \tau)\}\, ds\, d\tau \end{aligned} \tag{21}$$

In Eq. (21), the functions $\bar{\lambda}(M, \tau)$, $c\rho(M, \tau)$, and $\bar{\alpha}(M, \tau)$ are made symmetrical at the level $\tau = t$ by the rule of Eq. (20).

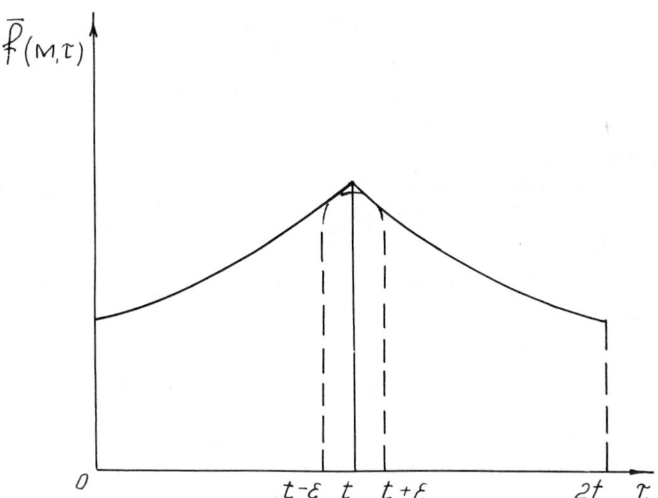

FIG. 2. Graphical representation of symmetrization of functions $f(M,\tau)$ at the level $\tau = t$.

Calculate the first variation of functional (21). This will give, with the omission of arguments in T, λ, $c\rho$, q_v, etc., except for the most necessary ones,

$$\delta J(T) = \int_0^{2t}\int_\Omega \left\{\frac{\partial}{\partial x}(\bar{\lambda}T_x) + \frac{\partial}{\partial y}(\bar{\lambda}T_y) + \frac{\partial}{\partial z}(\bar{\lambda}T_z) + \frac{1}{2}(\overline{c\rho_\tau})T - \overline{c\rho}T_\tau + 2q_v\right\}$$

$$+ \delta T(2t-\tau)\,d\Omega\,d\tau + \int_0^{2t}\int_\Omega \Big\{(\bar{\lambda}\,\delta T_x)_x + (\bar{\lambda}\,\delta T_y)_y + (\bar{\lambda}\,\delta T_z)_z$$

$$+ \frac{1}{2}\overline{(c\rho_\tau)}\,\delta T - \overline{c\rho}\,\delta T_\tau\Big\} T(2t-\tau)\,d\Omega\,d\tau$$

$$+ \int_\Omega \{\overline{c\rho}(M,0)\,\delta T(0)T(2t) + \overline{c\rho}(M,0)[T(0) - 2T_0]$$

$$\times\ \delta T(2t)\}\,d\Omega - \int_0^{2t}\int_{\Gamma_1}[2\bar{\lambda}T_w\,\delta\,\nabla T(2t-\tau)\vec{n}$$

$$- \bar{\lambda}\,\delta T(\tau)\,\nabla T(2t-\tau)\vec{n} - \bar{\lambda}T(\tau)\,\delta\,\nabla T(2t-\tau)\vec{n}]\,ds\,d\tau$$

$$- \int_0^{2t}\int_{\Gamma_2}\{2q\,\delta T(2t-\tau) + \bar{\lambda}\,\delta T(2t-\tau)\,\nabla T(\tau)\vec{n}$$

$$+ \bar{\lambda}T(2t-\tau)\,\delta\,\nabla T(\tau)\vec{n}]\,ds\,d\tau - \int_0^{2t}\int_{\Gamma_3}\{\bar{\lambda}\,\delta T(2t-\tau)$$

$$\times\ \nabla T(\tau)\vec{n} + \bar{\lambda}T(2t-\tau)\,\delta\,\nabla T(\tau)\vec{n} + \bar{\alpha}\,\delta T(\tau)$$

$$\times\ T(2t-\tau) + \bar{\alpha}[T(\tau) - 2T_c]\,\delta T(2t-\tau)\}\,ds\,d\tau \tag{22}$$

Applying Green's generalized equation twice

$$\int_\Omega u_x v \, d\Omega = -\int_\Omega u v_x \, d\Omega + \int_\Gamma uv \cos(n, x) \, ds$$

to the term from Eq. (22)

$$\int_0^{2t} \int_\Omega [(\bar{\lambda}\,\delta T_x)_x + (\bar{\lambda}\,\delta T_y)_y + (\bar{\lambda}\,\delta T_z)_z] T(2t - \tau) \, d\Omega \, d\tau$$

and accounting for the symmetry property of the convolution, obtain, instead of this term,

$$\int_0^{2t} \int_\Omega [(\bar{\lambda}T_x)_x + (\bar{\lambda}T_y)_y + (\bar{\lambda}T_z)_z] \delta T(2t - \tau) \, d\tau \, d\Omega$$

$$- \int_0^{2t} \int_\Gamma [\bar{\lambda}\,\delta T(2t - \tau) \nabla T \vec{n} - \bar{\lambda} T(2t - \tau)\, \delta\, \nabla T \vec{n}] \, ds \, d\tau$$

Moreover, making use of integration by parts in application to the term from Eq. (22),

$$\int_0^{2t} \int_\Omega \overline{c\rho}\, \delta T_\tau T(2t - \tau) \, d\Omega \, d\tau$$

will yield, instead of this term,

$$\int_\Omega [\overline{c\rho}(2t)\, \delta T(2t) T(0) - \overline{c\rho}(0)\, \delta T(0) T(2t)] \, d\Omega$$

$$- \int_0^{2t} \int_\Omega (\overline{c\rho})_\tau\, \delta T T(2t - \tau) \, d\Omega \, d\tau + \int_0^{2t} \int_\Omega \overline{c\rho} T T'(2t - \tau) \, d\Omega \, d\tau$$

Thus, the above transformations give the first variation of $J(T)$ (after the reduction of the similar terms) in the form

$$\begin{aligned}
\delta J(T) = {} & \int_0^{2t} \int_\Omega \{[2(\bar{\lambda}T_x)_x + 2(\bar{\lambda}T_y)_y + 2(\bar{\lambda}T_z)_z + 2q_v \\
& - 2\overline{c\rho}T_\tau] + [\overline{(c\rho_\tau)} - (\overline{c\rho})_\tau] T\}\, \delta T(2t - \tau) \, d\Omega \, d\tau \\
& + \int_\Omega \overline{c\rho}(0)[2T(0) - 2T_0]\, \delta T(2t) \, d\Omega + 2 \int_0^{2t} \int_{\Gamma_1} \{\bar{\lambda}[T(\tau) \\
& - T_w] \nabla \delta\, T(2t - \tau) \vec{n} \, ds \, d\tau - 2 \int_0^{2t} \int_{\Gamma_2} [\bar{\lambda} \nabla T(\tau) \vec{n} + q] \\
& \times \delta T(2t - \tau) \, ds \, d\tau - 2 \int_0^{2t} \int_{\Gamma_3} \{\bar{\lambda} \nabla T(\tau) \vec{n} + \bar{\alpha}[T(\tau) \\
& - T_c]\}\, \delta T(2t - \tau) \, ds \, d\tau
\end{aligned} \tag{23}$$

According to the main lemma of variational calculus [66], the first variation of functional (23) vanishes if and only if the cofactors in the variations $\delta T(2t-\tau)$ and $\delta T(2t)$ are equal to zero, that is, when the following relations are satisfied:

$$2(\bar{\lambda}T_x)_x + 2(\bar{\lambda}T_y)_y + 2(\bar{\lambda}T_z)_z + 2q_v - 2\overline{c\rho}T_\tau + [(\overline{c\rho_\tau}) - (\overline{c\rho})_\tau]T = 0, \qquad M \in \Omega, \qquad \tau > 0 \tag{24}$$

$$T(0) = T_0, \qquad M \in \Omega \tag{25}$$

$$T|_{\Gamma_1} = T_w, \qquad \tau > 0 \tag{26}$$

$$\left\{-\bar{\lambda}\frac{\partial T}{\partial n}\vec{n} = q\right\}\Bigg|_{\Gamma_2}, \qquad \tau > 0 \tag{27}$$

$$\left[-\bar{\lambda}\frac{\partial T}{\partial n}\vec{n} = \bar{\alpha}(T - T_c)\right]\Bigg|_{\Gamma_3}, \qquad \tau > 0 \tag{28}$$

It is not difficult to see that the set of Eqs. (24)–(28) does not coincide with the system, [Eqs. (15)–(19)] being solved because of the presence of an additional term in Eq. (24) of the form

$$[(\overline{c\rho_\tau}) - (\overline{c\rho})_\tau]T$$

However, since for $0 < \tau < t$, because of the symmetrization properties, the following equality holds:

$$\overline{\frac{\partial c\rho}{\partial \tau}} = \frac{\partial \overline{c\rho}}{\partial \tau}$$

then Eq. (24) coincides with Eq. (15) on the interval $(0, t)$.

Thus, Eqs. (15)–(19) can be solved by seeking the stationary point of functional (21) under conditions (16)–(19) on the interval $(0, t)$.

In order to construct the sequence that minimizes functional (21) (more precisely, to construct the sequence that provides the indicated functional with the stationary point) the nth approximation to the solution of Eqs. (15)–(19) will be presented, following Kantorovich and Krylov [28], in the form

$$T(M,\tau) = \sum_{i=1}^{n} \Psi_i(\tau)\chi_i(M,\tau) \tag{29}$$

where $\chi_i(M,\tau)$ is a complete system of the known functions of coordinates and time in the integrated region $\Omega \times [0, 2t)]$, which is selected to satisfy the boundary conditions of Eqs. (17)–(19), and $\Psi_i(\tau)$ is an unknown function of time.

In order to determine $\Psi_i(\tau)$, one must solve the system of Euler–Lagrange equations, which is equivalent to the condition for the vanishing of the first derivative of the functional ($\delta J = 0$). The system of Euler–Lagrange equations for $\Psi_i(\tau)$ is obtained by substituting the form of the solution of Eq. (29) into Eq. (23) for the first derivative of the functional and equating it to zero:

$$\delta J(T) = \int_0^{2t} \int_\Omega \Bigg\{ \sum_{i=1}^n [(\bar{\lambda}\chi_{i,x})_x + (\bar{\lambda}\chi_{i,y})_y + (\bar{\lambda}\chi_{i,z})_z] \times \Psi_i(\tau) - \overline{c\rho} \sum_{i=1}^n [\chi_i \Psi_i'(\tau) + \chi_{i,\tau}' \Psi_i(\tau)] + q_v + \frac{1}{2}\left(\frac{\overline{\partial c\rho}}{\partial \tau} - \frac{\partial \overline{c\rho}}{\partial \tau}\right) \sum_{i=1}^n \chi_i \Psi_i(\tau) \Bigg\} \sum_{j=1}^n \chi_j \delta\Psi_j(2t-\tau) \times d\Omega\, d\tau + \int_\Omega \overline{c\rho}(0) \left[\sum_{i=1}^n \chi_i(0)\Psi_i(0) - T_0\right] \sum_{j=1}^n \chi_j(2t) \times \delta\Psi_j(2t)\, d\Omega + \int_0^{2t} \int_{\Gamma_1} \left\{ \bar{\lambda} \left[\sum_{i=1}^n \chi_i \Psi_i(\tau) - T_w\right] \times \sum_{j=1}^n \nabla\chi_j \vec{n}\, \delta\Psi_j(2t-\tau)\, ds\, d\tau - \int_0^{2t} \int_{\Gamma_2} \left\{ \bar{\lambda} \sum_{i=1}^n \nabla\chi_i \vec{n} \Psi_i(\tau) + q \right] \times \sum_{j=1}^n \chi_j \delta\Psi_j(2t-\tau)\, ds\, d\tau - \int_0^{2t} \int_{\Gamma_3} \left\{ \bar{\lambda} \sum_{i=1}^n \nabla\chi_i \vec{n} \times \Psi_i(\tau) + \bar{\alpha} \left[\sum_{i=1}^n \chi_i \Psi_i(\tau) - T_c\right] \right\} \sum_{j=1}^n \chi_j \delta\Psi_j(2t-\tau)\, ds\, d\tau \tag{30}$$

Taking into account that the variations $\delta\Psi_j(2t-\tau)$ in $\Omega \times (0, 2t)$ and $\delta\Psi_j(2t)$ in Ω are independent at any $j = 1, 2, \ldots, n$ and that on the interval $(0, t)$ the equality

$$\frac{\overline{\partial c\rho}}{\partial \tau} = \frac{\partial \overline{c\rho}}{\partial \tau}$$

holds, the system of Euler–Lagrange equations is obtained on the range $(0, t)$ in the form

$$\int_\Omega \Bigg\{ \sum_{i=1}^n [(\bar{\lambda}\chi_{i,x})_x + (\bar{\lambda}\chi_{i,y})_y + (\bar{\lambda}\chi_{i,z})_z] \Psi_i(\tau) - \overline{c\rho} \sum_{i=1}^n [\chi_i \Psi_i'(\tau) + \chi_{i,\tau}' \Psi_i(\tau)] + q_v \Bigg\} \chi_j\, d\Omega + \int_{\Gamma_1} \left\{ \bar{\lambda} \left[\sum_{i=1}^n \chi_i \Psi_i(\tau) - T_w\right] \nabla\chi_j \vec{n}\, ds - \int_{\Gamma_2} \left[\bar{\lambda} \sum_{i=1}^n \nabla\chi_i \times \vec{n} \Psi_i(\tau) + q \right] \chi_j\, ds - \int_{\Gamma_3} \left\{ \bar{\lambda} \sum_{i=1}^n \nabla\chi_i \vec{n} \Psi_i(\tau) + \bar{\alpha} \left[\sum_{i=1}^n \chi_i \Psi_i(\tau) - T_c\right] \right\} \chi_j\, ds = 0 \tag{31}$$

$$\int_{\Omega} \overline{c\rho}(0)\left[\sum_{i=1}^{n} \chi_i(0)\Psi_i(0) - T_0\right]\chi_j(2t)\,d\Omega = 0 \tag{32}$$

$$j = 1, 2, \ldots, n$$

It is not difficult to see that this set of equations (32) represents the assignment of initial conditions [the assignment of $\Psi_i(0)$] for the system of equations (31) for $\Psi_i(\tau)$, if it is kept in mind that, allowing for the symmetrization operation, $\chi_j(2t) = \chi_j(0)$.

Reducing similar terms in Eq. (31), obtain the system of Euler–Lagrange equations

$$\sum_{i=1}^{n} A_{ji}(\tau)\Psi'_i(\tau) = \sum_{i=1}^{n} a_{ji}(\tau)\Psi_i(\tau) + A_j(\tau) \tag{33}$$

$$\sum_{i=1}^{n} A_{ji}(0)\Psi_i(0) = \int_{\Omega} \overline{c\rho}(M,0)T_0(M)\chi_j\,d\Omega \tag{34}$$

$$j = 1, 2, \ldots, n$$

where

$$a_{ji}(\tau) = \int_{\Omega} L(\chi_i)\chi_j\,d\Omega - \int_{\Omega} \overline{c\rho}\chi'_{i,\tau}\,d\Omega$$

$$+ \int_{\Gamma_1} \bar{\lambda}(M,\tau)\chi_i\frac{\partial\chi_j}{\partial n}\vec{n}\,ds - \int_{\Gamma_2} \bar{\lambda}(M,\tau)\frac{\partial\chi_i}{\partial n}\vec{n}\chi_j\,ds$$

$$- \int_{\Gamma_3}\left[\bar{\lambda}(M,\tau)\frac{\partial\chi_i}{\partial n}\vec{n} + \bar{\alpha}(M,\tau)\chi_i\right]\chi_j\,ds$$

$$L(\chi_i) = \operatorname{div}[\bar{\lambda}(M,\tau)\operatorname{grad}\chi_i]$$

$$A_{ji}(\tau) = \int_{\Omega} \overline{c\rho}(M,\tau)\chi_i\chi_j\,d\Omega$$

$$A_j(\tau) = \int_{\Omega} q_v(M,\tau)\chi_j\,d\Omega - \int_{\Gamma_1} \bar{\lambda}(M,\tau)T_w(M,\tau)$$

$$\times\, \vec{n}\frac{\partial\chi_j}{\partial n}\,ds - \int_{\Gamma_2} q(M,\tau)\chi_j\,ds + \int_{\Gamma_3} \bar{\alpha}(M,\tau)T_c(\tau)\chi_j\,ds \tag{35}$$

In matrix form the set of equations (33) is given by

$$\mathbf{A}\Psi' = \mathbf{a}\Psi + A_0 \tag{33$'$}$$

where **A** and **a**, respectively, are the matrices of the nth order of the coefficients $A_{ji}(\tau)$ and $a_{ji}(\tau)$, and Ψ', Ψ, and A_0 are the matrices columns of the functions $\Psi'_i(\tau)$, $\Psi_i(\tau)$, and $A_j(\tau)$ of length $n(j, i = 1, \ldots, n)$.

The system of equations (33′), when the determinant of the matrix A is different from zero ($\Delta = \det A \neq 0$), is brought to the normal form in the usual way:

$$\Psi' = \mathbf{A}^{-1}\mathbf{a}\Psi + \mathbf{A}^{-1}A_0 \tag{33''}$$

Knowing the eigen-values (and their multiplicities) of the matrix $\mathbf{A}^{-1}$, one can seek the solution of the indicated set of equations with the initial conditions

$$\Psi(0) = \mathbf{A}^{-1}(0)\int_{\Omega} \overline{c\rho}(M,0)T_0(M)\chi_j(M)\,d\Omega \tag{32'}$$

VI. An Example of Temperature Field Determination in an Infinite Plate

An illustration of the effectiveness of the previously described variational method for determining the temperature field will be given first for the case where the thermophysical properties of the plate material are space dependent and when the equation of the process has the form

$$\widetilde{c\rho}(\xi)\frac{\partial \tilde{T}(\xi,\mathrm{Fo})}{\partial\,\mathrm{Fo}} = \frac{\partial}{\partial\xi}\left[\tilde{\lambda}(\xi)\frac{\partial \tilde{T}(\xi,\mathrm{Fo})}{\partial\xi}\right] + \tilde{q}_v(\xi,\mathrm{Fo})$$

$$0 < \xi < 1, \qquad \mathrm{Fo} > 0 \tag{36}$$

the initial condition is

$$\tilde{T}(\xi,0) = \tilde{T}_0(\xi), \qquad 0 < \xi < 1 \tag{37}$$

and the boundary conditions are formulated as follows.

1. Boundary conditions of the first kind:

$$\tilde{T}(0,\mathrm{Fo}) = \tilde{T}_{w,1}(\mathrm{Fo}), \qquad \mathrm{Fo} > 0 \tag{38$_{\mathrm{I}}$}$$ [1]

$$\tilde{T}(1,\mathrm{Fo}) = \tilde{T}_{w,2}(\mathrm{Fo}), \qquad \mathrm{Fo} > 0 \tag{39$_{\mathrm{I}}$}$$

2. Boundary conditions of the second kind:

$$\tilde{\lambda}(0)\frac{\partial \tilde{T}}{\partial\xi}\bigg|_{\xi=0} = \tilde{q}_1(\mathrm{Fo}), \qquad \mathrm{Fo} > 0 \tag{38$_{\mathrm{II}}$}$$

$$-\tilde{\lambda}(1)\frac{\partial \tilde{T}}{\partial\xi}\bigg|_{\xi=1} = \tilde{q}_2(\mathrm{Fo}), \qquad \mathrm{Fo} > 0 \tag{39$_{\mathrm{II}}$}$$

[1] The subscripts on the equation numbers refer to the boundary conditions: I, first kind; II, second kind; and III, third kind.

3. Boundary conditions of the third kind:

$$\tilde{\lambda}(0)\frac{\partial \tilde{T}}{\partial \xi}\bigg|_{\xi=0} = \mathrm{Bi}_1[\tilde{T}(0,\mathrm{Fo}) - T_{c,1}(\mathrm{Fo})], \qquad \mathrm{Fo} > 0 \qquad (38_{\mathrm{III}})$$

$$-\tilde{\lambda}(1)\frac{\partial \tilde{T}}{\partial \xi}\bigg|_{\xi=1} = \mathrm{Bi}_2[\tilde{T}(1,\mathrm{Fo}) - T_{c,2}(\mathrm{Fo})], \qquad \mathrm{Fo} > 0 \qquad (39_{\mathrm{III}})$$

In Eqs. (36)–(39), the special notation used is $\tilde{T}(\xi,\mathrm{Fo})$, $\tilde{T}_{w,1}$ and $\tilde{T}_{w,2}$, and $\tilde{T}_{c,1}$ and $\tilde{T}_{c,2}$ are the dimensionless instantaneous value of temperature, the temperatures on bounding surfaces, and the temperatures of the surrounding media based on the characteristic difference of temperatures ϑ_0; x and $\xi = x/l_0$ are the dimensional and dimensionless coordinates, l_0 is the plate thickness; τ and $\mathrm{Fo} = \lambda_0\tau/c_0\rho_0 l_0^2$ are the time and Fourier number; $\widetilde{c\rho}(\xi) = c\rho(\xi)/c_0\rho_0$ and $c\rho(\xi)$ are the dimensionless and dimensional volumetric heat capacity, while $c_0\rho_0$ is its characteristic value; $\tilde{\lambda}(\xi) = \tilde{\lambda}(\xi)/\lambda_0$ and $\lambda(\xi)$ are the dimensionless and dimensional coefficients of heat conduction, with λ_0 being its characteristic value; $q_v(\xi,\mathrm{Fo})$ and $\tilde{q}_v(\xi,\mathrm{Fo}) = q_v l_0^2/\lambda_0\vartheta_0$ are the dimensional and dimensionless power of the volumetric heat generation sources; $\tilde{q}_1(\mathrm{Fo}) = q_1(\mathrm{Fo})l_0/\vartheta_0\lambda_0$, $\tilde{q}_2(\mathrm{Fo}) = q_2(\mathrm{Fo})l_0/\vartheta_0\lambda_0$, and $q_1(\mathrm{Fo})$ and $q_2(\mathrm{Fo})$ are the dimensionless and dimensional densities of the heat flux into the wall; and $\mathrm{Bi}_1 = \alpha_1 l_0/\lambda_0$, $\mathrm{Bi}_2 = \alpha_2 l_0/\lambda_0$, and α_1 and α_2 are the Biot numbers and the coefficients of convective heat transfer.

The boundary conditions of the first, second, and third kinds in the boundary-value problem being considered are reduced to homogeneous conditions by introducing a new, unknown function $u(\xi,\mathrm{Fo})$ by the rule

$$\tilde{T}(\xi,\mathrm{Fo}) = u(\xi,\mathrm{Fo}) + \xi\tilde{T}_{w,2}(\mathrm{Fo}) + (1-\xi)\tilde{T}_{w,1}(\mathrm{Fo}) \qquad (40_{\mathrm{I}})$$

$$\tilde{T}(\xi,\mathrm{Fo}) = u(\xi,\mathrm{Fo}) - \frac{(1-\xi)^2}{2\tilde{\lambda}(0)}\tilde{q}_1(\mathrm{Fo}) - \frac{\xi^2}{2\tilde{\lambda}(1)}\tilde{q}_2(\mathrm{Fo}) \qquad (40_{\mathrm{II}})$$

$$\tilde{T}(\xi,\mathrm{Fo}) = u(\xi,\mathrm{Fo}) + (1-\xi)^2(2\xi+1)\tilde{T}_{c,1}(\mathrm{Fo}) + \xi^2(3-2\xi)\tilde{T}_{c,2}(\mathrm{Fo}) \qquad (40_{\mathrm{III}})$$

This yields for $u(\xi,\mathrm{Fo})$ the equation

$$\widetilde{c\rho}(\xi)\frac{\partial u}{\partial \mathrm{Fo}} = \frac{\partial}{\partial \xi}\left[\tilde{\lambda}(\xi)\frac{\partial u}{\partial \xi}\right] + Q_v(\xi,\mathrm{Fo}), \qquad 0 < \xi < 1 \qquad \mathrm{Fo} > 0 \quad (36')$$

with the initial conditions for $0 < \xi < 1$

$$u(\xi,0) = \tilde{T}_0(\xi) - \xi\tilde{T}_{w,2}(0) - (1-\xi)\tilde{T}_{w,1}(0) \qquad (37'_{\mathrm{I}})$$

$$u(\xi,0) = \tilde{T}_0(\xi) + (1-\xi)^2\tilde{q}_1(0)/2\tilde{\lambda}(0) + \xi^2\tilde{q}_2(0)/2\tilde{\lambda}(1) \qquad (37'_{\mathrm{II}})$$

$$u(\xi,0)=\tilde{T}_0(\xi)-(1-\xi)^2(2\xi+1)\tilde{T}_{c,1}(0)-\xi^2(3-2\xi)\tilde{T}_{c,2}(0) \qquad (37'_{\text{III}})$$

and the boundary conditions of

1. The first kind:

$$u(0,\text{Fo})=0, \qquad \text{Fo}>0 \qquad (38'_{\text{I}})$$

$$u(1,\text{Fo})=0, \qquad \text{Fo}>0 \qquad (39'_{\text{I}})$$

2. The second kind:

$$\left.\frac{\partial u}{\partial \xi}\right|_{\xi=0}=0, \qquad \text{Fo}>0 \qquad (38'_{\text{II}})$$

$$\left.\frac{\partial u}{\partial \xi}\right|_{\xi=1}=0, \qquad \text{Fo}>0 \qquad (39'_{\text{II}})$$

3. The third kind:

$$\tilde{\lambda}(0)\left.\frac{\partial u}{\partial \xi}\right|_{\xi=0}=\text{Bi}_1 u(0,\text{Fo}), \qquad \text{Fo}>0 \qquad (38'_{\text{III}})$$

$$-\tilde{\lambda}(1)\left.\frac{\partial u}{\partial \xi}\right|_{\xi=1}=\text{Bi}_2 u(1,\text{Fo}), \qquad \text{Fo}>0 \qquad (39'_{\text{III}})$$

In Eq. (36′), the modified magnitudes of the volumetric heat generation source $Q_v(\xi,\text{Fo})$ for the boundary conditions of the first, second, and third kinds, respectively, are

$$Q_v(\xi,\text{Fo})=\tilde{q}_v(\xi,\text{Fo})+[\tilde{T}_{w,2}(\text{Fo})-\tilde{T}_{w,1}(\text{Fo})] \times\frac{d\tilde{\lambda}(\xi)}{\partial\xi}-\widetilde{c\rho}(\xi)[\xi\tilde{T}'_{w,2}(\text{Fo})+(1-\xi)\tilde{T}'_{w,1}(\text{Fo})] \qquad (41_{\text{I}})$$

$$Q_v(\xi,\text{Fo})=\tilde{q}_v(\xi,\text{Fo})+(1-\xi)^2\widetilde{c\rho}(\xi)\tilde{q}'_1(\text{Fo})2\tilde{\lambda}(0) + \xi^2\widetilde{c\rho}(\xi)\tilde{q}'_2(\text{Fo})/2\tilde{\lambda}(1)+\frac{\partial}{\partial\xi}[(1-\xi)\tilde{\lambda}(\xi)]\tilde{q}_1(\text{Fo}) :\tilde{\lambda}(0)-\frac{\partial}{\partial\xi}[\xi\tilde{\lambda}(\xi)]\tilde{q}_2(\text{Fo})/\tilde{\lambda}(1) \qquad (41_{\text{II}})$$

$$Q_v(\xi,\text{Fo})=\tilde{q}_v(\xi,\text{Fo})-(1-\xi)^2(2\xi+1)\widetilde{c\rho}(\xi)\tilde{T}'_{c,1}(\text{Fo}) -\xi^2(3-2\xi)\widetilde{c\rho}(\xi)\tilde{T}_{c,2}(\text{Fo})+6[\tilde{T}_{c,1}(\text{Fo})-\tilde{T}_{c,2}(\text{Fo})] \times\frac{\partial}{\partial\xi}[(\xi^2-\xi)\tilde{\lambda}(\xi)] \qquad (41_{\text{III}})$$

Note once again that the subscripts I, II, and III in all of the previously given numbers of equations indicate their relevance to the boundary conditions of the first, second, and third kind, respectively.

The thermophysical properties that depend on the coordinate ξ are specified as follows:

$$\tilde{\lambda}(\xi) = 1 + \delta\xi + v\xi^2, \qquad \widetilde{c\rho}(\xi) = 1 + \sigma\xi + \mu\xi^2 \tag{42}$$

When constructing an approximate solution of the problem defined by Eqs. (36′)–(39′), the coordinate functions $\chi_i(M, \tau)$ are taken to be as given in the following.

1. For boundary conditions of the first kind:

$$\chi_i(M, \tau) = \chi_i(\xi) = \sin i\pi\xi, \qquad i = 1, \ldots, n \tag{43_{I}}$$

2. For boundary conditions of the second kind:

$$\chi_i(M, \tau) = \chi_i(\xi) = \cos(i - 1)\pi\xi, \qquad i = 1, \ldots, n \tag{43_{II}}$$

3. For boundary conditions of the third kind:

$$\chi_i(M, \tau) = \chi_i(\xi) = a_i + b_i\xi + c_i\xi^{i+1}, \qquad i = 1, \ldots, n \tag{43_{III}}$$

where

$$a_i = \tilde{\lambda}(0)/\mathrm{Bi}_1, \qquad b_i = 1$$
$$c_i = -[\mathrm{Bi}_2(a_i + b_i) + \tilde{\lambda}(1)b_i]/[\tilde{\lambda}(1)(i + 1) + \mathrm{Bi}_2]$$

The selected systems of coordinate functions (43_{I}) and (43_{II}) are well defined over the segment [0 to 1], continuous, and differentiable within the region studied. Further the approximations to the solution constructed with $\chi_i(\xi)$ in the form

$$u(\xi, \mathrm{Fo}) = \sum_{i=1}^{n} \chi_i(\xi)\Psi_i(\mathrm{Fo}) \tag{44}$$

satisfy, as is easy to see, the boundary conditions (38′) and (39′).

In Tables I–IV, the components of the first and second approximations to the solution that enter into the equation for $\Psi_1(\mathrm{Fo})$ are presented. The specific forms of the equation $u(\xi, \mathrm{Fo})$ for the considered boundary conditions are also shown.

Now, approximations to the solution will be given for the case of identical initial temperature of the plate ($\tilde{T}_0(\xi) = \tilde{T}_0$) and in the absence of volumetric heat generation sources ($\tilde{q}_v(\xi, \mathrm{Fo}) = 0$.

1. For the boundary conditions of the first kind with constant temperature

on the boundary surfaces of the plate, the first approximation has the form

$$\tilde{T}(\xi, \mathrm{Fo}) = \xi \tilde{T}_{w,2} + (1 - \xi)\tilde{T}_{w,1} + \Bigg\{ 2(\tilde{T}_{w,2} - \tilde{T}_{w,1}) \times \frac{\delta + \nu}{\pi} \frac{1/2 + \sigma/4 + \mu(1/6 - 1/4\pi^2)}{\pi^2(1 + \delta/2 + \nu/3)/2 + \nu/4} \Bigg\{ \exp\Bigg\{ \Bigg[\frac{\pi^2}{2} \times \left(1 + \frac{\delta}{2} + \frac{\nu}{3}\right) + \frac{\nu}{4} \Bigg] \mathrm{Fo}/[1/2 + \sigma/4 + \mu(1/6 - 1/4\pi^2)] \Bigg\} - 1 \Bigg\} + \tilde{T}_0 \left[\frac{\sigma + 2}{\pi} + \frac{\mu(\pi^2 - 4)}{\pi^3} \right] - \tilde{T}_{w,2} \left[1/\pi + \frac{\sigma(\pi^2 - 4)}{\pi^3} + \frac{\mu(\pi^2 - 6)}{\pi^3} \right] - \tilde{T}_{w,1}(1/\pi + 4\sigma/\pi^3 + 2\mu/\pi^3) \Bigg\} \times \exp\{[-\pi^2/2(1 + \delta/2 + \nu/3) - \nu/4]\,\mathrm{Fo}/[1/2 + \sigma/4 + \mu(1/6 - 1/4\pi^2)]\} \sin \pi\xi/[1/2 + \sigma/4 + \mu(1/6 - 1/4\pi^2)] \quad (45)$$

2. For the boundary conditions of the second kind with constant heat flux into the plate wall, the first approximation is written as

$$\tilde{T}(\xi, \mathrm{Fo}) = (1 - \xi)^2 \tilde{q}_1/2 - \xi^2 \tilde{q}_2/2(1 + \delta + \nu) - [\tilde{q}_1 + \tilde{q}_2)\,\mathrm{Fo} + T_0(1 + \sigma/2 + \mu/3) + (1/3 + \sigma/12 + \mu/30\tilde{q}_1/2 + (1/3 + \sigma/4 + \mu/5)\tilde{q}_2/2(1 + \delta + \nu)]/(1 + \sigma/2 + \mu/3) \quad (46)$$

and the second approximation is

$$\tilde{T}(\xi, \mathrm{Fo}) = \Psi_1(\mathrm{Fo}) + \Psi_2(\mathrm{Fo}) \cos \pi\xi - \frac{(1 - \xi)^2 \tilde{q}_1}{2} - \frac{\xi^2 \tilde{q}_2}{2(1 + \delta + \nu)} \quad (47)$$

where

$$\Psi_1(\mathrm{Fo}) = \frac{A_{12}}{A_{11}^2 a_{22}} \{(A_{12}A_1 - A_{11}A_2)[1 - \exp(-A_{11}a_{22} \times \mathrm{Fo}/\Delta)] - \Psi_2(0)A_{11}a_{22}\} \exp(A_{11}a_{22}\,\mathrm{Fo}/\Delta) + [A_1\,\mathrm{Fo} + A_{11}\Psi_1(0) + A_{12}\Psi_2(0)]/A_{11}$$

$$\Psi_2(\mathrm{Fo}) = \exp(A_{11}a_{22}\,\mathrm{Fo}/\Delta)\{(A_{11}A_{12} - A_{12}A_1)[1 - \exp(-A_{11} \times a_{22}\,\mathrm{Fo}/\Delta)] + \Psi_2(0)A_{11}a_{22}\}/A_{11}a_{22}$$

$$\Psi_1(0) = (A_{22}F_1 - A_{12}F_2)/\Delta, \qquad \Psi_2(0) = (A_{11}F_2 - A_{12}F_1)/\Delta$$

$$\Delta = A_{11}A_{22} - A_{12}^2, \qquad F_1 = a_0 + a_1/2 + a_2/3 + a_3/4 + a_4/5$$

$$F_2 = -2(a_1 + a_2)/\pi^2 + 3a_3(4/\pi^4 - 1/\pi^2) + 4a_4(6/\pi^4 - 1/\pi^2)$$

TABLE I

COMPONENTS OF THE FIRST APPROXIMATION FOR BOUNDARY CONDITIONS OF THE FIRST KIND

Equation for determination	Resulting calculation equation
$a_{11} = \int_0^1 \frac{\partial}{\partial \xi}\left[(1 + \delta\xi + \nu\xi^2)\frac{\partial \sin \pi\xi}{\partial \xi}\right] \sin \pi\xi \, d\xi$	$-\pi^2(1 + \delta/2 + \nu/3)/2 - \nu/4$
$A_{11} = \int_0^1 (1 + \sigma\xi + \mu\xi^2) \sin^2 \pi\xi \, d\xi$	$1/2 + \sigma/4 + \mu(1/6 - 1/4\pi^2)$
$A_1(\mathrm{Fo}) = \int_0^1 Q_v(\xi, \mathrm{Fo}) \sin \pi\xi \, d\xi = \int_0^1 \tilde{q}_v(\xi, \mathrm{Fo}) \times \sin \pi\xi \, d\xi + \int_0^1 (\tilde{T}_{w,2} - \tilde{T}_{w,1}) \frac{d\tilde{\lambda}(\xi)}{d\xi} \sin \pi\xi \, d\xi - \int_0^1 \widetilde{c\rho}(\xi)[\xi \tilde{T}'_{w,2}(\mathrm{Fo}) + (1 - \xi)\tilde{T}'_{w,1}(\mathrm{Fo})] \sin \pi\xi \, d\xi$	$\int_0^1 \tilde{q}_v(\xi, \mathrm{Fo}) \sin \pi\xi \, d\xi + 2(\tilde{T}_{w,2} - \tilde{T}_{w,1})(\delta + \nu)/\pi - \tilde{T}'_{w,2}(\mathrm{Fo}) \times [1/\pi + \sigma(\pi^2 - 4)/\pi^3 + \mu(\pi^2 - 6)/\pi^3] - \tilde{T}'_{w,1}(\mathrm{Fo}) \times (1/\pi + 4\sigma/\pi^3 + 2\mu/\pi^3)$
$\psi_1(\mathrm{Fo}) = \left[\int_0^{\mathrm{Fo}} A_1(\mathrm{Fo}) \exp(-a_{11}\widetilde{\mathrm{Fo}}/A_{11}) \, d\widetilde{\mathrm{Fo}} + \int_0^1 \widetilde{c\rho}(\xi) u(\xi, 0) \sin \pi\xi \, d\xi\right]/A_{11}$	$\left\{\int_0^{\mathrm{Fo}} [A_1(\widetilde{\mathrm{Fo}}) \exp(-a_{11}\widetilde{\mathrm{Fo}}/A_{11}) \, d\widetilde{\mathrm{Fo}} + \int_0^1 (1 + \sigma\xi + \mu\xi^2) \times [\tilde{T}_0(\xi) - \xi\tilde{T}_{w,2}(0) - (1 - \xi)\tilde{T}_{w,1}(0)] \sin \pi\xi \, d\xi\right\} \exp(a_{11}\mathrm{Fo}/A_{11})/A_{11}$

TABLE II

COMPONENTS OF THE FIRST APPROXIMATION FOR BOUNDARY CONDITIONS OF THE SECOND KIND

Equation for determination	Resulting calculation equation
$a_{11} = \int_0^1 \frac{\partial}{\partial \xi}\left[(1 + \delta\xi + \nu\xi^2)\frac{\partial \cos 0}{\partial \xi}\right]\cos 0\, d\xi$	0
$A_{11} = \int_0^1 (1 + \sigma\xi + \mu\xi^2)\cos^2 0\, d\xi$	$1 + \sigma/2 + \mu/3$
$A_1(\mathrm{Fo}) = \int_0^1 Q_v(\xi, \mathrm{Fo})\cos 0\, d\xi = \int_0^1 \tilde{q}_v(\xi, \mathrm{Fo})\, d\xi$ $+ \tilde{q}'_1(\mathrm{Fo})\int_0^1 (1-\xi)^2 \widetilde{c\rho}(\xi)\, d\xi/2\tilde{\lambda}(0) + \tilde{q}'_2(\mathrm{Fo})$ $\times \int_0^1 \xi^2 \widetilde{c\rho}(\xi)\, d\xi/2\tilde{\lambda}(1) + \tilde{q}_1(\mathrm{Fo})\int_0^1 \frac{\partial}{\partial\xi}[(1-\xi)$ $\times \tilde{\lambda}(\xi)]\, d\xi/\tilde{\lambda}(0) - \tilde{q}_2(\mathrm{Fo})\int_0^1 \frac{\partial}{\partial \xi}[\xi\tilde{\lambda}(\xi)]\, d\xi/\tilde{\lambda}(1)$	$\int_0^1 \tilde{q}_v(\xi, \mathrm{Fo}) + \tilde{q}'_1(\mathrm{Fo})(1/3 + \sigma/12 + \mu/15)/2 + \tilde{q}'_2(\mathrm{Fo})(1/3 + \sigma/4$ $+ \mu/5)/2(1 + \delta + \nu) - \tilde{q}_1(\mathrm{Fo}) - \tilde{q}_2(\mathrm{Fo})$
$\psi_1(\mathrm{Fo}) = \left[\int_0^{\mathrm{Fo}} A_1(\widetilde{\mathrm{Fo}})\, d\widetilde{\mathrm{Fo}} + \int_0^1 \widetilde{c\rho}(\xi) u(\xi, 0)\cos 0\, d\xi\right]/A_{11}$	$\left\{\int_0^{\mathrm{Fo}} A_1(\widetilde{\mathrm{Fo}})\, d\widetilde{\mathrm{Fo}} + \int_0^1 (1 + \sigma\xi + \mu\xi^2) \times [\tilde{T}_0(\xi) + (1-\xi)^2\tilde{q}_1(0)/2\tilde{\lambda}(0)\right.$ $\left. + \xi^2\tilde{q}_2(0){:}\ 2\tilde{\lambda}(1)]\, d\xi\right\}/A_{11}$

TABLE III

COMPONENTS OF THE SECOND APPROXIMATION TO THE SOLUTION WITH THE BOUNDARY CONDITIONS OF THE SECOND KIND[a]

Equations for determination	Resulting calculation equation
$a_{12} = \int_0^1 \frac{\partial}{\partial\xi}\left[(1 + \delta\xi + v\xi^2)\frac{\partial \cos \pi\xi}{\partial\xi}\right] \cos 0 \, d\xi$	0
$a_{21} = \int_0^1 \frac{\partial}{\partial\xi}\left[(1 + \delta\xi + v\xi^2)\frac{\partial \cos 0}{\partial\xi}\right] \cos \pi\xi \, d\xi$	0
$a_{22} = \int_0^1 \frac{\partial}{\partial\xi}\left[1 + \delta\xi + v\xi^2)\frac{\partial \cos \pi\xi}{\partial}\right] \cos \pi\xi \, d\xi$	$-\pi^2(1 + \delta/2 + v/3)/^2 - v/4$
$A_{12} = A_{21} = \int_0^1 (1 + \sigma\xi + \mu\xi^2) \cos \pi\xi \cos 0 \, d\xi$	$-2(\sigma + \mu)/\pi^2$
$A_{22} = \int_0^1 (1 + \sigma\xi + \mu\xi^2) \cos^2 \pi\xi \, d\xi$	$1/2 + \sigma 4 + \mu/6 - \mu/4\pi^2$
$A_2(\mathrm{Fo}) = \int_0^1 Q_v(\xi, \mathrm{Fo}) \cos \pi\xi \, d\xi = \int_0^1 \tilde{q}_v(\xi, \mathrm{Fo}) \cos \pi\xi \, d\xi$	$\int_0^1 \tilde{q}_v(\xi, \mathrm{Fo}) \cos \pi\xi \, d\xi + \tilde{q}'_1(\mathrm{Fo})[2(\sigma - \mu + 1)/\pi^2 + (12\sigma + 3\pi^2$
$+ \tilde{q}'_1(\mathrm{Fo}) \int_0^1 \widetilde{c\rho}(\xi)(1 - \xi)^2 \cos \pi\xi \, d\xi / 2\tilde{\lambda}(0)$	$- 30\mu + 7\mu\pi^2)\pi^4]/2 + \tilde{q}'_2(\mathrm{Fo})[(12\sigma + 3\pi^2\sigma - 4\mu\pi^2$
$+ \tilde{q}'_2(\mathrm{Fo}) \int_0^1 \widetilde{c\rho}(\xi)\xi^2 \cos \pi\xi \, d\xi : 2\tilde{\lambda}(1)$	$+ 24\mu)/\pi^4 - 2/\pi^2]/2(1 + \delta + v) + \tilde{q}_1(\mathrm{Fo})(4\delta + 2v)$
$+ \tilde{q}_1(\mathrm{Fo}) \int_0^1 \frac{\partial}{\partial\xi}[(1 - \xi)\tilde{\lambda}(\xi)] \cos \pi\xi \, d\xi$	$: \pi^2 + \tilde{q}_2(\mathrm{Fo})(4\delta + 6v)/\pi^2(1 + \delta + v)$
$: \tilde{\lambda}(0) - \tilde{q}_2(\mathrm{Fo}) \int_0^1 \frac{\partial}{\partial\xi}[\xi\tilde{\lambda}(\xi)] \cos \pi\xi \, d\xi / \tilde{\lambda}(1)$	

[a] a_{11}, A_{11}, and $A_1(\mathrm{Fo})$ are presented in Table II.

TABLE IV

COMPONENTS OF THE FIRST APPROXIMATION WITH THE BOUNDARY CONDITIONS OF THE THIRD KIND

Equations for determination	Resulting calculation equation
$a_{11} = \int_0^1 \frac{\partial}{\partial \xi}\left[(1 + \delta\xi + \nu\xi^2)\frac{\partial(a_1 + b_1\xi + c_1\xi^2)}{\partial \xi}\right](a_1 + b_1\xi + c_1\xi^2)\,d\xi$	$2a_1c_1 + a_1b_1\delta + 2a_1c_1\delta + a_1b_1\nu + 2a_1c_1\nu + b_1c_1 + b_1^2\delta/2 + 5b_1c_1\delta/3$ $+ 2b_1^2\nu/3 + 2b_1c_1\nu + 2c_1^2/3 + c_1^2\delta + 6c_1^2\nu/5$
$A_{11} = \int_0^1 (1 + \sigma\xi + \mu\xi^2)(a_1 + b_1\xi^2 + c_1\xi^2)^2\,d\xi$	$a_1^2 + a_1b_1 + a_1^2\sigma/2 + (2a_1c_1 + b_1^2 + 2a_1b_1\sigma + a_1^2\mu)/3 + (2b_1c_1$ $+ 2a_1c_1\sigma + b_1^2\sigma + 2a_1b_1\mu)/4 + (c_1^2 + 2b_1c_1\sigma$ $+ 2a_1c_1\mu + \mu b_1^2)/5 + (c_1^2\sigma + 2b_1c_1\mu)/6 + c_1^2\mu/7$
$A_1(\mathrm{Fo}) = \int_0^1 Q_v(\xi, \mathrm{Fo})(a_1 + b_1\xi + c_1\xi^2)\,d\xi = \int_0^1 \tilde{q}_v(\xi, \mathrm{Fo})(a_1 + b_1\xi$ $+ c_1\xi^2)\,d\xi - \tilde{T}'_{c,1}(\mathrm{Fo})\int_0^1 (1 + \sigma\xi + \mu\xi^2)(1 - \xi)^2(2\xi + 1)(a_1$ $+ b_1\xi + c_1\xi^2)\,d\xi - \tilde{T}'_{c,2}(\mathrm{Fo})\int_0^1 (1 + \sigma\xi + \mu\xi^2)\xi^2(3 - 2\xi)(a_1$	$\int_0^1 \tilde{q}_v(\xi, \mathrm{Fo})(a_1 + b_1\xi + c_1\xi^2)\,d\xi - \tilde{T}'_{c,1}(\mathrm{Fo})(a_1/2 + 3b_1/20$ $+ c_1/15 + 3a_1\sigma/20 + b_1\sigma/15 + c_1\sigma/28 + a_1\mu/15 + b_1\mu/28$ $+ 3c_1\mu/140) - \tilde{T}'_{c,2}(\mathrm{Fo})(a_1/2 + 7b_1/20 + 4c_1/15 + 7a_1\sigma/20$

$$+ b_1\xi + c_1\xi^2)\, d\xi + 6[\tilde{T}_{c,1}(\mathrm{Fo}) - \tilde{T}_{c,2}(\mathrm{Fo})] \int_0^1 \frac{\partial}{\partial \xi}$$

$$\times [(\xi^2 - \xi)(1 + \delta\xi + \nu\xi^2)](a_1 + b_1\xi + c_1\xi^2)\, d\xi$$

$$\psi_1(\mathrm{Fo}) = \Big[\int_0^{\mathrm{Fo}} A_1(\widetilde{\mathrm{Fo}}) \exp(-a_{11}\, \widetilde{\mathrm{Fo}}/A_{11})\, d\,\widetilde{\mathrm{Fo}}$$

$$+ \int_0^1 \widetilde{c\rho}(\xi) u(\xi, 0)(a_1 + b_1\xi + c_1\xi^2)\, d\xi \Big] \exp(a_{11}\,\mathrm{Fo}/A_{11})/A_{11}$$

$$+ 4b_1 \sigma/15 + 3c_1\sigma/14 + 4a_1\mu/15 + 3b_1\mu/14 + 5c_1\mu/28)$$

$$+ 6[\tilde{T}_{c,1}(\mathrm{Fo}) - \tilde{T}_{c,2}(\mathrm{Fo})](b_1/6 + c_1/6 + b_1\sigma/12 + b_1\nu/20$$

$$+ c_1\sigma/10 + c_1\nu/15); \quad a_1 = \tilde{\lambda}(0)/\mathrm{Bi}_1; \quad b_1 = 1; \quad c_1 = -[Bi_2(a_1 + b_1)$$

$$+ \tilde{\lambda}(1)]/[2\tilde{\lambda}(1) + \mathrm{Bi}_2]$$

$$\Big[\int_0^{\mathrm{Fo}} A_1(\widetilde{\mathrm{Fo}}) \exp(-a_{11}\, \widetilde{\mathrm{Fo}}/A_{11})\, d\,\widetilde{\mathrm{Fo}} + \int_0^1 (1 + \sigma\xi + \mu\xi^2)(a_1 + b_1\xi + c_1\xi^2)\tilde{T}_0(\xi)$$

$$\times d\xi - \tilde{T}_{c,1}(0)(a_1/2 + 3b_1/20 + c_1/15 + 3a_1\sigma/20 + b_1\sigma/15 + c_1\sigma/28$$

$$+ a_1\mu/15 + b_1\mu/28 + 3c_1\mu/140) - \tilde{T}_{c,2}(0)(a_1/2 + 7b_1/20 + 4c_1/15$$

$$+ 7a_1\sigma/20 + 4b_1\sigma/15 + 3c_1\sigma/14 + 4a_1\mu/15 + 3b_1\mu/14 + 5c_1\mu/28)\Big]$$

$$\times \exp(a_{11}\,\mathrm{Fo}/A_{11})/A_{11}$$

With $\tilde{T}_0(\xi) = \tilde{T}_0 = \mathrm{const}$:

$$\int_0^1 (1 + \sigma\xi + \mu\xi^2)(a_1 + b_1\xi + c_1\xi^2)\tilde{T}_0(\xi)\, d\xi = \tilde{T}_0[a_1 + (a_1\sigma + b_1)/2$$

$$+ (a_1\mu + b_1\sigma + c_1)/3 + (b_1\mu + c_1\sigma)/4 + c_1\mu/5]$$

$$a_0 = b_1 + b_2, \quad a_1 = \sigma(b_1 + b_2) - 2b_2, \quad a_2 = \mu(b_1 + b_2) + b_2 + b_3 - 2b_2\sigma, \quad a_3 = \sigma(b_2 + b_3) - 2\mu b_2, \quad a_4 = \mu(b_2 + b_3)$$

$$b_1 = \tilde{T}_0, \quad b_2 = \tilde{q}_1/2, \quad b_3 = \tilde{q}_2/2(1 + \delta + v)$$

3. For the boundary conditions of the third kind with the constant temperature of the media surrounding the plate, the first approximation is

$$\tilde{T}(\xi, \text{Fo}) = \tilde{T}_{c,1}(1 + \xi)^2(2\xi + 1) + \tilde{T}_{c,2}\xi^2(3 - 2\xi) + \Psi_1(\text{Fo})(a_1 + b_1\xi + c_1\xi^2) \tag{48}$$

where

$$\begin{aligned}\Psi_1(\text{Fo}) = \{&-6(\tilde{T}_{c,1} - \tilde{T}_{c,2})[(b_1 + c_1)/6 + b_1\delta/12 \\ &+ b_1 v/20 + c_1\delta/10 + c_1 v/15]A_{11}[\exp(-a_{11}\,\text{Fo}/A_1) - 1] \\ &: a_{11} + \tilde{T}_0[a_1 + (a_1\sigma + b_1)/2 + (a_1\mu + b_1\sigma + c_1)/3 \\ &+ (b_1\mu + c_1\sigma)/4 + c_1\mu/5] - \tilde{T}_{c,1}(a_1/2 + 3b_1/20 + c_1/15 \\ &+ 3a_1\sigma/20 + b_1\sigma/15 + c_1\sigma/28 + a_1\mu/15 + b_1\mu/28 \\ &+ 3c_1/140) - \tilde{T}_{c,2}(a_1/2 + 7b_1/20 + 4c_1/15 + 7a_1\sigma/20 \\ &+ 4b_1\sigma/15 + 3c_1\sigma/14 + 4a_1\mu/15 + 3b_1\mu/14 \\ &+ 5c_1\mu/28\}\exp(a_{11}\,\text{Fo}/A_{11})/A_{11}\end{aligned}$$

(the values of a_{11} and A_{11} are given in Table IV).

In Figs. 3, 4, and 5, graphical interpretations of the solutions for Eqs. (45), (46), and (47), respectively, are presented for $v = \delta = \mu = \sigma = 0.5$ and $\tilde{T}_0 = 0$. In this case, it was assumed for the boundary conditions of the first kind that $\tilde{T}_{w,1} = 0.5$ and $\tilde{T}_{w,2} = 1$; for the boundary conditions of the second kind that $\tilde{q}_1 = -2.5$ and $\tilde{q}_2 = -0.5$; and for the boundary conditions of the third kind that $\text{Bi}_1 = 5$, $\text{Bi}_2 = 15$, $\tilde{T}_{c,1} = 0.5$, and $\tilde{T}_{c,2} = 1$. The same problems were solved on a computer by the finite-difference method. Comparison of the analytical and numerical results indicates their practical coincidence for all the types of boundary conditions starting from $\text{Fo} \geq 0.05$.

At constant thermophysical properties of the plate material ($\tilde{\lambda} = 1$, $\widetilde{c\rho} = 1$, or $v = \delta = \mu = \sigma = 0$), it is possible to indicate the nth approximation to the solution of the boundary-value problems [Eqs. (15)–(19)] with the chosen system of coordinate functions [Eq. (43)]. For the case of the boundary conditions of the first kind, Eqs. (33) (34), and (35) give

$$a_{ji} = -j^2\frac{\pi^2}{2}\delta_{ji}, \qquad A_{ji} = \frac{1}{2}\delta_{ji}$$

$$A_j(\text{Fo}) = \int_0^1 [\tilde{q}_v(\xi, \text{Fo}) - \xi\tilde{T}_{w,2}(\text{Fo}) - (1 - \xi)\tilde{T}_{w,1}(\text{Fo})] \sin j\pi\xi\, d\xi$$

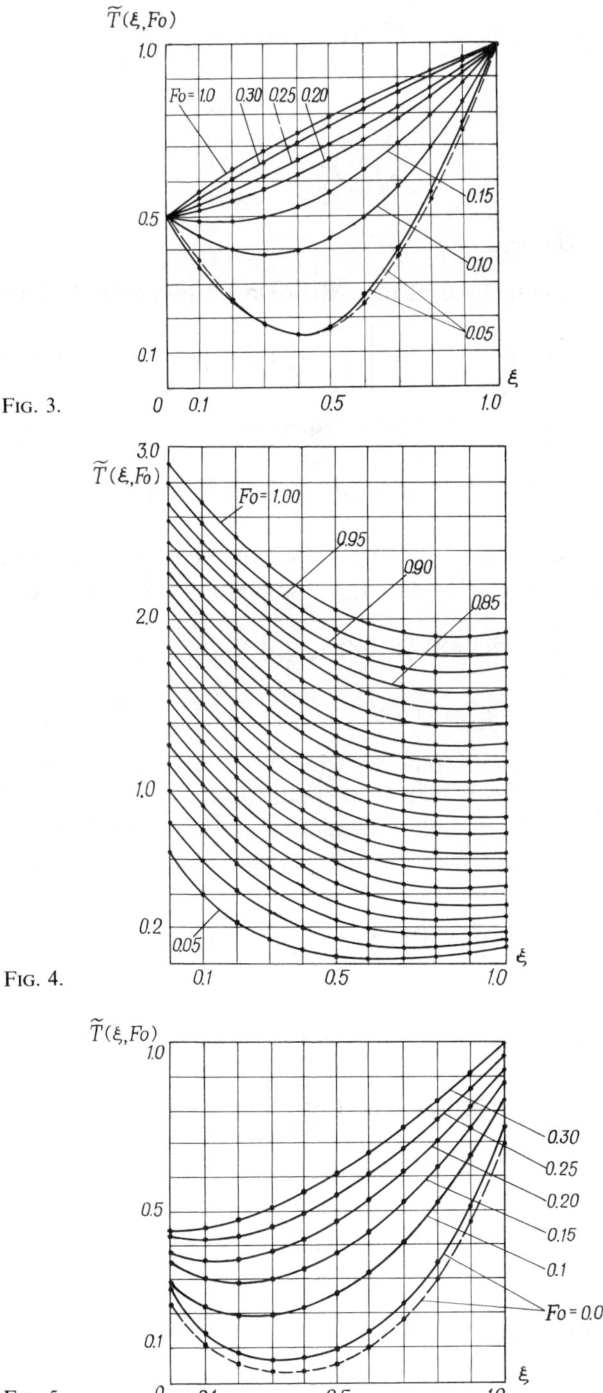

FIGS. 3, 4, and 5. Dependence of temperature $\widetilde{T}$ on coordinate ξ and time Fo in a plate for boundary conditions of the first, second, and third kinds, respectively: (——), analytical solution; (– – –), numerical solution by the grid method.

where

$$\delta_{ji} = \begin{cases} 1, & i = j \\ 0, & i \neq j \end{cases}$$

is the Kronecker delta.

The unknown functions of time $\Psi_i(\text{Fo})$ in Eq. (44) have the form

$$\Psi_i(\text{Fo}) = 2\exp(-i^2\pi^2\,\text{Fo})\left[\int_0^{Fo} A_i(\text{Fo})\exp(i^2\pi^2\tilde{\text{F}}\text{o})\,d\tilde{\text{F}}\text{o} + \Psi_i(0)\right] \quad (49)$$

where the initial values of $\Psi_i(0)$ are equal to

$$\Psi_i(0) = 2\left[\int_0^1 \tilde{T}_0(\xi)\sin i\pi\xi\,d\xi + (-1)^i\,\tilde{T}_{w,2}(0)/i\pi - \tilde{T}_{w,1}(0)/i\pi\right]$$

In the absence of internal sources of volumetric heat generation and at constant temperatures $\tilde{T}_{w,1}$ and $\tilde{T}_{w,2}$, the solution is simplified to the form

$$\tilde{T}(\xi, \text{Fo}) = \xi\tilde{T}_{w,2} + (1-\xi)\tilde{T}_{w,1} + 2\sum_{i=1}^{n}\exp(-i^2\pi^2\,\text{Fo})$$
$$\times\,[\tilde{T}_0 + (-1)^{i+1}\tilde{T}_0 + (-1)^i\tilde{T}_{w,2} - \tilde{T}_{w,1}]\sin i\pi\xi/i\pi \quad (50)$$

When $n \to \infty$, Eq. (50) coincides with the exact solution of the corresponding problem of unsteady-state heat conduction.

In the case of the boundary conditions of the second kind and constant thermophysical properties, the use of Eqs. (33), (34), and (35) will give successively

$$a_{ji} = -\frac{(j-1)^2\pi^2}{2}\delta_{ji}, \qquad A_{ji} = \begin{cases} 1, & i = j = 1, \\ \delta_{ji}/2, & i, j \neq 1 \end{cases}$$

$$A_j(\text{Fo}) = \int_0^1 [\tilde{q}_v(\xi, \text{Fo}) + \tilde{q}_1'(\text{Fo})(1-\xi)^2/2 + \tilde{q}_2'(\text{Fo})\xi^2/2$$
$$+\,\tilde{q}_1(\text{Fo}) - \tilde{q}_2(\text{Fo})]\cos(j-1)\pi\xi\,d\xi$$

$$\Psi_i(\text{Fo}) = 2\exp[-(i-1)^2\pi^2\,\text{Fo}]\left\{\int_0^{\text{Fo}} A_i(\text{Fo})\exp(i-1)^2\pi^2\right.$$
$$\left.\times\,\tilde{\text{F}}\text{o}]\,d\tilde{\text{F}}\text{o} + \Psi_i(0)\right\}$$

$$\Psi_i(0) = 2\left[\int_0^1 \tilde{T}_0(\xi)\cos(i-1)\pi\xi\,d\xi + \tilde{q}_1(0)(i-1)^2\pi^2\right.$$
$$\left. + (-1)^{i-1}\tilde{q}_2(0)/(i-1)^2\pi^2\right], \qquad i > 1$$

At $\tilde{q}_v(\xi, \text{Fo}) = 0$ and constant $\tilde{q}_1$ and $\tilde{q}_2$, the following relation is specifically obtained

$$\tilde{T}(\xi, \text{Fo}) = -(1-\xi)^2\tilde{q}_1/2 - \xi^2\tilde{q}_2/2 - (\tilde{q}_1 + q_2) \times (\text{Fo} - 1/6) + \int_0^1 \tilde{T}_0(\xi)d\xi + \sum_{i=2}^{n} \exp[-(i-1)^2\pi^2 \text{Fo}] \times \Psi_i(0)\cos \pi\xi. \tag{51}$$

The approximate solution [Eq. (51)] for $n \to \infty$ coincides with the exact solution of the problem.

VII. One-Dimensional Temperature Field of an Infinite Plate with a Time-Dependent Rate of Convective Heat Transfer

Another application of the variational method will be demonstrated using the practically important example of the nonstationary temperature field in a plate with an arbitrarily time-dependent temperature of the surrounding media, $T_{c,1}(\tau)$ and $T_{c,2}(\tau)$, and with the coefficients of convective heat transfer, $\alpha_1(\tau)$, $\alpha_2(\tau)$, on the bounding surfaces.

This problem with constant thermophysical properties of the body material, using a dimensionless representation of the quantities entering into the problem, takes the form

$$\frac{\partial \tilde{T}(\xi, \text{Fo})}{\partial \text{Fo})} = \frac{\partial^2 \tilde{T}(\xi, \text{Fo})}{\partial \xi^2} + \tilde{q}_v(\xi, \text{Fo}), \qquad 0 < \xi < 1, \qquad \text{Fo} > 0 \tag{52}$$

$$\tilde{T}(\xi, 0) = \tilde{T}_0(\xi), \qquad 0 < \xi < 1 \tag{53}$$

$$\left.\frac{\partial \tilde{T}}{\partial \xi}\right|_{\xi=0} = \text{Bi}_1(\text{Fo})[\tilde{T}(0, \text{Fo}) - \tilde{T}_{c,1}(\text{Fo})], \qquad \text{Fo} > 0 \tag{54}$$

$$-\left.\frac{\partial \tilde{T}}{\partial \xi}\right|_{\xi=1} = \text{Bi}_2(\text{Fo})[\tilde{T}(1, \text{Fo}) - \tilde{T}_{c,2}(\text{Fo})], \qquad \text{Fo} > 0 \tag{55}$$

Note that the solution of this problem [Eqs. (52)–(55)] by traditional methods meets with great difficulties.

Introduce a new unknown function $u(\xi, \text{Fo})$ so that

$$\tilde{T}(\xi, \text{Fo}) = u(\xi, \text{Fo}) + \tilde{T}_{c,1}(\text{Fo})(1-\xi)^2(2\xi + 1) + \tilde{T}_{c,2}(\text{Fo})\xi^2(3 - 2\xi) \tag{56}$$

and transform Eqs. (52)–(55) to the homogeneous boundary conditions

$$\frac{\partial u(\xi, \text{Fo})}{\partial \text{Fo}} = \frac{\partial^2 u(\xi, \text{Fo})}{\partial \xi^2} + Q_v(\xi, \text{Fo}) \tag{52'}$$

$$u(\xi,0) = \tilde{T}_0(\xi) - \tilde{T}_{c,1}(0)(1-\xi)^2(2\xi+1) - \tilde{T}_{c,2}(0)\xi^2(3-2\xi) \quad (53')$$

$$\left.\frac{\partial u(\xi,\mathrm{Fo})}{\partial \xi}\right|_{\xi=0} = \mathrm{Bi}_1(\mathrm{Fo})u(0,\mathrm{Fo}) \quad (54')$$

$$-\left.\frac{\partial u(\xi,\mathrm{Fo})}{\partial \xi}\right|_{\xi=1} = \mathrm{Bi}_2(\mathrm{Fo})u(1,\mathrm{Fo}) \quad (55')$$

The function $Q_v(\xi, \mathrm{Fo})$ is equal to

$$Q_v(\xi,\mathrm{Fo}) = \tilde{q}_v(\xi,\mathrm{Fo}) - (1-\xi)^2(2\xi+1)\tilde{T}'_{c,1}(\mathrm{Fo}) - \xi^2(3-2\xi) \times \tilde{T}'_{c,2}(\mathrm{Fo}) + 6[\tilde{T}_{c,1}(\mathrm{Fo}) - \tilde{T}_{c,2}(\mathrm{Fo})](2\xi-1)$$

According to the Kantorovich method, approximations to the stationary point of the functional (21) for Eqs. (52′)–(55′) have the form.

$$u(\xi,\mathrm{Fo}) = \sum_{i=1}^{n} \chi_i(\xi,\mathrm{Fo})\Psi_i(\mathrm{Fo}) \quad (57)$$

in which the functions $\chi_i(\xi, \mathrm{Fo})$ are selected as

$$\chi_i(\xi,\mathrm{Fo}) = a_i(\mathrm{Fo}) + b_i(\mathrm{Fo})\xi + c_i(\mathrm{Fo})\xi^{i+1}, \qquad i = 1,\ldots,n \quad (58)$$

When constructing the first approximation to the solution of the problem set in the form

$$u(\xi,\mathrm{Fo}) = \chi_1(\xi,\mathrm{Fo})\Psi_1(\mathrm{Fo}) \quad (59)$$

the components of the solution that enter into the equation for $\Psi_1(\mathrm{Fo})$

$$\Psi_1(\mathrm{Fo}) = \left\{\int_0^{\mathrm{Fo}} [A_1(\tilde{\mathrm{Fo}})/A_{11}(\tilde{\mathrm{Fo}})] \exp\left\{-\int_0^{\mathrm{Fo}} [a_{11}(\Theta)/A_{11}(\Theta) \times d\Theta\right\} d\,\tilde{\mathrm{Fo}} + \Psi_1(0)\right\} \exp\left\{\int_0^{\mathrm{Fo}} [a_{11}(\tilde{\mathrm{Fo}})/A_{11}(\tilde{\mathrm{Fo}})]\, d\,\tilde{\mathrm{Fo}}\right\} \quad (60)$$

are, according to Eq. (35), as follows:

$$\begin{aligned} a_{11}(\mathrm{Fo}) &= \int_0^1 \left[\frac{\partial^2}{\partial \xi^2}(a_1 + b_1\xi + c_1\xi^2)\right](a_1 + b_1\xi + c_1\xi^2)\,d\xi \\ &\quad - \int_0^1 \left[\frac{\partial}{\partial \mathrm{Fo}}(a_1 + b_1\xi + c_1\xi^2)\right](a_1 + b_1\xi + c_1\xi^2)\,d\xi \\ &= 2c_1(a_1 + b_1/2 + c_1/3) - a_1a'_1 - a_1{b'_1}^2/2 \\ &\quad - a_1c'_1/3 - a'_1b_1/2 - b_1b'_1/3 - b_1c'_1/4 - a'_1c_1/3 - b'_1c_1/4 \\ &\quad - c_1c'_1/5 \end{aligned}$$

$\left(a_1' = \dfrac{da_1}{d\,\mathrm{Fo}}, \text{ and so forth}\right)$,

$$A_{11}(\mathrm{Fo}) = \int_0 \chi_1^2\, d\xi = a_1^2 + b_1^2/3 + c_1^2/5 + a_1 b_1 + 2a_1 c_1/3 + b_1 c_1/2$$

The function $A_1(\mathrm{Fo})$, which is equal to $\int_0^1 Q_{\mathrm{v}}\chi_1\, d\xi$, can be calculated when $\tilde{q}_{\mathrm{v}}(\xi, \mathrm{Fo})$, $\tilde{T}_{c,1}(\mathrm{Fo})$, and $\tilde{T}_{c,2}(\mathrm{Fo})$ are assigned. In the particular case when $\tilde{q}_{\mathrm{v}}(\xi, \mathrm{Fo}) = 0$, $T_{c,1} = \text{const}$, and $\tilde{T}_{c,2} = \text{const}$,

$$A_1(\mathrm{Fo}) = \int_0^1 Q_{\mathrm{v}}\chi_1\, d\xi = (\tilde{T}_{c,1} - \tilde{T}_{c,2})(b_1 + c_1)$$

Moreover, according to Eq. (32′) at a constant initial temperature $\tilde{T}_0$, in Eq. (60),

$$\Psi_1(0) = \left[a_1\left(\tilde{T}_0 - \frac{\tilde{T}_{c,1}}{2} - \frac{\tilde{T}_{c,2}}{2}\right) + b_1(\tilde{T}_0/2 - 3\tilde{T}_{c,1}/20 \right.$$
$$\left. - 7\tilde{T}_{c,2}/20) + c_1(\tilde{T}_0/3 - T_{c,1}/15 - 4\tilde{T}_{c,2}/15)\right] \Big/ A_{11}(0)$$

Then, as a result of all the calculations, we obtain the unknown first approximation to the solution in the form

$$\tilde{T}(\xi, \mathrm{Fo}) = \tilde{T}_{c,1}(1 - \xi^2)(2\xi + 1) + \tilde{T}_{c,2}\xi^2(3 - 2\xi)$$
$$+ [a_1(\mathrm{Fo}) = b_1(\mathrm{Fo})\xi + c_1(\mathrm{Fo})\xi^2]\psi_1(\mathrm{Fo}) \tag{61}$$

In Fig. 6, the results of calculations by Eq. (61) are compared with the grid solution of Eqs. (52)–(55) obtained on a computer. It was assumed in the

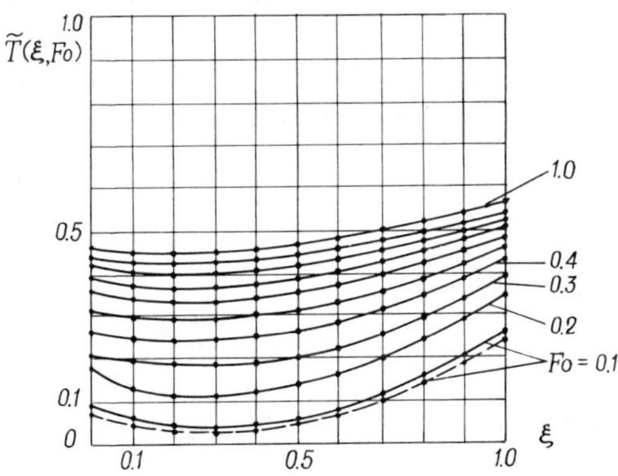

FIG. 6. Dependence of $\tilde{T}$ on ξ and Fo in a plate for time-dependent boundary conditions of the third kind: (——), analytical solutions; (– – –), numerical solution by the grid method.

calculation that $\tilde{T}_0 = 0$, $\tilde{T}_{c,1} = 0.5$, and $\tilde{T}_{c,2} = 1$; $\mathrm{Bi}_1(\mathrm{Fo}) = 0.5\exp(-0.5\,\mathrm{Fo})$; and $\mathrm{Bi}_2 = \exp(-\mathrm{Fo})$. Figure 6 indicates that good agreement between the analytical first approximation to the solution and the finite-difference (i.e., grid) values of $\tilde{T}(\xi, \mathrm{Fo})$ is obtained starting from $\mathrm{Fo} \geq 0.1$. Our latest investigations show that substitution $u(\xi, \mathrm{Fo}) = v(\xi, \mathrm{Fo})\exp[-(\mathrm{I} - \xi)^k \mathrm{Bi}_1(\mathrm{Fo})/k\tilde{\lambda}(0) - \xi^p \mathrm{Bi}_2(\mathrm{Fo})\colon p\tilde{\lambda}(\mathrm{I})]$, where k and p are integers, leads to boundary conditions $(38'_{\mathrm{II}})$, $(39'_{\mathrm{II}})$ and to the basis (43_{II}).

VIII. Temperature Field of Complex-Shaped Bodies

As an example of the application of the variational description using the convolution-type functional for complex-shaped regions with space-dependent thermophysical properties, consider a two-dimensional problem, which for the unknown temperature $\tilde{T}(\mathrm{M}, \mathrm{Fo})$, has the form

$$c\rho(\xi,\eta)\frac{\partial \tilde{T}(\xi,\eta,\mathrm{Fo})}{\partial \mathrm{Fo}} = \frac{\partial}{\partial \xi}\left[\tilde{\lambda}(\xi,\eta)\frac{\partial \tilde{T}(\xi,\eta,\mathrm{Fo})}{\partial \xi}\right] + \frac{\partial}{\partial \eta}\left[\tilde{\lambda}(\xi,\eta)\frac{\partial \tilde{T}(\xi,\eta,\mathrm{Fo})}{\partial \eta}\right] + \tilde{q}_v(\xi,\eta,\mathrm{Fo}), \qquad \mathrm{Fo} > 0, \qquad (\xi,\eta) \in \Omega \tag{62}$$

$$\tilde{T}(\xi,\eta,0) = \tilde{T}_0(\xi,\eta), \qquad (\xi,\eta) \in \Omega \tag{63}$$

$$\tilde{T}(\xi,\eta,\mathrm{Fo})\Big|_{\Gamma} = \tilde{T}_w(\xi,\eta,\mathrm{Fo}), \qquad \mathrm{Fo} > 0, \qquad (\xi,\eta) \in \Gamma \tag{64}$$

The substitution

$$\tilde{T}(\xi,\eta,\mathrm{Fo}) = u(\xi,\eta,\mathrm{Fo}) + \tilde{T}_w(\xi,\eta,\mathrm{Fo}) \tag{65}$$

reduces Eqs. (62)–(64) to the one-dimensional boundary conditions

$$\widetilde{c\rho}(\xi,\eta)\frac{\partial u}{\partial \mathrm{Fo}} = \frac{\partial}{\partial \xi}\left[\tilde{\lambda}(\xi,\eta)\frac{\partial u}{\partial \xi}\right] + \frac{\partial}{\partial \eta}\left[\tilde{\lambda}(\xi,\eta)\frac{\partial u}{\partial \eta}\right] + Q_v(\xi,\eta,\mathrm{Fo}) \tag{62'}$$

$$u(\xi,\eta,0) = \tilde{T}_0(\xi,\eta) - \tilde{T}_w(\xi,\eta,0) \tag{63'}$$

$$u(\xi,\eta,\mathrm{Fo})\Big|_{\Gamma} = 0 \tag{64'}$$

In Eq. (62′), there is a modified source function

$$Q_v(\xi,\eta,\mathrm{Fo}) = \tilde{q}_v(\xi,\eta,\mathrm{Fo}) - c\rho(\xi,\eta,\mathrm{Fo})\frac{\partial \tilde{T}_w(\xi,\eta,\mathrm{Fo})}{\partial \mathrm{Fo}} + \frac{\partial}{\partial \xi}\left[\tilde{\lambda}(\xi,\eta)\frac{\partial \tilde{T}_w(\xi,\eta,\mathrm{Fo})}{\partial \xi}\right] + \frac{\partial}{\partial \eta}\left[\tilde{\lambda}(\xi,\eta)\frac{\partial \tilde{T}_w(\xi,\eta,\mathrm{Fo})}{\partial \eta}\right] \tag{66}$$

Following Kantorovich [28], the nth approximation to the solution of Eqs. (62′)–(64′) will be sought in the form

$$u(\xi, \eta, \mathrm{Fo}) = \sum_{i=1}^{n} \chi_i(\xi, \eta)\Psi_i(\mathrm{Fo}) \tag{67}$$

where $\chi_i(\xi, \eta)$ is the system of the known coordinate functions that satisfy the condition $\chi_i|_\Gamma = 0$ and is complete in the region Ω, and $\Psi_i(\mathrm{Fo})$ are the unknown functions of time.

According to Ref. 28, the system of functions $\chi_i(\xi, \eta)$ should be selected in the given case as $\chi_i = \omega(\xi, \eta)$, $\chi_2 = \xi\omega(\xi, \eta)$, $\chi_3 = \eta\omega(\xi, \eta)$, $\chi_4 = \xi^2\omega(\xi, \eta)$, $\chi_5 = \xi\eta\omega(\xi, \eta), \ldots$, when $\omega(\xi, \eta) > 0$ in the region Ω, $\omega(\xi, \eta)|_\Gamma \equiv 0$, and $\omega(\xi, \eta)$ is continuous together with the second-order derivatives in the region Ω. It is clear that these requirements are first satisfied by the functions $\omega(\xi, \eta)$, which are the explicit parts of the equation for the line Γ bounding the region Ω.

Consider now the practically important case of determining a temperature field in an infinite prism of triangular cross section that is formed by the planes $\xi = 1, \eta = k_1\xi$, and $\eta = k_2\xi$ (Fig. 7). When seeking the first approximation to the solution of Eqs. (62)–(64), the coordinate function $\chi_1(\xi, \eta)$ is selected in the following way to satisfy the previously mentioned requirements for $\chi_i(\xi, \eta)$:

$$\begin{aligned}\chi_1(\xi, \eta) = \omega(\xi, \eta) = (1 - \xi)(\eta - k_1\xi)(k_2\xi - \eta) = k_1k_2\xi^3 \\ + (k_1 + k_2)\xi^2\eta + \xi\eta^2 - k_1k_2\xi^2 + (k_1 + k_2)\xi\eta - \eta^2\end{aligned} \tag{68}$$

Assume the temperature of the prism surface to be constant over time $\tilde{T}_w = \mathrm{const}$, the initial temperature in the prism $\tilde{T}_0 = \mathrm{const}$, and $\tilde{q}_v(\xi, \eta, \mathrm{Fo}) = 0$. Then, when seeking the first approximation to the solution in the form

$$\tilde{T}(\xi, \eta, \mathrm{Fo}) = u(\xi, \eta, \mathrm{Fo}) + \tilde{T}_w = \Psi_1(\mathrm{Fo})\chi_1(\xi, \eta) + \tilde{T}_w \tag{65′}$$

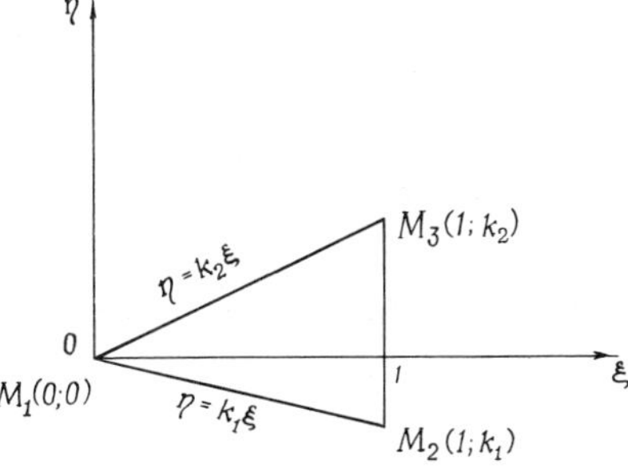

FIG. 7. Two-dimensional region of the process of unsteady-state heat conduction in a triangular prism.

the solution of the set of Eqs. (33)–(34) for $\psi_1(\text{Fo})$ will be

$$\Psi_1(\text{Fo}) = \left[\int_0^{\text{Fo}} A_1(\tilde{\text{F}}\text{o}) \exp(-a_{11}\tilde{\text{F}}\text{o}/A_{11})\, d\tilde{\text{F}}\text{o} + \int_\Omega \widetilde{c\rho}(\xi,\eta) u_0(\xi,\eta) \times \chi_1\, d\Omega\right] \exp(a_{11}\,\text{Fo}/A_{11})/A_{11} \tag{69}$$

where, according to Eq. (35)

$$a_{11} = \int_0^1 \int_{k_1\xi}^{k_2\xi} \left\{\left\{\frac{\partial}{\partial\xi}\left[\tilde{\lambda}(\xi,\eta)\frac{\partial\chi_1(\xi,\eta)}{\partial\xi}\right] + \frac{\partial}{\partial\eta}\left[\tilde{\lambda}(\xi,\eta)\frac{\partial\chi_1(\xi,\eta)}{\partial\eta}\right]\right\} \times \chi_1\right\} d\eta\, d\xi$$

$$A_{11} = \int_0^1 \int_{k_1\xi}^{k_2\xi} \widetilde{c\rho}(\xi,\eta)\chi_1^2\, d\eta\, d\xi, \qquad A_1(\text{Fo}) = \int_0^1 \int_{k1\xi}^{k_2\xi} Q_{\text{v}}(\xi,\eta,\text{Fo})\chi_1\, d\eta\, d\xi \tag{70}$$

at

$$u_0(\xi,\eta) = \tilde{T}_0 - \tilde{T}_w, \qquad Q_{\text{v}}(\xi,\eta,\text{Fo}) = 0$$

Bearing in mind that the solution of the set problem applies to both the determination of the temperature field in a solid prism and to the calculation of heat transfer in a fluid flowing in a prismatic channel of triangular cross section, the following relations will be assigned for the thermophysical characteristics of the medium:

$$\widetilde{c\rho}(\xi,\eta) = a_1 + b_1\xi^2 + c_1\eta^2 + d_1\xi^2\eta + e_1\xi\eta^2 + f_1\xi^3 \tag{71}$$

$$\tilde{\lambda}(\xi,\eta) = a_2 + b_2\xi^2 + c_2\eta^2 + d_2\xi^2\eta + e_2\xi\eta^2 \tag{72}$$

The use of Eqs. (70) (71), and (72) allows one to obtain the unknown functions a_{11}, A_{11}, and A_1:

$$\begin{aligned} a_{11} = {} & (k_2 - k_1)k_1k_2[-d_2(k_1 + k_2)/252 - (a_2/10 + b_2/14) \\ & \times k_1k_2 + a_2/30 + b_2/84] + (k_2^2 - k_1^2)[(a_2/6 + 11b_2/84 \\ & - c_2/84 - e_2/126)(k_1 + k_2)k_1k_2 + d_2(k_1 + k_2)^2/252 - (a_2/30 \\ & + b_2/84)(k_1 + k_2) - 5d_2k_1^2k_2^2/84 + d_2k_1k_2/63]/2 \\ & + (k_2^3 - k_1^3)[3d_2(k_1 + k_2)k_1k_2/28 + (-a_2/15 - 5b_2/84 \\ & + c_2/84 + e_2/126)(k_1 + k_2)^2 - 5d_2(k_1 + k_2)/252 - (5c_2/84 \\ & + 3e_2/56)k_2^2k_2^2 + (-a_2/10 - 3b_2/28 + c_2/28 + e_2/42)k_1k_2 \\ & + a_2/30 + b_2/84]/3 + (k_2^4 - k_1^4)[(2c_2/21 + 23e_2/252)(k_1 + k_2) \end{aligned}$$

$$\times k_1k_2 - d_2(k_1+k_2)^2/21 + (a_2/15 + 2b_2/21 - c_2/21 - 2e_2/63)$$
$$\times (k_1+k_2) - 11d_2k_1k_2/126 + d_2/63]/4 + (k_2^5 - k_1^5)[(-c_2/28$$
$$- 19e_2/504)(k_1+k_2)^2 + 19d_2(k_1+k_2)/252 - (5c_2/84$$
$$+ 17e_2/252)k_1k_2 + (-b_2/28 + c_2/28 + e_2/42)]/5$$
$$+ (k_2^6 - k_1^6)[(c_2/28 + 13e_2/252)(k_1+k_2) - d_2/36]/6$$
$$- (k_2^7 - k_1^7)e_2/504 \tag{73}$$

$$A_{11} = (k_2 - k_1)\alpha k_1^2k_2^2 + (k_2^2 - k_1^2)k_1k_2[\beta k_1k_2 - 2\alpha(k_1+k_2)]$$
$$: 2 + (k_2^3 - k_1^3)\{\gamma k_1^2k_2^2 - 2\beta k_1k_2(k_1+k_2) + \alpha[(k_1+k_2)^2 + 2k_1k_2]\}$$
$$: 3 + (k_2^4 - k_1^4)\{-2\gamma k_1k_2(k_1+k_2) + \beta[(k_1+k_2)^2 + 2k_1k_2]$$
$$- 2\alpha(k_1+k_2)\}/4 + (k_2^5 - k_2^5)\{\gamma[k_1+k_2)^2 + 2k_1k_2] - 2\beta(k_1+k_2)$$
$$+ \alpha\}/5 + (k_2^6 - k_1^6)[-2\gamma(k_1+k_2) + \beta]/6 + (k_2^7 - k_1^7)\gamma/7 \tag{74}$$

where

$$\alpha = a_1/168 + b_1/360 + f_1/495; \qquad \beta = d_1/495, \qquad \gamma = c_1/360 + e_1/495$$

$$A_1(\text{Fo}) = 0 \tag{75}$$

The unknown function $\Psi_1(\text{Fo})$ will take the form

$$\psi_1(\text{Fo}) = (\tilde{T}_0 - \tilde{T}_w)\exp(a_{11}\,\text{Fo}/A_{11})\{(k_2 - k_1)k_1k_2p + (k_2^2 - k_1^2)$$
$$\times [rk_1k_2 - p(k_1+k_2)]/2 + (k_2^3 - k_1^3)[qk_1k_2 - r(k_1+k_2) + p]/3$$
$$- (k_2^4 - k_1^4)[q(k_1+k_2) - r)]/4 + (k_2^5 - k_1^5)q/5\}/A_{11} \tag{76}$$

where

$$p = -(a_1/20 + b_1/42 + f_1/56), \qquad q = -(c_1/42 + e_1/56), \qquad r = -d_1/56$$

In the case of a prism of elliptical cross section bounded by a curve with the canonical equation

$$\xi^2 + \eta^2 = 1, \qquad \xi = x/r_1, \qquad \eta = y/r_2 \tag{77}$$

(r_1 and r_2 are the major and minor semiaxes of the ellipse), when seeking the first approximation to the solution of Eqs. (62)–(64), the following coordinate function $\chi_1(\xi,\eta)$ from Eq. (65′) will be selected in order to satisfy the requirements [28] for $\chi_i(\xi,\eta)$:

$$\chi_1(\xi,\eta) = \omega(\xi,\eta) = 1 - \xi^2 - \eta^2 \tag{78}$$

The functions a_{11}, A_{11}, and $A_1(\text{Fo})$ in Eq. (69) are determined according to Eq. (35) when reducing the calculations of double integrals to repeated calculations. In the case considered, it is necessary to take into account the substitutions of the variables x by $\xi = x/r_1$ and y by $\eta = y/r_2$ with the aid of the familiar Jacobian transform which, in our case, is equal to

$$J(\xi, \eta) = r_1 r_2$$

and, moreover, pass to the cylindrical coordinate system by the rule

$$\xi = \rho \cos \phi, \qquad \eta = \rho \sin \phi$$

so that

$$\begin{aligned}\int_\Omega \int F(x, y)\, d\Omega &= \int_\Omega \int F[(\Psi_1(\xi, \eta), \Psi_2(\xi, \eta))] |J(\xi, \eta)|\, d\xi\, d\eta \\ &= \int_0^{2\pi} d\phi \int_0^1 F[\Psi_1(\rho, \phi), \Psi_2(\rho, \phi)]|J(\xi, \eta)|\, d\rho \qquad (79)\end{aligned}$$

Consider the case of the development in an elliptic prism of a temperature field that is symmetric about the abscissa and ordinate axes; for this purpose select the medium thermophysical properties $\widetilde{c\rho}$ and $\tilde{\lambda}$ as even functions of ξ and η

$$\widetilde{c\rho}(\xi, \eta) = a_1 + b_1 \xi^2 + c_1 \eta^2 \qquad (80)$$

$$\tilde{\lambda}(\xi, \eta) = a_2 + b_2 \xi^2 + c_2 \eta^2 \qquad (81)$$

At $\tilde{T}_w = \text{const}$, $\tilde{T}_0 = \text{const}$, and $\tilde{q}_v(\xi, \eta, \text{Fo}) = 0$, the function in Eq. (69) will successively be equal to

$$\begin{aligned}a_{11} &= -\frac{\pi}{r_1 r_2}[(r_1^2 + r_2^2)a_2 + (r_1^2 + 3r_2^2)b_2/6 + (3r_1^2 + r_2^2)c_2/6] \\ A_{11} &= \pi r_1 r_2 (a_1/3 + b_1/24 + c_1/24 \\ A_1(\text{Fo}) &= 0 \qquad (82)\end{aligned}$$

and, finally, the unknown time function is

$$\Psi_1(\text{Fo}) = (\tilde{T}_0 - \tilde{T}_w)\pi r_1 r_2 \exp(a_{11}\,\text{Fo}/A_{11})(a_1/2 + b_1/12 + c_1/12)/A_{11} \qquad (83)$$

Other approximations to the solution are obtained similarly.

IX. Use of the Computer for Analytic Transformations in Variational Solutions

The material presented in Sections V–VIII shows that the main stage in the method of obtaining the first and second analytic approximations to the solution of unsteady-state heat conduction problems by the convolution-type functional is the determination of analytic relations for the coefficients and free terms $a_{ji}(\mathrm{Fo})$, $A_{ji}(\mathrm{Fo})$, and $A_j(\mathrm{Fo})$ and the initial values of the functions $\Psi_i(0)$ in the set of Eqs. (33)–(34) from Eqs. (35) and (32′). To save computation time, the same analytic relations for $a_{ji}(\mathrm{Fo})$, $A_{ji}(\mathrm{Fo})$, $A_j(\mathrm{Fo})$, and $\Psi_i(0)$ should be established prior to numerical solution of the set of eqs. (33)–(34) when seeking the nth ($n > 2$) nonanalytic approximation to the solution of the problem set.

The determination of $a_{ji}(\mathrm{Fo})$, $A_{ji}(\mathrm{Fo})$, $A_j(\mathrm{Fo})$, and $\Psi_i(0)$ is not difficult in principle: it only amounts to computation of certain integrals in the geometric region of the problem and its bounding surface, and this is the indisputable advantage of the method developed. A number of such computations were made by the author "manually" for solving the problems of practical importance, and they are presented in Sections V–VII.

At the same time, it is possible to get rid of the routine manual calculation of the previously mentioned integrals. This makes the variational method even more attractive for application to thermophysical computations by performing analytic computations on a computer. This possibility was pointed out by academician L. V. Kantorovich as early as 1957.

For this purpose, a program for the analytic determination of the functions $a_{ji}(\mathrm{Fo})$, $A_{ji}(\mathrm{Fo})$, $A_j(\mathrm{Fo})$, and $\Psi_i(0)$ was developed and written in the language of symbolic programming PL/1-FORMAC [67] called "Integration in a polygonal region." The program provides for analytic calculation of these functions, according to Eqs. (35), and (32′): of the double integrals from the plane region Ω and of the curvilinear integrals of the first kind from the curve Γ bounding the region Ω (and from its portions Γ_1, Γ_2, and Γ_3).

In the program, all the integrands are assumed to be polynomials in the rectangular coordinates x, and y and time τ. Thus, polynomials are assigned to the aforementioned functions $C\rho(x, y, \tau)$, $\lambda(x, y, \tau)$, $q_v(x, y, \tau)$, $T_0(x, y)$, $T_w(x, y, \tau)$, $q(x, y, \tau)$, $T_c(\tau)$, and $\alpha(x, y, \tau)$, which are known when the problem is stated, and to the coordinate functions $\chi_i(x, y, \tau)$, $\chi_j(x, y, \tau)$. It is envisaged that the coefficients of the polynomials can be represented by the numbers, identifiers or their combinations. Each polynomial in the program is given a name (designation) and, when introduced into the computer, the quantity of the variables used in the polynomial and the number of its terms (monomials) are specified and they are introduced in succession as $x^0y^0, x, y, x^2, xy, y^2, x^3, x^2y$, xy^2, y^3, x^4, x^3y, x^2y^2, xy^3, y^4, x^5, x^4y, x^3y^2, x^2y^3, xy^4, y^5, etc. If some monomial is absent in the polynomial, the zero coefficient connected with it is introduced.

Integration in a plane singly connected region Ω bounded by the curve Γ was replaced by integration in the polygonal region Ω_ε, that is, in a polygon. In this case, the polygonal region is divided into triangular elements, which, just as the vertices of the broken line Γ_ε that bounds the region Ω_ε, are numbered counterclockwise. The following information on the characteristics of Ω_ε and

Γ_ε is fed to the computer memory: the magnitude of the sections of the broken line Γ_ε and the coordinates of its nodal points $M_i(x_i, y_i)$ in the form of numbers and symbols or their combinations. Then, the lengths of the polygon sides are determined on the computer:

$$l_i = \sqrt{(x_{i+1} - x_i)^2 + (y_{i+1} - y_i)^2} \tag{84}$$

as well as the explicit parts of the equations for the polygon boundaries (i.e., the sections of the broken line) by the formula

$$\omega_i = (y_{i+1} - y_i)(x - x_i) - (x_{i+1} - x_i)(y - y_i), \qquad i = 1, \ldots, N + 1 \tag{85}$$

Then, the coordinate functions $\chi_i(x, y)$ are calculated or entered (or displayed). For the case of homogeneous boundary equations of the first kind, it can be shown, for example, that

$$\chi_1 = (-1) \prod_{i=1}^{N} \omega_i, \qquad \chi_2 = x\chi_1, \qquad \chi_3 = y\chi_1,$$
$$\chi_4 = x^2\chi_1, \qquad \chi_5 = xy\chi_1, \ldots \tag{86}$$

The direction cosines to each section of the boundary Γ_ε are found from the equations

$$\cos(n_i, x) l_i = \frac{\partial}{\partial x}\omega_i = y_{i+1} - y_i$$
$$\cos(n_i, y) l_i = \frac{\partial}{\partial y}\omega_i = x_i - x_{i+1} \tag{87}$$

Since the boundary Γ_ε is traversed counterclockwise, then the normal to the ith section with the nodal points $M_i(x_i, y_i)$, $M_{i+1}(x_{i+1}, y_{i+1})$ represents the rotation of the vector $\mathbf{M_iM_{i+1}}$ through the angle $\pi/2$.

The program provides for the input of the boundary conditions of the first, second, or third kind on each section of the broken line Γ_ε.

Making use of the information entered, successive calculation is performed of the integrals that enter into $a_{ji}(\text{Fo})$, $A_{ji}(\text{Fo})$, $A_j(\text{Fo})$, and $\Psi_i(0)$.

The efficiency of the program was verified by obtaining an analytic first approximation to the determination of the temperature field in a prism whose cross section is an isosceles triangle with the vertices $M_1(0;0)$, $M_2(H_1 - A)$, and $M_3(H, A)$ at $T_w = \text{const}$, $T_0 = \text{const}$, and $q_v = 0$ and with the functions $c\rho$ and λ defined by Eqs. (71) and (72). As the second example, the same problem was solved for a prism with the cross section in the form of a nonisosceles triangle, which had the vertices $M_1(0;0)$, $M_2(1;-0.5)$, and $M_3(1;0.8)$ in order to reveal the effect of thermophysical characteristics on the unknown temperature field.

The program not only envisages obtaining an analytic approximation to the solution but also the substitution into it of numerical values of the arguments (coordinates of the points of a body and time) and numerical values of the coefficients in the equations for λ, $c\rho$, etc., for determining numerical values of temperature at any point of the region Ω and at time instants of interest.

X. Estimation of Approximations when Using the Convolution-Type Functional

For unsteady states, the variational description in the classical sense seems to be nonexistent. For linear problems of heat conduction and convective heat transfer, it is possible to construct the functionals $J(T)$ that have, at best, a "stationary" point that corresponds to the solution of the problems considered previously. On the contrary, it seems impossible that functionals having extreme properties (necessary and sufficient) could be obtained forunsteady-state problems. This makes an exact estimation of approximation to the solution of corresponding problems difficult. This remark pertains completely to the convolution-type functional used in the present paper.

To show this, consider the simplest boundary-value problem of unsteady-state heat conduction over the segment $[0, 1]$ with constant thermophysical properties of the medium material that has the form

$$\frac{\partial u(\xi,\tau)}{\partial \tau} = \frac{\partial^2 u(\xi,\tau)}{\partial \xi^2} + Q_v(\xi,\tau), \qquad 0 < \xi < 1, \qquad \tau > 0 \tag{88}$$

$$u(\xi, 0) = 0, \qquad 0 < \xi < 1 \tag{89}$$

$$\left.\frac{\partial u(\xi,\tau)}{\partial \xi}\right|_{\xi=1} + u(1,\tau) = 0, \qquad \tau > 0 \tag{90}$$

$$\left.\frac{\partial u(\xi,\tau)}{\partial \xi}\right|_{\xi=0} = 0, \qquad \tau > 0 \tag{91}$$

In the class of functions that satisfy conditions (89)–(90), the aforegoing shows that the convolution-type functional for the problem under consideration is

$$J(u) = \int_0^1 \int_0^{2t} \left[\frac{\partial^u}{\partial \mathrm{Fo}} - \frac{\partial^2 u}{\partial \xi^2} - 2Q_v\right] u(\xi, 2t - \tau)\, d\tau\, d\xi \tag{92}$$

Having denoted $u(\xi,\tau)$ by $\delta u(\xi,\tau) = \theta(\xi,\tau)$, calculate the increment of functional (92) as

$$\Delta J(u) = J(u + \delta u) - J(u) = J(u + \theta) - J(u)$$

to obtain

$$\Delta J(u) = \int_0^{2t}\int_0^1 \left[\frac{\partial\theta(\xi,\tau)}{\partial\tau} - \frac{\partial^2\theta(\xi,\tau)}{\partial\xi^2}\right] u(\xi, 2t-\tau)\,d\xi\,d\tau$$

$$+ \int_0^{2t}\int_0^1 \left[\frac{\partial\theta(\xi,\tau)}{\partial\tau} - \frac{\partial^2\theta(\xi,\tau)}{\partial\xi^2}\right] \theta(\xi, 2t-\tau)\,d\xi\,d\tau$$

$$+ \int_0^{2t}\int_0^1 \left[\frac{\partial u(\xi,\tau)}{\partial\tau} - \frac{\partial^2 u(\xi,\tau)}{\partial\xi^2} - 2Q_v(\xi,\tau)\right] \theta(\xi, 2t-\tau)\,d\xi\,d\tau \quad (93)$$

Making use of integration by parts, the commutability of convolution, and conditions (89)–(90), transform the first term on the right-hand side (RHS) of Eq. (93) as

$$\int_0^{2t}\int_0^1 \left[\frac{\partial\theta(\xi,\tau)}{\partial\tau} - \frac{\partial^2\theta(\xi,\tau)}{\partial\xi^2}\right] u(\xi, 2t-\tau)\,d\xi\,d\tau$$

$$= \int_0^{2t}\int_0^1 \frac{\partial\theta(\xi,\tau)}{\partial\tau} u(\xi, 2t-\tau)\,d\xi\,d\tau - \int_0^{2t}\int_0^1 \frac{\partial^2\theta(\xi,\tau)}{\partial\xi^2} u(\xi, 2t-\tau)\,d\xi\,d\tau$$

$$= \int_0^1 \left[\theta(\xi,\tau)u(\xi, 2t-\tau)\Big|_0^{2t} - \int_0^{2t} \frac{\partial u(\xi, 2t-\tau)}{\partial\tau}\theta(\xi,\tau)\,d\tau\right] d\xi$$

$$- \int_0^{2t}\left[\frac{\partial\theta(\xi,\tau)}{\partial\xi} u(\xi, 2t-\tau)\Big|_0^1 - \int_0^1 \frac{\partial\theta(\xi,\tau)}{\partial\xi}\frac{\partial u(\xi, 2t-\tau)}{\partial\xi}\,d\xi\right] d\tau$$

$$= \int_0^{2t}\int_0^1 \frac{\partial u(\xi, 2t-\tau)}{\partial(2t-\tau)}\theta(\xi,\tau)\,d\xi\,d\tau$$

$$+ \int_0^{2t}\int_0^1 \frac{\partial\theta(\xi,\tau)}{\partial\xi}\frac{\partial u(\xi, 2t-\tau)}{\partial\xi}\,d\xi\,d\tau - \int_0^{2t} \frac{\partial\theta(\xi,\tau)}{\partial\xi}\bigg|_{\xi=1} u(1, 2t-\tau)\,d\tau$$

$$= \int_0^{2t}\int_0^1 \frac{\partial u(\xi,\tau)}{\partial\tau}\theta(\xi, 2t-\tau)\,d\xi\,d\tau$$

$$+ \int_0^{2t}\left[\theta(\xi,\tau)\frac{\partial u(\xi, 2t-\tau)}{\partial\tau}\Big|_0^1 - \int_0^1 \frac{\partial^2 u(\xi, 2t-\tau)}{\partial\xi^2}\theta(\xi,\tau)\,d\xi\right] d\tau$$

$$- \int_0^{2t} \frac{\partial\theta(\xi,\tau)}{\partial\xi}\bigg|_{\xi=1} u(1, 2t-\tau)\,d\tau$$

$$= \int_0^{2t}\int_0^1 \left[\frac{\partial u(\xi,\tau)}{\partial\tau} - \frac{\partial^2 u(\xi,\tau)}{\partial\xi^2}\right]\theta(\xi, 2t-\tau)\,d\xi\,d\tau$$

$$+ \int_0^{2t}\left[\theta(1,\tau)\frac{\partial u(\xi, 2t-\tau)}{\partial\xi}\bigg|_{\xi=1} - \frac{\partial\theta(\xi,\tau)}{\partial\xi}\bigg|_{\xi=1} u(1, 2t-\tau)\right] d\tau \quad (94)$$

It follows from the boundary condition [Eq. (90)] that the second term on the RHS of Eq. (94) is equal to zero, so that instead of Eq. (93), it is possible to obtain

$$\Delta J(u) = 2\int_0^{2t}\int_0^1\left[\frac{\partial u(\xi,\tau)}{\partial \tau} - \frac{\partial^2 u(\xi,\tau)}{\partial \xi^2} - Q_v(\xi,\tau)\right]\theta(\xi, 2t-\tau)\,d\xi\,d\tau + \int_0^{2t}\int_0^1\left[\frac{\partial \theta(\xi,\tau)}{\partial \tau} - \frac{\partial^2 \theta(\xi,\tau)}{\partial \xi^2}\right]\theta(\xi, 2t-\tau)\,d\xi\,d\tau \tag{93'}$$

In Eq. (93′), the first term on the RHS is the first derivative of functional (92), which vanishes because of the validity of Eq. (88):

$$\delta J(u) = \int_0^{2t}\int_0^1\left[\frac{\partial u(\xi,\tau)}{\partial \tau} - \frac{\partial^2 u(\xi,\tau)}{\partial \xi^2} - Q_v(\xi,\tau)\right]\theta(\xi, 2t-\tau)\,d\xi\,d\tau = 0 \tag{95}$$

These calculations ultimately yield the following relation for $\Delta J(u)$ in the class of functions that satisfy the boundary-value problem of Eqs. (88)–(91):

$$\Delta J(u) = \int_0^{2t}\int_0^1\left[\frac{\partial \theta(\xi,\tau)}{\partial \tau} - \frac{\partial^2 \theta(\xi,\tau)}{\partial \xi^2}\right]\theta(\xi, 2t-\tau)\,d\xi\,d\tau \tag{96}$$

Determine the sign of the increment of the functional $\Delta J(u)$ by having selected first as $\theta(\xi,\tau)$ the function

$$\theta(\xi,\tau) = \varepsilon(\xi^2 - 3\xi^3/4)\sin\tau \tag{97}$$

where ε is an arbitrary constant.

For function (97), which satisfies conditions (88)–(91), calculate the increment of the functional $\Delta J(u)$ at $2t = 2\pi$.

$$\begin{aligned}\Delta J &= \int_0^{2t}\int_0^1 [\varepsilon(\xi^2 - 3\xi^3/4)\cos\tau - \varepsilon\sin\tau(2 - 9\xi/2)] \\ &\quad\times \varepsilon(\xi^2 - 3\xi^3/4)\sin(2\pi-\tau)\,d\xi\,d\tau = -\varepsilon^2\int_0^{2\pi}[(\xi^2 - 3\xi^3/4)^2 \\ &\quad\times \sin\tau\cos\tau - (2 - 9\xi/2)(\xi^2 - 3\xi^3/4)\sin^2\tau)]\,d\xi\,d\tau \\ &= -19\varepsilon^2\pi/120 < 0\end{aligned} \tag{98}$$

Consequently, in any ε-vicinity of the function $u(\xi,\tau)$, there is the function $w(\xi,\tau) = u(\xi,\tau) + \theta(\xi,\tau)$, where $\theta(\xi,\tau)$ is found according to Eq. (97) so that

$$\Delta J(u) < 0 \tag{99}$$

It follows from the inequality of Eq. (99) that the solution of Eqs (88)–(91) is not a local minimum of the functional (92).

Similarly, considering as $\theta(\xi, \tau)$ the function

$$\theta(\xi, \tau) = \varepsilon\tau(\xi^2 - 3\xi^3/4) \tag{100}$$

and assuming $2t = 1$, it can be verified that $u(\xi, \tau)$ is not a local maximum of functional (92).

Thus, in the case of the functional (92) there is no problem at the extremum, but only a problem at the critical (stationary) "point" of this functional. It follows from the theory of the existence and uniqueness of solutions for the boundary-value problem that this point exists and is unique. The problem of the convergence of approximations to the critical point is a difficult one, because even in the case of the convergence of functionals of approximations, that is, when the following equality is satisfied

$$\lim_{n\to\infty} J(u_n) = J(u) \tag{101}$$

it is impossible to extract from it the convergence of the sequence u_n in any form. The calculation of Eq. (101) is the most difficult problem, since the expression for $\Delta J(u)$ is of variable sign.

It is just for this reason that the first, second, or nth approximation to the solution of the problems analyzed was compared with the well-known exact or numerical solutions.

In the author's opinion, the general evidence for the impossibility of constructing a classical variational description of unsteady states in the case of a linear model of the phenomenon could be reduced to the problem of the impossibility to construct an elliptic analog for a parabolic problem as regards unsteady-state heat conduction.

Really, it is not difficult to construct a functional which has the extreme properties for the problems of steady-state heat conduction with the elliptic equation of the process. Thus, for a plane region Ω inside of which there is a heat source $q_v(M)$, and on the outer boundary Γ, the temperature $T_w(M)$ is prescribed, there is the following boundary-value problem:

$$\frac{\partial^2 T(x, y)}{\partial x^2} + \frac{\partial^2 T(x, y)}{\partial y^2} = -q_v(x, y), \qquad M(x, y) \in \Omega \tag{102}$$

$$T|_\Gamma = T_w(x, y), \qquad M(x, y) \in \Gamma \tag{103}$$

whose solution can be reduced to finding the functional extremum

$$J(T) = \int_\Omega \left[\left(\frac{\partial T}{\partial x}\right)^2 + \left(\frac{\partial T}{\partial y}\right)^2 - 2q_v T \right] dx\, dy \tag{104}$$

in the class of doubly continuous-differentiated functions in the region Ω, which on the boundary Γ attain the prescribed values [Eq. (103)].

By calculating the first derivative of functional (104) it is possible to verify that it is Eq. (102) that makes the variation vanish and consequently, is the Euler equation for functional (104). The constant sign of the functional (104) increment proves that the unknown functions $T(M)$ provide it with the extremum.

For problems with an elliptic equation describing the process, it is possible to construct the proof of the convergence of the approximations that were obtained using the variational description [22] to the exact solution. Moreover, the variational formulation of the steady-state heat conduction problem, as shown in Ref. 68, allows the use of dual evaluations to control the error that appears in the approximate determination of the process parameters and to find the limits of the range within which the actual values of each of them can exist.

Consider now a "parabolic" problem that for $T(x, \tau)$ has the form

$$\frac{\partial}{\partial x}\left[\lambda(x,\tau)\frac{\partial T(x,\tau)}{\partial x}\right] - c\rho(x,\tau)\frac{\partial T(x,\tau)}{\partial \tau} = -q_v(x,\tau) \qquad a < x < b, \qquad \tau > 0 \tag{105}$$

$$T(x, 0) = 0 \tag{106}$$

$$-\lambda(a,\tau)\left.\frac{\partial T(x,\tau)}{\partial x}\right|_{x=a} + \alpha_1(\tau)T(a,\tau) = 0, \qquad \tau > 0 \tag{107}$$

$$\lambda(b,\tau)\left.\frac{\partial T(x,\tau)}{\partial x}\right|_{x=b} + \alpha_2(\tau)T(b,\tau) = 0, \qquad \tau > 0 \tag{108}$$

Thus, consider the boundary-value problem of unsteady-state heat conduction on the segment $[a, b]$ when the boundary conditions of the third kind are prescribed at its ends at the known rate of convective heat transfer $\alpha_1(\tau)$ and $\alpha_2(\tau)$ and with zero temperature of the medium around the body (infinite plate). The thermal conductivity of the body material λ and its volumetric specific heat $c\rho$, as well as the volumetric heat generation source q_v, are assumed to be known functions of the problem arguments—of the coordinate x and time τ.

Replace the initial problem, Eqs. (105)–(108), by the problem of the form

$$\frac{\partial}{\partial x}\left[\lambda\frac{\partial T(x,\tau)}{\partial x}\right] + \frac{\partial}{\partial \tau}\left[A\frac{\partial T(x,\tau)}{\partial \tau}\right] = -q_v(x,\tau) \tag{109}$$

$$a < x < b, \qquad 0 < \tau < t, \qquad A > 0$$

$$[N(T) + \sigma(M,\tau)T]\,|_S = 0, \qquad M = M(x,\tau), \qquad \sigma \geq 0 \tag{110}$$

Here, S is the boundary of the rectangle $[a, b] \times [0, t]$ and

$$N(T) = \lambda(M,\tau)\frac{\partial T}{\partial x}\cos(n, x) + A(M,\tau)\frac{\partial T}{\partial \tau}\cos(n,\tau)$$

When evolved, condition (110) is written as

$$-\lambda(a,\tau)\frac{\partial T(x,\tau)}{\partial x}\bigg|_{x=a} + \sigma(a,\tau)T(a,\tau) = 0 \tag{111}$$

$$\lambda(b,\tau)\frac{\partial T(x,\tau)}{\partial x}\bigg|_{x=b} + \sigma(b,\tau)T(b,\tau) = 0 \tag{112}$$

$$-A(x,0)\frac{\partial T(x,\tau)}{\partial \tau}\bigg|_{\tau=0} + \sigma(x,0)T(x,0) = 0 \tag{113}$$

$$A(x,t)\frac{\partial T(x,\tau)}{\partial \tau}\bigg|_{\tau=t} + \sigma(x,t)T(x,t) = 0 \tag{114}$$

In Fig. 8, the region of the existence of the solution for Eqs. (109) and (110) is shown.

Now, the energy method [22] for solving the Eqs. (109) and (110) will be given. Let L denote the operator that is defined by the equality

$$LT = -\frac{\partial}{\partial x}\left(\lambda\frac{\partial T}{\partial x}\right) - \frac{\partial}{\partial \tau}\left(A\frac{\partial T}{\partial \tau}\right) = q_v(x,\tau) \tag{115}$$

Show that this operator is symmetric.

In fact, Green's formula gives for v from the class of functions, which are doubly continuously differentiated in $(a,b)x(o,t)$ and which satisfy Eqs. (111)–(114), that

$$\int_\Omega (vLT - TLv)\,dx\,d\tau = -\int_S [vN(T) - TN(v)]\,ds \tag{116}$$

Then, the equality

$$[N(T) = \sigma T]\,|_S = [N(v) + \sigma v]\,|_S = 0 \tag{117}$$

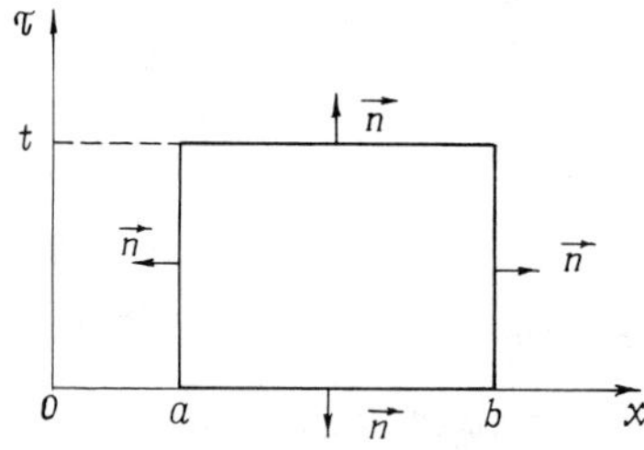

FIG. 8. Region of the existence of the parabolic-type problem solution.

yields that

$$[vN(T) - TN(v)]\,|_S = 0 \tag{118}$$

and then Eqs. (117) and (118) give

$$\int_\Omega (vLT - TLv)\,dx\,d\tau = (LT, v) - (T, Lv) = 0$$

where the scalar products are defined as

$$(LT, v) = \int_a^b \int_0^t LT \cdot v\,d\tau\,dx$$

and this means that the operator L is symmetric on the corresponding sets of functions.

It is assumed that there exists such a number $\mu_0 > 0$ that

$$\lambda k^2 + Al^2 \geq \mu_0(k^2 + l^2)$$

for any point $M \in \bar{\Omega} = \Omega US(k = T_x$ and $l = T_\tau$ are real numbers). This condition means that the operator L is elliptic in the region Ω.

Determine the conditions that provide a positive definiteness of the operator L.

The Green's formula yields

$$\begin{aligned}(LT, T) &= \int_\Omega TLT\,dx\,d\tau = \int_\Omega \left[\lambda\left(\frac{\partial T}{\partial \tau}\right)^2 + A\left(\frac{\partial T}{\partial \tau}\right)^2\right] dx\,d\tau \\ &\quad - \int_S TN(T)\,ds = \int_\Omega \left[\lambda\left(\frac{\partial T}{\partial x}\right)^2 + A\left(\frac{\partial T}{\partial \tau}\right)^2\right] dx\,d\tau + \int_S \sigma\; T^2\,ds \\ &\geq \mu_0 \int_\Omega (T_x^2 + T_\tau^2)\,d\Omega \end{aligned} \tag{119}$$

It will be assumed that the function $\sigma(M, \tau)$ is limited from below by a certain positive number σ_0, so that $\sigma(M, \tau) \geq \sigma_0 \geq 0$. Then, making use of Fridrichs's inequality

$$\int_\Omega (T_x^2 + T_\tau^2)\,dx\,d\tau \geq k \int_\Omega T^2\,dx\,d\tau$$

it is possible to obtain from Eq. (119) that

$$(LT, T) \geq \gamma^2 \|T\|^2 = \gamma^2(T, T)$$

which means a positive definiteness of the operator L.

Thus, the problem of the solution of Eq. (109) with condition (110) is equivalent to the problem of finding the minimum of the functional [22].

$$J(T) = (LT, T) - 2(T, q_v) \tag{120}$$

Under the condition of Eq. (110), there exists the minimum of the functional (120), and the minimizing sequences, constructed by the methods of Ritz, Currant, and Kantorovich, converge to this minimum [21].

Consider now the interrelation between Eqs. (105)–(108), (109), and (110). If the function $A = A(x, \tau, \varepsilon)$, which satisfies the conditions

1. $A \to 0$ as $\varepsilon \to 0$
2. $\dfrac{\partial A}{\partial \tau} \to -c\rho$ as $\varepsilon \to 0$
3. $A > 0$ at $\varepsilon > 0$

then at $\sigma(a, \tau) = \alpha_1(\tau)$, $\sigma(b, \tau) = \alpha_2(\tau)$, $\sigma(x, 0) = 1$, Eqs. (109) and (110) with $\varepsilon \to 0$ would formally be converted into Eqs. (105)–(108).

However, there are no continuously-differentiated functions that satisfy conditions 1–3.

Lemma. In the class of continuously-differentiated functions, there is no function $\Psi(\varepsilon, \tau)$ that satisfies the conditions

$$\lim_{\varepsilon \to 0} \Psi(\varepsilon, \tau) = 0, \qquad \tau \in [0, t] \tag{121}$$

$$\lim_{\varepsilon \to 0} \frac{\partial \Psi(\varepsilon, \tau)}{\partial \tau} = -c\rho(\tau) = B(\tau) < 0, \qquad \tau \in [0, t] \tag{122}$$

Proof. The equality

$$\Psi(\varepsilon, \tau) - \Psi(\varepsilon, 0) = \int_0^\tau \frac{\partial \Psi(\varepsilon, \tilde{\tau})}{\partial \tilde{\tau}} d\tilde{\tau} \tag{123}$$

is evident. It will be assumed that property (122) is fulfilled in the space $L_2[0, \tau]$.

It follows from Eq. (123) that

$$\left| \Psi(\varepsilon, \tau) - \Psi(\varepsilon, 0) - \int_0^\tau B(\tilde{\tau}) d\tilde{\tau} \right| = \left| \int_0^\tau \left[\frac{\partial \Psi(\varepsilon, \tilde{\tau})}{\partial \tilde{\tau}} - B(\tilde{\tau}) \right] d\tilde{\tau} \right| \le \int_0^\tau \left| \frac{\partial \Psi}{\partial \tilde{\tau}} - B(\tilde{\tau}) \right| d\tilde{\tau} \le \tau^{1/2} \left\{ \int_0^\tau \left| \frac{\partial \Psi}{\partial \tau} - B(\tilde{\tau}) \right|^2 d\tau \right\}^{1/2} \tag{124}$$

Passing in Eq. (124) to the limit with $\varepsilon \to 0$, obtain

$$\left| \int_0^\tau B(\tilde{\tau}) d\tilde{\tau} \right| = 0, \tag{125}$$

which is impossible according to Eq. (122).

Now, let $\Psi(\varepsilon, \tau)$ be from the class of generalized functions. Then, for the finite function $\phi(\tau)$,

$$\int_0^t \left[\frac{\partial\Psi}{\partial\tilde{\tau}} - B(\tau)\right]\phi(\tilde{\tau})\,d\tilde{\tau} = -\int_0^\tau B(\tilde{\tau})\phi(\tilde{\tau})\,d\tilde{\tau} - \int_0^\tau \Psi\phi\,d\tilde{\tau} \qquad (126)$$

Again, from Eq. (126) follows the inadmissible value

$$\int_0^\tau B(\tilde{\tau})\,d\tilde{\tau} = 0.$$

To summarize, there are no functions $\Psi(\varepsilon,\tau)$ that would exactly satisfy conditions (121) and (122) of the lemma.

Finally, consider the problem of the existence of functions for which conditions (121) and (122) are fulfilled in an approximate sense, that is, consider the problem of finding the functions $\Psi(\varepsilon,\tau)$ for which the following evaluations are fulfilled:

$$|\Psi(\varepsilon,\tau)| \leq \beta \qquad (127)$$

$$\left|\frac{\partial\Psi(\varepsilon,\tau)}{\partial\tau} - B(\tau)\right| \leq \beta \qquad (128)$$

at sufficiently small ε [here β is a small number and $-B(\tau) \geq M > 0$].

It follows from inequality (124) that

$$\left|\int_0^\tau B(\tilde{\tau})\,d\tilde{\tau}\right| \leq \tau^{1/2}\left\{\int_0^\tau \left|\frac{\partial\Psi}{\partial\tilde{\tau}} - B(\tilde{\tau})\right|^2 d\tilde{\tau}\right\}^{1/2} + \left|\Psi(\varepsilon,\tau) - \Psi(\varepsilon,0)\right|$$

Then, Eqs. (127) and (128) give the equality

$$\left|\int_0^\tau B(\tilde{\tau})\,d\tilde{\tau}\right| = -\int_0^\tau B(\tilde{\tau})\,d\tilde{\tau} \leq \tau^{1/2}\left(\int_0^\tau \beta^2\,d\tilde{\tau}\right)^{1/2} + \beta = \beta\tau + \beta \qquad (129)$$

Since M is a prescribed positive number, the latter result yields

$$M \leq \beta(1 + 1/\tau)$$

that is, the function $B(\tau)$ in the norm of the space L_1 is small, namely,

$$\int_0^\tau \left|B(\tilde{\tau})\right| d\tilde{\tau} \leq \beta(1 + \tau)$$

In fact, $c\rho(x,\tau) = -B(\tau)$ in Eq. (105) is not small, and, consequently, the functions $\Psi(\varepsilon,\tau)$, which would satisfy Eqs. (127), and (128), do not exist in an approximate sense.

XI. Conclusion

It can be concluded that a variational description using a convolution-type functional, while providing the mathematical equivalence to the initial boundary-value and the corresponding variational problems, is very effective for solving complex linear problems of heat conduction. The procedure of obtaining an analytical solution, when constructing approximations according to the Kantorovich scheme, is so simplified in this case that its solution becomes possible on a computer. The invalidity of some variational descriptions for nonlinear problems and the impossibility to construct a functional for linear evolutional problems described by a parabolic equation, which would be identical to that used in a variational description of the steady-state heat conduction process are shown.

NOMENCLATURE

$q(T)$ or $q(M, \tau)$	density of heat flux through the bounding surface of a body	$T_c(\tau)$	temperature of surrounding medium
$q_v(T)$ or $q_v(M, \tau)$	power of volumetric heat generation sources	$T_0(M)$	function of initial temperature distribution
$T(M, \tau)$	temperature at point M with coordinates x, y, z at time instant τ	$T_w(M, \tau)$	temperature on the bounding surface of a body

Greek Symbols

$\alpha(M, \tau)$	coefficient of convective heat transfer	τ_r	relaxation time
$\lambda(T)$ and $c\rho(T)$	or $\lambda(M, \tau)$ and $c\rho(M, \tau)$ coefficient of thermal conductivity and volumetric specific heat of the body material	Ω	geometric region of process progress bounded by the surface Γ(or its segments, Γ_1, Γ_2, and Γ_3)

REFERENCES

1. A. V. Luikov, "Teoriya teploprovodnosti" (Heat Conduction Theory). Vysshaya Shkola Press, Moscow, 1967.
2. A. V. Luikov and Yu. A. Mikhailov, "Teoriya teplo- i massoperenosa" (Heat and Mass Transfer Theory). Gosenergoizdat Press, Moscow-Leningrad, 1963.
3. E. M. Goldfarb, "Teplotekhnika metallurgicheskikh protsessov" (Thermal Technology of Metallurgical Processes). Metallurgiya Press, Moscow, 1967.
4. B. S. Petukhov, "Teploobmen i soprotivleniye pri laminarnom techenii zhidkosti v trubakh" (Heat Exchange and Resistance in Laminar Fluid Flow in Tubes). Energiya Press, Moscow, 1967.

5. A. G. Tyomkin, "Obratnye metody teploprovodnosti" (Inverse Methods of Heat Conduction). Energiya Press, Moscow, 1973.
6. E. M. Kartashov, "Analiticheskiye metody v teploprovodnosti tvyordykh tel" (Analytical Methods in Heat Conduction of Solid Bodies). Vysshaya Shkola Press, Moscow, 1979.
7. P. V. Tsoi, "Metody raschyota zadach teplomassoperenosa" (Methods of Calculation of Heat and Mass Transfer Problems). Energoatomizdat Press, Moscow, 1984.
8. H. S. Carlslaw and J. C. Jaeger, "Conduction of Heat in Solids," 2nd ed., Oxford Univ. Press (Clarendon), London and New York.
9. P. J. Schneider, "Conduction Heat Transfer." Addison-Wesley, Cambridge, Massachusetts, 1955.
10. M. D. Mikhailov, "Nestatsionarnyi teplo- i massoobmen v odnomernykh telakh" (Non-stationary Heat and Mass Transfer in One-dimensional Bodies). ITMO AN BSSR Press, Minsk, 1969.
11. A. F. Chudnovskiy, "Teplofizicheskiye kharakteristiki dispersnykh materialov" (Thermophysical Characteristics of Dispersed Materials). Fizmatgiz Press, Moscow, 1962.
12. A. V. Luikov, Methods for solving linear equations of nonstationary heat conduction. *Rev. Izv. AN SSSR, Ser. Energetika Transp.* No. 2, 3 (1969).
13. A. N. Gordov, Temperature field of bodies under the conditions of the variable ambient temperature and alternating thermal conductivity. *Trudy VNIIM* **35,** 129 (1965).
14. G. A. Grinberg, On temperature or concentration fields established inside an infinite or finite region by moving surfaces with time-dependent temperature or concentration being set on them. *Prikl. Mat. Mekh.* **33,** 1051 (1969).
15. E. M. Kartashov and V. M. Nechayev, The method of Green's functions for boundary-value problems of heat conduction equation in non-cylindrical regions. *ZAMM* **58,** 199 (1978).
16. A. V. Luikov, Methods for solving non-linear equations of non-stationary heat conduction. *Rev. Izv. AN SSSR, Ser. Energetika Transp.* No. 5, 109 (1970).
17. L. A. Kozdoba, "Metody resheniya nelineynykh zadach teploprovodnosti" (Methods for Solving Non-linear Problems of Heat Conduction). Nauka Press, Moscow, 1975.
18. V. N. Kozlov, Temperature field of an infinite plate in the case of variable heat-exchange coefficient. *J. Eng. Phys.* **16,** 125 (1969).
19. E. M. Kartashov and B. Ya. Lyubov, Analytical methods for solving boundary-value problems of heat conduction equation in a region with moving boundaries. *Rev. Izv. AN SSSR, Ser. Energetika Transp.* No. 6, 33 (1974).
20. L. A. Kozdoba, Methods for solving the problems of solidification. *Rev. Fiz. Khim. Obrabotki Materialov* No. 2, 4 (1979).
21. V. S. Novikov, Heat conduction in complex shapes of bodies. *Rev. Izv. AN SSSR, Ser. Energetika Transp.* No. 4, 134 (1982).
22. S. G. Mikhlin, "Variatsionnye metody v matematicheskoy fizike" (Variational Methods in Mathematical Physics). GITTL Press, 1957.
23. T. R. Goodman, The heat balance integral and its application to problems involving a change of phase. *Trans. ASME Ser. C* **10,** 335 (1958).
24. T. Goodman, Application of integral methods in non-linear non-stationary heat conduction problems. *In* "Problemy teploobmena" (Problems of Heat Transfer). Atomizdat Press, Moscow, 1967.
25. A. I. Veinik, "Priblizhyennyy raschyot protsessov teploprovodnosti" (Approximate Calculation of Heat Conduction Processes). Gosenergoizdat Press, Moscow-Leningrad, 1959.
26. B. A. Finlayson, "The Method of Weighted Residuals and Variational Principles." Academic Press, New York, 1972.
27. N. M. Belyayev, A. A. Kochubey, A. A. Ryadno *et al.*, "Nestatsionarnyi teploobmen v trubakh" (Non-stationary Heat Transfer in Tubes). Vysshaya Shkola Press, Kiev, Donetsk, 1980.

28. L. V. Kantorovich and V. I. Krylov, "Priblizhonnye metody vysshego analiza" (Approximate Methods of Higher Analysis). Fizmatgiz Press, Moscow-Leningrad, 1962.
29. R. S. Anderssen, The numerical solution of parabolic differential equations using variational methods. *Numer. Math.* **13,** 129 (1969).
30. R. S. Anderssen, A class of densely invertible parabolic operator equations. *Bull. Aust. Math. Soc.* **1,** 363 (1969).
31. M. A. Biot, "Variational Principles in Heat Transfer." Oxford Univ. Press, London and New York, 1970.
32. P. M. Kolesnikov, Variational methods in the theory of transfer. *In* "Metody issledovaniya i optimizatsii protsessov perenosa" (Methods of Study and Optimization of Transfer Processes). ITMO AN BSSR Press, Minsk, 1979.
33. T. J. Lardner, Biot's variational principle in heat conduction. *AIAA J.* **1,** 196 (1963).
34. G. F. Muchnik and Yu. A. Polyakov, The Biot variational principle in the problems of heat conduction with variable boundary conditions. *Teplofiz. Vysok. Temp.* **2,** 424 (1964).
35. Yu. A. Samoylovich, Application of the Biot variational principle for solving the Stephan problem. *Teplofiz. Vysok. Temp.* **4,** 832 (1966).
36. B. A. Finlayson and L. E. Scriven, On the search for variational principles. *Int. J. Heat Mass Transfer* **19,** 799 (1967).
37. B. A. Finlayson and L. E. Scriven, The method of weighted residuals and its relation to certain variational principles for the analysis of transport processes. *Chem. Eng. Sci.* **20,** 395 (1965).
38. Yu. T. Glazunov, Additional form of the variational principle of non-linear phenomena of interrelated transfer. *Izv. AN Latv. SSR* No. 5, 69 (1978).
39. Yu. T. Glazunov, Variational solutions of problems of interrelated heat and mass transfer. *Izv. AN Latv. SSR* No. 2, 60 (1979).
40. Yu. T. Glazunov, Variational principles of the phenomena of non-linear anisotropic mutually-related transfer. *Izv. AN Latv. SSR* No. 2, 51 (1979).
41. Yu. T. Glazunov, Variational principle for the phenomena of interrelated heat and mass transfer which takes into account the finite velocity of the propagation of perturbations. *J. Eng. Phys.* **40,** 134 (1981).
42. Yu. T. Glazunov, Variational principle of the phenomena of interrelated heat and mass transfer which takes into account the finite velocity of the propagation of perturbations. *J.* Eng. Phys. **40,** 134 (1981).
43. R. Schechter, "The Variational Method in Engineering." McGraw Hill, New York, 1967.
44. R. Rosen, On variational principles for irreversible processes. *J. Chem. Phys.* **21,** 1220 (1953).
45. R. Rosen, Use of restricted variational principles for solution of differential equations. *J. Mech. Appl. Math.* **9,** 869 (1956).
46. L. G. Chambers, A variational principle for the conduction of heat. *Q. J. Mech. Appl. Math.* **9,** 234 (1956).
47. Dj. Djukic and B. Vujanovic, On a new variational principle of Hamiltonian type for classical field theory. *ZAMM* **51,** 611 (1971).
48. B. Vujanovic, An approach to linear and non-linear heat transfer problems using a lagrangian. *AIAA J.* **9,** 131 (1971).
49. B. Vujanovic and A. M. Strauss, Heat transfer with non-linear boundary conditions via a variational principle. *AIAA J.* **9,** 327 (1971).
50. G. Lebon and J. Lambermont, Some variational principles pertaining to non-stationary heat conduction and coupled thermoelasticity. *Acta Mech.* **15,** 121 (1972).
51. G. Lebon and Ph. Mathieu, A numerical calculation of nonlinear transient heat conduction in the fuel elements of a nuclear reactor. *Int. J. Heat Mass Transfer* **22,** 1187 (1979).
52. B. Krajewski, On a direct variational method for non-linear heat tranfer. *Int. J. Heat Mass Transfer* **18,** 495 (1975).

53. G. Lebon and J. Lambermont, Comment on the paper "On a Direct Variational Method for Non-linear Heat Transfer" by B. Krayewski. *Int. J. Heat Mass Transfer* **19,** 1340 (1976).
54. B. Krayewski, Rejoinder. *Int. J. Heat Mass Transfer* **19,** 1341 (1976).
55. L. E. Elsgolts, "Differentsialnye uravneniya i variatsionnoe ischesleniye" (Differential Equations and Variational Calculus). Nauka Press, Moscow, 1969.
56. R. W. Atherton and G. M. Homsy, On the existence and formulation of variational principles for non-linear differential equations. *Stud. Appl. Math.* **34,** 31 (1975).
57. L. N. Tao, Variational analysis of forced heat convection in a duct of arbitrary cross section. *Proc. Int. Heat Transfer Conf., 3rd, Chicago* 56 (1966).
58. L. Ya. Ainola, Variational principles for nonstationary heat conduction problems. *J. Eng. Phys.* **12,** 465 (1967).
59. A. V. Tokarenko and V. F. Kravchenko, On a numerical method for calculating non-stationary temperature fields in complex-shaped bodies. *In* "Energeticheskoe Mashinostroeniye, vyp. 14." Kharkov Univ. Press, Kharkov, 1972.
60. N. M. Tsirelman, "Variatsionnye metody v teorii teploobmena" (Variational Methods in the Heat Transfer Theory). Ufimsk. Aviats. Inst. Press, Ufa, 1981.
61. N. M. Tsirelman, Variational solution of heat and mass transfer problems for multi-dimensional regions. *In* "Metody i algoritmy parametricheskogo analiza lineynykh i nelineynykh modeley perenosa" (Methods and Algorithms of Parametric Analysis of Linear and Non-linear Models of Transfer), Vyp. 2, Moscow, 1986.
62. N. M. Tsirelman, Variational solution of multi-dimensional problems of non-stationary heat conduction. *In* "Teploprovodnost i diffuziya" (Heat Conduction and Diffusion), Rizhsk. Politekhn. Inst. Press, Riga, 1987.
63. N. M. Tsirelman, Construction of the functional for variational solution of the non-stationary heat transfer problem. *In* "Voprosy teorii i raschyota rabochikh protsessov teplovykh dvigateley" (Problems of Theory and Calculation of Working Processes of Heat Engines), Vyp. 6. Ufimsk. Aviats. Inst. Press, Ufa, 1982.
64. N. M. Tsirelman, Variational solution of non-stationary heat conduction problem for regions with a movable boundary. *Teplofiz. Vysok. Temp.* **18,** 886 (1980).
65. N. M. Tsirelman and E. M. Bronstein, Variational solution of the third boundary-value problem of heat transfer in channel fluid flow. *Teplofiz. Vysok. Temp.* **13,** 1003 (1975).
66. N. M. Tsirelman and E. M. Bronstein, "Variatsionnoye resheniye zadachi teploobmena pri laminarnom techenii zhidkosti v slozhnykh sistemakh" (Variational Solution of Problem of Heat Transfer in Laminar Fluid Flow in Complex Systems). ITMO AN BSSR Press, Minsk, 1974.
67. M. B. Zaks, "Analiticheskoe preobrazovaniye na ES EVM" (Analytical Transformation on an ES Electronic Computer). Saratovsk. Univ. Press, Saratov, 1981.
68. V. S. Zarubin, "Inzhenernye metody resheniya zadach teploprovodnosti" (Engineering Methods of Solving Heat Conduction Problems). Energoatomizdat Press, Moscow, 1983.

ADVANCES IN HEAT TRANSFER, VOLUME 19

Heat Transfer to Newtonian and Non-Newtonian Fluids in Rectangular Ducts

JAMES P. HARTNETT AND MILIVOJE KOSTIC*

Energy Resources Center,
University of Illinois at Chicago,
Chicago, Illinois 60680

I. Introduction

A. Scope and Overview

The fluid dynamics and heat transfer behavior of laminar and turbulent flow through noncircular ducts is of special interest because of the wide application of such geometries in compact heat exchangers. Consequently, extensive analytical and experimental studies have been carried out on such geometries. The analysis of the hydrodynamics and heat transfer of noncircular duct flows is generally more complicated than in the case of circular pipe flow. For example, the determination of the fully developed friction factor and the fully developed heat transfer in noncircular ducts requires a two-dimensional analysis in constrast to the usual one-dimensional anaylsis for fully established circular pipe flow. For developing flows in noncircular channels the analysis becomes three-dimensional.

The usual boundary condition on the velocity for a fluid flowing through a noncircular channel is the relatively simple nonslip condition (i.e., the velocity goes to zero on the boundaries), the same as for circular pipe flow. However, the thermal boundary conditions encountered in noncircular duct flows are generally more complex than in the case of pipe flow. In the case of circular pipe flow, under conditions where forced convection dominates, the temperature and velocity profiles are usually symmetric. Unfortunately, noncircular

* Present address: Department of Mechanical Engineering, Northern Illinois University, DeKalb, Illinois 60115.

ducts do not possess such symmetry and as a result there often exist peripheral variations in shear rate and shear stress as well as in temperature and heat flux, requiring simultaneous analysis of the conduction problem in the wall and the convection problem in the fluid. To avoid this complication of considering conduction in the wall, several limiting cases dealing only with the boundary conditions on the fluid are often used. Such solutions may provide an engineering estimate of the heat transfer performance in actual practice if they are judiciously applied.

Since the most common engineering application of noncircular geometries involve Newtonian fluids such as air, water, and oils, rather extensive analytical and experimental studies have been carried out for the friction factor and heat transfer behavior of Newtonian fluids in laminar and turbulent flow through noncircular ducts. These studies have been reviewed by Shah and London [1] and Shah and Bhatti [2] for laminar flow and by Bhatti and Shah [3] for turbulent flow.

However, in many industries, such as the chemical, pharmaceutical, biological, and food industries, it is not uncommon to encounter non-Newtonian fluids. Thus, it is important to develop an understanding of the heat transfer behavior of such non-Newtonian fluids flowing through non-circular ducts. Among noncircular geometries the rectangular duct, because of its technical importance, has received more attention than other noncircular geometries.

In the course of preparing this article two reviews have appeared which treat non-Newtonian fluids [4, 5]. Irvine and Karni [4] provide a general overview of the subject, discussing rheological property measurements and treating pressure drop and heat transfer for fully established thermal conditions. Lawal and Majumdar [5] present an overview of laminar duct flow and heat transfer to purely viscous non-Newtonian fluids, taking into account the influence of variable properties and viscous dissipation.

This review article will be restricted to the rectangular duct geometry, with special attention given to the friction factor and heat transfer behavior of non-Newtonian fluids. It is recognized that non-Newtonian behavior is generally more complicated than Newtonian flow. As a result, most of the available analytical solutions are restricted to special limiting cases. Experimental data are even more scarce. Consequently, the practicing engineer has limited information on which to base design decisions when dealing with non-Newtonian fluids. An added complication is the fact that the nomenclature varies from study to study, making comparisons difficult. It is the objective of this article to recast the available analytical studies in the framework of a consistent nomenclature. Variable properties and viscous dissipation effects are not taken into account here, and the reader is referred to ref. [5] for a treatment of these effects. The resulting baseline constant property solutions should yield lower limits of the heat transfer performance if properly applied.

The relatively few experimental studies of laminar heat transfer to non-

Newtonian fluids in rectangular passages are covered in some detail. Experimental Nusselt numbers are generally higher than the analytical predictions. Taken together, the analytical and experimental studies should provide a basis for making engineering estimates of heat transfer to non-Newtonian fluids in laminar flow through rectangular geometries.

B. CLASSIFICATION OF NON-NEWTONIAN FLUIDS

The main characteristic of non-Newtonian fluids is a nonlinear dependency of shear stress on shear rate. Metzner [6] classified non-Newtonian fluids into three broad categories: (1) purely viscous fluids, (2) time-dependent fluids and (3) viscoelastic fluids. Newtonian fluids are a subclass of purely viscous fluids. Shenoy and Mashelkar [7] combined purely viscous and time-dependent fluids into the broad category of inelastic fluids.

Purely viscous fluids include pseudoplastics that are shear thinning (that is, the viscosity decreases with increasing shear rate) and dilatant fluids that are shear thickening (that is, the viscosity increases with increasing shear rate). If a fluid requires a certain imposed stress level (i.e., yield stress) before it begins to flow, it is called a Bingham plastic.

Time-dependent fluids are those for which the shear stress is a function of both magnitude and duration of shear rate. These fluids are usually classified as thixotropic and rheopectic fluids, depending on whether the shear stress decreases or increases with time at a given shear rate, respectively. Thixotropic and rheopectic behavior is common to slurries and suspension of solids or colloidal aggregates in liquids.

A more complicated class is the viscoelastic fluid, which exhibits properties that lie between those of a Hookean solid and a purely viscous non-Newtonian fluid. These fluids exhibit differences in the stresses acting in the three orthogonal directions, resulting in the so-called normal stress differences. The unusual characteristics of viscoelastic fluids may be demonstrated experimentally by the Weissenberg effect (e.g., rod climbing effect), die swell, fluid recoil, and the tubeless siphon. Visscoelastic fluids may also exhibit drag and heat transfer reduction under turbulent flow conditions.

Recent experiments with aqueous solutions of 1000 wppm Carbopol have called into question the above classification, which has been widely used in the study of non-Newtonian fluids. It has been experimentally determined that such Carbopol solutions are viscoelastic on the basis of oscillatory viscometric studies (i.e., they show a phase shift between the input shear rate and fluid stress response of less than $\pi/2$ radians). However, to the present authors' knowledge, all turbulent flow studies of pressure drop and heat transfer of aqueous Carbopol solutions (both in circular and noncircular channels) reveal the same behavior as so-called purely viscous nonelastic fluids, such as aqueous Attagel solutions. That is, aqueous Carbopol solutions do not show drag reduction as found for aqueous polyethylene oxide or polyacrylamide solutions. At the same time, there is strong experimental evidence that aqueous Carbopol solutions experience strong secondary motions in laminar flow in noncircular channels, a behavior that is expected in viscoelastic fluids. On the basis of such evidence, it seems clear that not all viscoelastic fluids show drag reduction. Whether all drag-reducing fluids are viscoelastic is also an open question. Furthermore, there is a question as to whether all non-Newtonian fluids, purely viscous and viscoelastic, show secondary flows in steady laminar flow in noncircular channels. As a result of these considerations, it may be necessary in the future to modify the classification of non-Newtonian fluids.

C. Governing Equations

1. *Conservation Equations*

To solve a problem related to momentum and heat transfer in a fluid flowing through a channel, it is convenient to start with the conservation equations. These equations include (1) the continuity equation, (2) the momentum equation, and (3) the energy equation:

$$D\rho/Dt = -\rho(\nabla \cdot \mathbf{V}) \tag{1.1}$$

$$\rho(D\mathbf{V}/Dt) = -\nabla p + \mathbf{F}_b + \nabla \cdot \bar{\bar{\tau}} \tag{1.2}$$

$$\rho(De/Dt) = -\nabla \cdot \mathbf{q} - p\nabla \cdot \mathbf{V} + \nabla\mathbf{V} : \bar{\bar{\tau}} + S \tag{1.3}$$

where the material (substantial) derivative in rectangular coordinates is given by

$$D/Dt = \partial/\partial t + u\partial/\partial x + v\,\partial/\partial y + w\,\partial/\partial z \tag{1.4}$$

These equations describe the changes of velocity and temperature with respect to time and position in the system.

For steady flow of an incompressible fluid with negligible viscous dissipation and without heat generation, the equations reduce to the following in a rectangular coordinate system.

$$\partial u/\partial x + \partial v/\partial y + \partial w/\partial z = 0 \tag{1.5}$$

$$u\frac{\partial u}{\partial x} + v\frac{\partial u}{\partial y} + w\frac{\partial u}{\partial z} = -\frac{1}{\rho}\frac{\partial p}{\partial x} + \frac{1}{\rho}\left(\frac{\partial \tau_{xx}}{\partial x} + \frac{\partial \tau_{yx}}{\partial y} + \frac{\partial \tau_{zx}}{\partial z}\right) + g_x \tag{1.6}$$

$$u\frac{\partial v}{\partial x} + v\frac{\partial v}{\partial y} + w\frac{\partial v}{\partial z} = -\frac{1}{\rho}\frac{\partial p}{\partial y} + \frac{1}{\rho}\left(\frac{\partial \tau_{xy}}{\partial x} + \frac{\partial \tau_{yy}}{\partial y} + \frac{\partial \tau_{zy}}{\partial z}\right) + g_y \tag{1.7}$$

$$u\frac{\partial w}{\partial x} + v\frac{\partial w}{\partial y} + w\frac{\partial w}{\partial z} = -\frac{1}{\rho}\frac{\partial p}{\partial z} + \frac{1}{\rho}\left(\frac{\partial \tau_{xz}}{\partial x} + \frac{\partial \tau_{yz}}{\partial y} + \frac{\partial \tau_{zz}}{\partial z}\right) + g_z \tag{1.8}$$

$$u\frac{\partial T}{\partial x}+v\frac{\partial T}{\partial y}+w\frac{\partial T}{\partial z}=\frac{1}{\rho C_p}\left[\frac{\partial}{\partial x}\left(k\frac{\partial T}{\partial x}\right)+\frac{\partial}{\partial y}\left(k\frac{\partial T}{\partial y}\right)+\frac{\partial}{\partial z}\left(k\frac{\partial T}{\partial z}\right)\right] \quad (1.9)$$

2. *Constitutive Equations*

While the continuity and energy equations are the same for both Newtonian and non-Newtonian fluids, the momentum equation depends on the relationship between the shear stress and the shear rate (velocity gradients). For Newtonian and purely viscous non-Newtonian fluids, the following simple relation has been used:

$$\tau_{ij}=\eta d_{ij}=\eta(I,II,III)\left(\frac{\partial u_i}{\partial x_j}+\frac{\partial u_j}{\partial x_i}\right) \quad (1.10)$$

where η is the apparent viscosity of the fluid.

The viscosity is a scalar function (if we assume isotropicity) of the three invariants of the rate of deformation tensor d_{ij} for purely viscous non-Newtonian fluids (and for a viscoelastic fluid as well). It reduces to a constant for the limiting case of a Newtonian fluid. For an incompressible fluid, the first invariant vanishes and for a simple shear flow even the third invariant vanishes. It is customary to assume, based on experimental evidence [8–10], that the viscosity is a function of the second invariant even for nonviscometric flows. Thus, the viscosity dependence on shear rate may be given in the invariant functional notation as

$$\eta=\eta(\sqrt{II/2}) \quad (1.11)$$

A simple analytical model that allows for a nonlinear dependence of shear stress on shear rate is the power law or Ostwaald–de Waele model ($\tau_w=K\dot{\gamma}^n$). It is the most widely used model and may be generally used for purely viscous fluids at least for a narrow range of shear rate. For such power law fluids, the viscosity may be presented by the following equation:

$$\eta=K(II/2)^{(n-1)/2} \quad (1.12)$$

$$\sqrt{\tfrac{1}{2}II}=\sqrt{\tfrac{1}{2}(d_{ij}:d_{ij})}=\left\{2\left[\left(\frac{\partial u}{\partial x}\right)^2+\left(\frac{\partial v}{\partial y}\right)^2+\left(\frac{\partial w}{\partial z}\right)^2\right]+\left(\frac{\partial u}{\partial y}+\frac{\partial v}{\partial x}\right)^2+\left(\frac{\partial u}{\partial z}+\frac{\partial w}{\partial x}\right)^2+\left(\frac{\partial v}{\partial z}+\frac{\partial w}{\partial y}\right)^2\right\}^{1/2} \quad (1.13)$$

Where u, v, and w are, respectively, the velocities in the x, y, and z directions. For fully developed flow, it is plausible to assume the following conditions:

$$\frac{\partial u}{\partial x}=0, \qquad v=w=0, \qquad p=p(x), \qquad u=u(y,z) \quad (1.14)$$

This yields for the second invariant:

Then
$$II = 2\left[\left(\frac{\partial u}{\partial y}\right)^2 + \left(\frac{\partial u}{\partial z}\right)^2\right] \tag{1.15}$$

$$\tau_{yx} = K\left[\left(\frac{\partial u}{\partial y}\right)^2 + \left(\frac{\partial u}{\partial z}\right)^2\right]^{(n-1)/2} \frac{\partial u}{\partial y} \tag{1.16}$$

$$\tau_{zx} = K\left[\left(\frac{\partial u}{\partial y}\right)^2 + \left(\frac{\partial u}{\partial z}\right)^2\right]^{(n-1)/2} \frac{\partial u}{\partial z} \tag{1.17}$$

Thus, neglecting body forces, the momentum equation for a power law fluid in fully developed flow through a rectangular channel becomes

$$\frac{1}{K}\frac{dp}{dx} = \frac{\partial}{\partial y}\left\{\left[\left(\frac{\partial u}{\partial y}\right)^2 + \left(\frac{\partial u}{\partial z}\right)^2\right]^{(n-1)/2} \frac{\partial u}{\partial y}\right\} + \frac{\partial}{\partial z}\left\{\left[\left(\frac{\partial u}{\partial y}\right)^2 + \left(\frac{\partial u}{\partial z}\right)^2\right]^{(n-1)/2} \frac{\partial u}{\partial z}\right\} \tag{1.18}$$

subject to the condition that the velocity, u, goes to zero on the boundaries of the flow.

In the current review, the power law model will be used for the analytical hydrodynamic and heat transfer studies of purely viscous non-Newtonian fluids in laminar flow through a rectangular channel. Such predictions will be compared with the available experimental data. In the case of turbulent flow, emphasis will be placed on the available experimental results.

A viscoelastic fluid also needs a constitutive equation to describe stress relaxation and normal stress development. Since Eq. (1.11) may not be adequate for this purpose, two general classes of equations have been introduced for viscoelastic fluids with some success: (1) rate equations (differential equations) and (2) integral equations. The details of these constitutive equations can be found elsewhere [10]. Given the uncertainty and complexity of the analytical formulation of the constitutive equations for viscoelastic fluids, this review will concentrate on the simplest model and available experimental information, both for laminar and turbulent flow.

3. *Boundary Conditions*

In order to solve the conservation equations, once the constitutive relations are given, the appropriate boundary and initial conditions must be prescribed. For the velocity,

$$u, v, w = 0 \qquad \text{at wall} \tag{1.19}$$

$$u(0, y, z) = \text{given} \tag{1.20}$$

There are an infinite number of possible thermal boundary conditions describing the temperature and the heat flux that can be imposed on the boundaries of the fluid flowing through a rectangular duct. The heat transfer is strongly dependent on the thermal boundary conditions in the laminar flow regime while much less dependent in the turbulent flow regime, particularly for fluids with a Prandtl number larger than unity. This survey will be restricted mainly to three classes of thermal boundary conditions:

1. constant temperature imposed on the boundary of the fluid, the so-called T condition;
2. constant axial heat flux with constant local peripheral wall temperature imposed on the boundary of the fluid, the $H1$ condition; and
3. constant heat flux imposed both axially and peripherally on the boundary of the fluid, the $H2$ condition.

If not all boundary walls are heated then the usual nomenclature (i.e., T, $H1$, and $H2$) must be modified. Consideration is given herein to the thermal boundary conditions: (1) constant temperature imposed on one or more bounding walls with the remaining walls adiabatic; (2) constant heat input per unit length imposed on one or more walls with the associated peripheral wall temperature being constant, while the remaining unheated walls are adiabatic; (3) constant heat input per unit area imposed on one or more walls while the remaining walls are adiabatic. The designations of such thermal boundary conditions are given in Table I. The following examples illustrate the use of the definition:

$H1(3\mathrm{L})$ = thermal boundary condition of the $H1$ type imposed on three walls (longer version), while one shorter wall is adiabatic (1.21)

$T(2\mathrm{S})$ = two opposite shorter walls held at constant temperature, while two longer walls are adiabatic

If these terms are used in a subscript, such as $\mathrm{Nu}_{xH1(3\mathrm{L})}$, it is obvious that this relates to the axially local Nusselt number for the $H1(3\mathrm{L})$ thermal boundary condition. In general, when T, $H1$, and $H2$ appear alone this corresponds to the case where all bounding walls are heated, that is

$$H1 = H1(4) \tag{1.22}$$

4. *Dimensionless Groups and Generalized Solutions*

In the previous sections, the appropriate differential equations and boundary conditions are outlined for the flow of a non-Newtonian fluid through a rectangular channel. Solutions to such problems have been generally presented in terms of dimensionless groups, which can be determined by making

TABLE I

DEFINITION OF DIFFERENT COMBINATIONS OF THERMAL BOUNDARY CONDITIONS: *T*, *H*, *H*1, OR *H*2

Symbol	Description
4	 Four (all) walls heated
3L	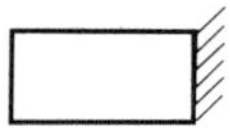 Three walls (longer version) heated
3S	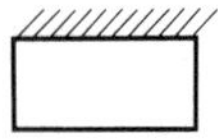 Three walls (shorter version) heated
2L	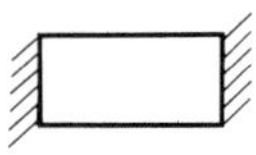 Two walls (longer version) heated
2S	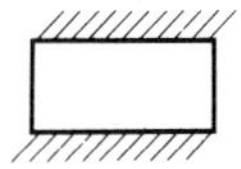 Two walls (shorter version) heated
2C	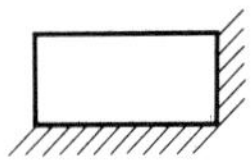 Two walls (corner version) heated
1L	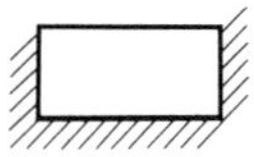 One wall (longer version) heated
1S	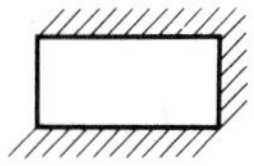 One wall (shorter version) heated
	Adiabatic (unheated) wall

the governing equations and the corresponding boundary conditions dimensionless. In this section, some important dimensionless groups that result from the solutions of the momentum and energy equations are described.

The major dimensionless groups for friction factor and heat transfer to Newtonian and non-Newtonian fluids are now given.

1. Dimensionless peripheral coordinates:

$$Y^* = y/D_h \qquad Z^* = z/D_h \tag{1.23}$$

2. Dimensionless hydraulic axial distance based on Re^+:

$$x_{\mathrm{hy}}^+ = x/(D_{\mathrm{h}}\,\mathrm{Re}^+) \tag{1.24}$$

3. Dimensionless hydraulic axial distance based on Re^*:

$$x_{\mathrm{hy}}^* = x/(D_{\mathrm{h}}\,\mathrm{Re}^*) \tag{1.25}$$

4. Dimensionless thermal axial distance:

$$x_{\mathrm{th}}^+ = x/(D_{\mathrm{h}}\,\mathrm{Re}^+\,\mathrm{Pr}^+) \tag{1.26}$$

5. Dimensionless thermal axial distance based on Re^*:

$$x_{\mathrm{th}}^* = x/(D_{\mathrm{h}}\,\mathrm{Re}^*\,\mathrm{Pr}^*) = x_{\mathrm{th}}^+ \tag{1.27}$$

6. Graetz number based on hydraulic diameter:

$$\mathrm{Gz} = \frac{1}{x_{\mathrm{th}}^*} = \frac{1}{x_{\mathrm{th}}^+} = \mathrm{Pe}/(x/D_{\mathrm{h}}) = \mathrm{Gz}_{\dot{m}}(D_{\mathrm{h}}^2/A) \tag{1.28}$$

7. Graetz number based on mass flow rate:

$$\mathrm{Gz}_{\mathrm{m}} = \frac{\dot{m}C_{\mathrm{p}}}{xk} = \frac{(1+\alpha^*)^2}{4\alpha^*}\frac{1}{x_{\mathrm{th}}^+} \tag{1.29}$$

8. Dimensionless velocities:

$$u^* = u/U, \qquad v^* = v/U, \qquad w^* = w/U \tag{1.30}$$

9. Dimensionless complex velocity gradient:

$$\eta^* = \left[\left(\frac{\partial u^*}{\partial Y^*}\right)\right]^2 + \left(\frac{\partial u^*}{\partial Z^*}\right)^2\Bigg]^{(n-1)/2} \tag{1.31}$$

10. Fanning friction factor:

$$f = \tau_{\mathrm{w}}/(\rho U^2/2) = (-dp/dx)(D_{\mathrm{h}}/4)/(\rho U^2/2) \tag{1.32}$$

11. Incremental pressure drop number:

$$K^*(x_{\mathrm{hy}}^+) = \frac{\Delta p}{\frac{1}{2}\rho U^2} - 4x_{\mathrm{hy}}^+(f_{\mathrm{fd}}\,\mathrm{Re}) \tag{1.33}$$

$$K^*(x_{\mathrm{hy}}^*) = \frac{\Delta p}{\frac{1}{2}\rho U^2} - 64x_{\mathrm{hy}}^* \tag{1.34}$$

12. Dimensionless temperature:

$$T^* = (T - T_{\mathrm{w}})/(T_{\mathrm{in}} - T_{\mathrm{w}}) \qquad \text{for } T \text{ boundary condition} \tag{1.35}$$

$$T^* = (T - T_{\mathrm{in}})/(q'/k) \qquad \text{for } H1 \text{ boundary condition} \tag{1.36}$$

$$T^* = (T - T_{\mathrm{in}})/(q''D_{\mathrm{h}}/k) \qquad \text{for } H2 \text{ boundary condition} \tag{1.37}$$

13. Nusselt number:

$$\mathrm{Nu} = hD_{\mathrm{h}}/k \tag{1.38}$$

14. Heat transfer j factor:

$$j_{\mathrm{H}} = \mathrm{Nu}/(\mathrm{Re}\,\mathrm{Pr}^{1/3}) \tag{1.39}$$

15. Reynolds number based on apparent viscosity, also Newtonian Reynolds number:

$$\mathrm{Re} = \rho U D_{\mathrm{h}}/\eta \tag{1.40}$$

16. Generalized Reynolds number:

$$\mathrm{Re}^{+} = \rho U^{2-n} D_{\mathrm{h}}^{n}/K \tag{1.41}$$

17. Kozicki generalized Reynolds number:

$$\mathrm{Re}^{*} = \rho U^{2-n} D_{\mathrm{h}}^{n} \Bigg/ \left[8^{n-1} K \left(b^{*} + \frac{a^{*}}{n} \right)^{n} \right] \tag{1.42}$$

18. Peclet number:

$$\mathrm{Pe} = \rho C_{\mathrm{p}} U D_{\mathrm{h}}/k = \mathrm{Re}\,\mathrm{Pr} = \mathrm{Re}^{+}\,\mathrm{Pr}^{+} = \mathrm{Re}^{*}\,\mathrm{Pr}^{*} \tag{1.43}$$

19. Prandtl number:

$$\mathrm{Pr} = \eta C_{\mathrm{p}}/k = \mathrm{Pe}/\mathrm{Re} \tag{1.44}$$

$$\mathrm{Pr}^{+} = \mathrm{Pe}/\mathrm{Re}^{+} \tag{1.45}$$

$$\mathrm{Pr}^{*} = \mathrm{Pe}/\mathrm{Re}^{*} \tag{1.46}$$

20. Grashof number:

$$\mathrm{Gr} = g\beta\,\Delta T\, D_{\mathrm{h}}^{3}\rho^{2}/\eta^{2} \tag{1.47}$$

$$\mathrm{Gr}_{q} = g\beta q'' D_{\mathrm{h}}^{4}\rho^{2}/(\eta^{2}k) \tag{1.48}$$

21. Rayleigh number:

$$\mathrm{Ra} = \mathrm{Gr}\,\mathrm{Pr} \tag{1.49}$$

$$\mathrm{Ra}_{q} = \mathrm{Gr}_{q}\,\mathrm{Pr} \tag{1.50}$$

22. Weissenberg or Deborah number, Ws or De:

$$\mathrm{Ws} = \lambda U/D_{\mathrm{h}} \tag{1.51}$$

$$\mathrm{De} = \lambda/\bar{t} \tag{1.52}$$

where λ is a characteristic time of the fluid and $\bar{t}$ is the characteristic time of the flow system. If n is equal to unity, these dimensionless groups become those for Newtonian fluids. The Weissenberg and Deborah numbers, which take account of the elasticity of viscoelastic fluids, are zero for Newtonian fluids. The friction factor and the dimensionless heat transfer behavior of Newtonian and non-Newtonian fluids in flow through rectangular channels will be shown to be functions of one or more of the above dimensionless groups, depending on the particular problem.

As an example, the momentum equation [Eq. (1.18)] and the energy equation [Eq. (1.9)] under assumption of constant fluid properties (k, ρ, C_p, K, n), fully developed flow, and neglecting axial thermal conduction, may be written as

$$\frac{\partial}{\partial Y^*}\left(\eta^*\frac{\partial u^*}{\partial Y^*}\right) + \frac{\partial}{\partial Z^*}\left(\eta^*\frac{\partial u^*}{\partial Z^*}\right) + 2f\,\mathrm{Re}^+ = 0 \tag{1.53}$$

and

$$u^*\frac{\partial T^*}{\partial x^*_{\mathrm{th}}} = \frac{\partial^2 T^*}{\partial Y^{*^2}} + \frac{\partial^2 T^*}{\partial Z^{*^2}} \tag{1.54}$$

Also, the boundary conditions will change accordingly:

$$u^* = v^* = w^* = 0 \qquad \text{on perimeter, } P \tag{1.55}$$

$$T^* = 0 \qquad \text{on } P \text{ for } T \text{ boundary condition} \tag{1.56}$$

$$\int_{P^*_h} (\partial T^*/\partial n^*)\, dP^*_h = -1 \qquad \text{for } H1 \text{ boundary condition} \tag{1.57}$$

$$(\partial T^*/\partial n^*)_P = -1 \qquad \text{for } H2 \text{ boundary condition} \tag{1.58}$$

where n^* is the coordinate normal to the perimeter (P), either Z^* or Y^*; P_{h} is the heated part of the perimeter P; $P^*_{\mathrm{h}} = P_{\mathrm{h}}/D_{\mathrm{h}}$; and dP^*_{h} is the differential of the heated perimeter, either dY^* or dZ^*, along the heated perimeter. For the purpose of general presentation or numerical calculation, it is much more convenient to make the governing equations and boundary conditions dimensionless.

II. Laminar Flow

A. Hydrodynamically Developed Flow

1. *Velocity Profile*

a. *Newtonian fluids*

The fully developed laminar velocity profile and friction factor of a Newtonian fluid in a parallel plate geometry (Fig. 1a) can be easily obtained by solving the one-dimensional momentum equation and can be found in many textbooks.

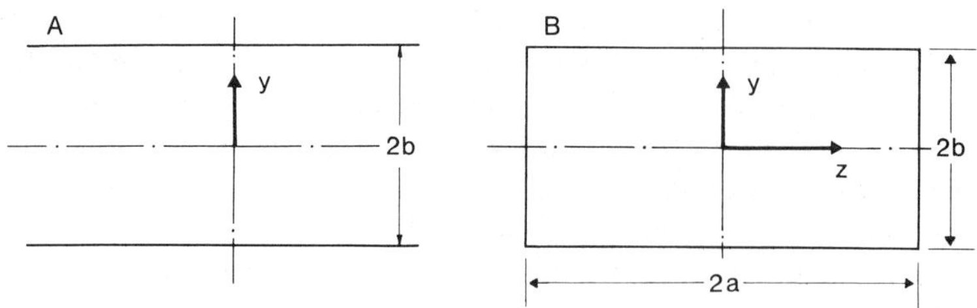

FIG. 1. Parallel plate (A) and rectangular duct (B) geometries.

$$\frac{u}{U} = \frac{3}{2}\left[1 - \left(\frac{y}{b}\right)^2\right] \tag{2.1}$$

The analytical solution is also available for the fully developed laminar velocity profile of a Newtonian fluid in rectangular ducts (Fig. 1b). It has been determined using the similarity to the well-known Lagrange's stress function for the small deflection of a thin rectangular plate subjected to a uniform load [11, 12]. The velocity profile is

$$u(y, z) = \frac{16a^2}{\pi^3}\left(\frac{-dp/dx}{\eta}\right)\sum_{i=1}^{\infty}\left[\frac{(-1)^{i-1}}{(2i-1)^3}\right] \times \left\{1 - \frac{\cosh[(2i-1)\pi y/2a]}{\cosh[(2i-1)\pi b/2a]}\right\}\cos\frac{(2i-1)\pi z}{2a} \tag{2.2}$$

This prediction has been confirmed by the experimental measurements of Holmes and Vermeulen [13], Goldstein and Kreid [14], Sparrow *et al.* [15], and Muchnik *et al.* [16]. The aspect ratio range and experimental methods are described in Table II.

Because of the computational complexity of Eq. (2.2) it is more convenient to use an approximate solution suggested by Purday [17] to describe the laminar velocity profile in rectangular ducts.

$$\frac{u}{u_{\max}} = \left[1 - \left(\frac{y}{b}\right)^r\right]\left[1 - \left(\frac{z}{a}\right)^s\right] \tag{2.3}$$

where $r = 2$ and s was obtained by applying the principle of minimum energy dissipation:

$$s \cong 1.54/\alpha^* \qquad \text{for} \qquad 0 < \alpha^* \leq 2/3 \tag{2.4}$$

$$s \cong 2.3 \qquad \text{for} \qquad 2/3 < \alpha^* \leq 1 \tag{2.5}$$

This approach was evaluated by Holmes and Vermeulen [13] and the values of exponents r and s were determined based on their experimental measurements in two different ways: (1) by measuring the velocity gradient at the wall

TABLE II

EXPERIMENTAL MEASUREMENTS OF NEWTONIAN VELOCITY PROFILES IN RECTANGULAR DUCTS

Investigator	Working fluid	Aspect ratio	Method
Holmes and Vermeulen [13]	Water	0.10	Flow visualization
		0.15	
		0.25	
		1.00	
Goldstein and Kried [14]	Water	1.0	LDV
Sparrow *et al.* [15]	Air	0.2	Pitot tube
		0.5	
Muchnik *et al.* [16]	Water	0.10	Flow visualization
		0.25	
		0.50	

and (2) by measuring the area under the velocity profile curve. They reported good agreement with the values suggested by Purday.

Natarajan and Lakshmanan [18] solved the continuity and momentum equations by a finite-difference method. They matched the velocity profile to the empirical equation [Eq. (2.3)] and proposed the following equations for exponents r and s:

$$r = \begin{cases} 2 & \text{for} \quad \alpha^* \leq 1/3 \\ 2 + 0.3(\alpha^* - 1/3) & \text{for} \quad \alpha^* > 1/3 \end{cases} \tag{2.6}$$

$$s = 1.7 + 0.5(\alpha^*)^{-1.4} \qquad \text{for} \qquad 0 \leq \alpha^* \leq 1 \tag{2.7}$$

They also noted that

$$u_{\max}/U = \left(\frac{r+1}{r}\right)\left(\frac{s+1}{s}\right) \tag{2.8}$$

Agreement between the approximate equation [Eq. (2.3)] and the exact series solution [Eq. (2.2)] is within 2% over most of the cross section; however in regions close to the corner, the differences may be above 35%, being highest for the square duct.

b. *Non-Newtonian fluids*

Turning to non-Newtonian fluids, a simple analytical solution is available for a power law fluid in a parallel plate channel [19]:

$$\frac{u}{U} = \frac{2n+1}{n+1}\left[1 - \left(\frac{y}{b}\right)^{(n+1)/n}\right] \tag{2.9}$$

Schechter [20] also obtained the velocity distribution of a power law fluid in a parallel plate channel using a variational principle:

$$\frac{u}{U} = \tfrac{1}{4}\left[1 - \left(\frac{y}{b}\right)^2\right]\left[a_1 + a_2\left(\frac{y}{2b}\right)^2\right] \tag{2.10}$$

The values of a_1 and a_2 are functions of the power law index n as shown in Table III. Since this velocity profile does not explicitly contain the power law index, n, it is particularly convenient to use in the energy equation to analytically calculate the developing temperature profiles in a parallel plate under hydrodynamically fully established conditions [21].

For a rectangular duct flow of a power law fluid, an exact analytical solution for the velocity profile does not appear to be available. However, Schechter [20] did solve the momentum equation for a power law fluid in laminar flow through a rectangular duct having aspect ratios of 0.25, 0.5, 0.75, and 1.0 using a variational principle involving six terms:

$$\frac{u(y,z)}{U} = \sum_{i=1}^{6} A_i \sin\left[\alpha_i \pi \frac{(z/a)+1}{2}\right] \sin\left[\beta_i \pi \frac{(y/b)+1}{2}\right] \tag{2.11}$$

where the values of α_i and β_i are shown in Table IV, and the values of the

TABLE III

CONSTANTS PERTINENT TO FLOW BETWEEN FLAT PLATES IN EQ. (2.10)

n	a_1	a_2	$f\,\mathrm{Re}^+$
1.0	6.0	0.00	24.00
0.75	5.770	4.598	13.98
0.50	5.3673	12.683	8.01

TABLE IV

VALUES OF CONSTANTS IN EQ. (2.11)

i	α_i	β_i
1	1	1
2	3	1
3	1	3
4	3	3
5	5	1
6	1	1

coefficients, A_i are shown in Table V. Subsequently, Wheeler and Wissler [22] applied an overrelaxation method to obtain more accurate velocity profiles as well as the friction factor for a power law fluid in fully established laminar flow through rectangular ducts of aspect ratios of 0.5, 0.75, and 1.0. Later Chadrupatla [23] numerically solved the same problem for a square duct.

The fully established velocity profiles of a power law fluid in a parallel plate and along the centerline of a square duct are shown in Fig. 2 for n values of 0, 0.5, 0.75, and 1.0 [22, 23]. Fluids with low n values approach slug flow over a substantial central portion of the flow field.

2. *Friction Factor*

a. *Newtonian fluids*

The friction factors of a Newtonian fluid in fully established laminar flow through rectangular ducts were calculated accurately by Shah and London [1] using the analytical solution given by Eq. (2.2). They also proposed the following simple equation, which is within 0.05% of the analytical solution.

$$f\,\mathrm{Re} = 24(1 - 1.3553\alpha^* + 1.9467\alpha^{*2} - 1.7012\alpha^{*3} + 0.9564\,\alpha^{*4} - 0.2537\alpha^{*5}) \tag{2.12}$$

TABLE V

COMPUTED RESULTS FOR FLOW IN RECTANGULAR DUCT[a]

n	α^*	A_1	A_2	A_3	A_4	A_5	A_6
1.00	1.00	2.346	0.156	0.156	0.0289	0.0360	0.0360
0.75	1.00	2.313	0.205	0.205	0.0007	0.0434	0.0434
0.50	1.00	2.263	0.278	0.278	−0.0285	0.0555	0.0555
1.00	0.75	2.341	0.204	0.119	0.0256	0.0498	0.0303
0.75	0.75	2.310	0.235	0.180	0.0001	0.0568	0.0364
0.50	0.75	2.263	0.286	0.267	−0.0277	0.0644	0.0505
1.00	0.50	2.311	0.296	0.104	0.0285	0.0795	0.0303
0.75	0.50	2.288	0.299	0.174	0.0120	0.0811	0.0364
0.50	0.50	2.249	0.312	0.274	−0.0101	0.0780	0.0501
1.00	0.25	2.227	0.503	0.0867	0.0274	0.184	0.0189
0.75	0.25	2.221	0.459	0.160	0.0312	0.160	0.0210
0.50	0.25	2.205	0.407	0.270	0.0257	0.131	0.0364

[a] See Eq. (2.11).

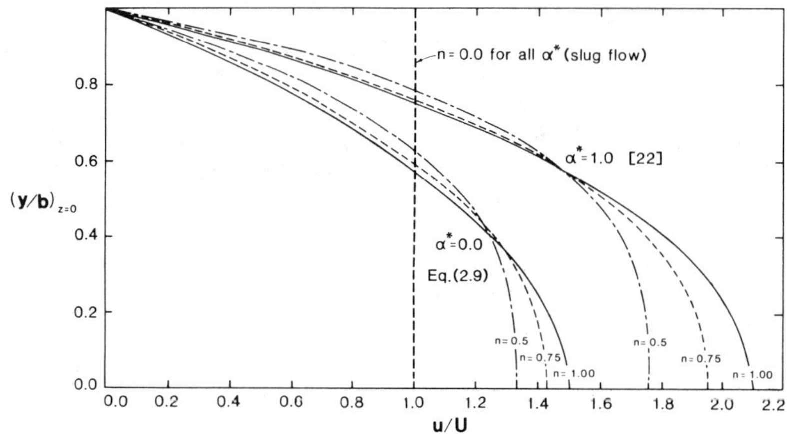

FIG. 2. Fully developed velocity profiles for power law fluids in laminar flow through rectangular channels, $\alpha^* = 0$ and 1.0.

The f Re values of Shih [24], determined by a point matching method, are in excellent agreement with Eq. (2.12). Natarajan and Lakshmanan [18] presented a simpler formula for f Re, which is within 5.6% of the results given by Eq. (2.12).

$$f\,\mathrm{Re} = 14.4(\alpha^*)^{-1/6} \qquad \text{for} \qquad \alpha^* \geq 0.05 \tag{2.13}$$

b. *Non-Newtonian fluids*

The friction factors of a power law fluid in fully established laminar flow through a rectangular channel were calculated by Schechter [20] using a variational principle and by Wheeler and Wissler [22] using a numerical method. Wheeler and Wissler also presented an approximate equation for the square duct geometry.

$$(4f)\,\mathrm{Re}^+ = 7.4942\left(\frac{1.7330}{n} + 5.8606\right)^n \tag{2.14}$$

where

$$\mathrm{Re}^+ = \rho U^{2-n} D_{\mathrm{h}}^n / K \tag{2.15}$$

Chandrupatla reported fully established friction factor results for flow in a square duct [23].

Perhaps the most elegant friction factor analysis has been provided by Kozicki *et al.* [25, 26], who generalized the Rabinowitsch–Mooney equation for non-Newtonian fluids including the special case of power law fluids in arbitrary ducts having a constant cross section. Kozicki *et al.* [25] introduced a new Reynolds number such that the friction factor for fully developed laminar flow of a power law fluid through noncircular geometries having a constant cross section area is given by a unique equation:

$$f = 16/\mathrm{Re}^* \tag{2.16}$$

where

$$\mathrm{Re}^* = \rho U^{2-n} D_{\mathrm{h}}^n \Bigg/ \left[8^{n-1} \left(b^* + \frac{a^*}{n} \right)^n K \right] \tag{2.17}$$

The values of a^* and b^* depend on the geometry of the duct. Table VI presents these values for a rectangular channel as a function of the aspect ratio, α^*. It is of interest to note that a^* and b^* are 0.25 and 0.75 for the circular

TABLE VI

GEOMETRIC CONSTANTS a^* AND b^* FOR RECTANGULAR DUCTS[a,b]

α^*	a^*	b^*	c
1.00	0.2121	0.6771	14.227
0.95	0.2123	0.6774	14.235
0.90	0.2129	0.6785	14.261
0.85	0.2139	0.6803	14.307
0.80	0.2155	0.6831	14.378
0.75	0.2178	0.6870	14.476
0.70	0.2208	0.6921	14.605
0.65	0.2248	0.6985	14.772
0.60	0.2297	0.7065	14.980
0.55	0.2360	0.7163	15.236
0.50	0.2439	0.7278	15.548
0.45	0.2538	0.7414	15.922
0.40	0.2659	0.7571	16.368
0.35	0.2809	0.7750	16.895
0.30	0.2991	0.7954	17.512
0.25	0.3212	0.8183	18.233
0.20	0.3475	0.8444	19.071
0.15	0.3781	0.8745	20.042
0.10	0.4132	0.9098	21.169
0.05	0.4535	0.9513	22.477
0.0	0.5000	1.0000	24.000

[a] See Eq. (2.17).

[b] Note: $c = 16(a^* + b^*) = f\,\mathrm{Re}$ for Newtonian fluid.

duct, and the generalized Reynolds number Re* becomes identical to that proposed by Metzner and Reed [27].

$$\mathrm{Re}' = \rho U^{2-n} D_{\mathrm{h}}^{n} \Bigg/ \left[8^{n-1} \left(0.75 + \frac{0.25}{n} \right)^{n} K \right] \tag{2.18}$$

Dividing Eq. (2.15) by Eq. (2.17) the following is obtained:

$$\mathrm{Re}^{+}/\mathrm{Re}^{*} = 8^{n-1} \left(b^{*} + \frac{a^{*}}{n} \right)^{n}$$

It is a straightforward exercise to reformulate Eq. (2.16) in terms of $f\ \mathrm{Re}^{+}$

$$f\ \mathrm{Re}^{+} = 2^{3n+1} \left(b^{*} + \frac{a^{*}}{n} \right)^{n} \tag{2.19}$$

The friction factor results of Schechter [20], Wheeler and Wissler [22], Chandrupatla [23], and Kozicki *et al.* [25] are given in Table VII, where it can be seen that the various predictions are in good agreement over a wide range of aspect ratio, α^{*}, and power law index n.

The results of the Kozicki analysis are also shown in Fig. 3 in the form of $f\ \mathrm{Re}^{+}$ as a function of α^{*} and n [Eq. (2.19)]. Inspection of the figure and of Table VII reveals that the value of $f\ \mathrm{Re}^{+}$ decreases with a decrease of the power law index at a fixed aspect ratio; it decreases with an increase in the aspect ratio at a fixed n value.

In a second paper, Kozicki and Tiu [26] presented an improved parametric method. An infinite number of geometric parameters are assumed to characterize the flow geometry. The actual number of the parameters depends on the rheological fluid model equation. For the power law (Ostwaald–de Waele) model, the number of geometric parameters required to represent the relationship between the flow rate and pressure drop is given by $(1 + 1/n)$ if $1/n$ is an integer, otherwise the number of parameters is infinite. For rectangular ducts, the use of these higher order parameters (ε_2 for $n = 1/2$, and ε_2 and ε_3 for $n = 1/3$) changes the f Re product by 1% and 2%, respectively, compared with the case when only ε_0 and ε_1 are taken into account. The ratio of f Re* using the new parameters ε_0 and ε_1 [26] to the value using the original parameters a^{*} and b^{*} [25] is presented in Table VIII. The ratio generally decreases with an increasing aspect ratio and a decreasing power law index n, showing a maximum difference between the two calculations of about 3%. This difference is of the order of the accuracy of the determination of the ε_i parameters. Accordingly, the use of the simpler two-parameter (a^{*} and b^{*}) method is recommended for rectangular ducts.

Experimental confirmation of the friction factor predictions has been reported by Wheeler and Wissler [22], Hartnett *et al.* [28], and Hartnett and Kostic [29]. The laminar friction factors of aqueous polyacrylic acid

TABLE VII

$f\,Re^+_{gen}$ AS A FUNCTION OF α^* AND POWER LAW INDEX n

	$f\,Re^+$ $\alpha^* = 1.00$				$f\,Re^+$ $\alpha^* = 0.75$			$f\,Re^+$ $\alpha^* = 0.50$			$f\,Re^+$ $\alpha^* = 0.25$	
n	Schechter [20]	Chandrupatla [23]	Wheeler [22]	Kozicki [25]	Schechter [20]	Wheeler [22]	Kozicki [25]	Schechter [20]	Wheeler [22]	Kozicki [25]	Schechter [20]	Kozicki [25]
1.0	14.27	14.228	14.228	14.219	14.30		14.470	15.60	15.544	15.546	18.38	18.23
0.9		11.910	11.905	11.965			12.158			12.981		15.03
0.8		9.923	9.915	10.061			10.210			10.830		12.37
0.75	9.085	9.060	9.055	9.222	9.069	9.117	9.349	[a]	9.693	9.889	11.11	11.22
0.7		8.270	8.268	8.451			8.561			9.026		10.18
0.6		6.885	6.883	7.089			7.170			7.511		8.352
0.5	5.755	5.733	5.723	5.935	5.724		5.993	6.033		6.237	6.641	6.837
0.4			4.743	4.954			4.993			5.162		5.574

[a] The value given by Schechter for $n = 0.75$ and $\alpha^* = 0.50$ is obviously in error and is omitted from the table.

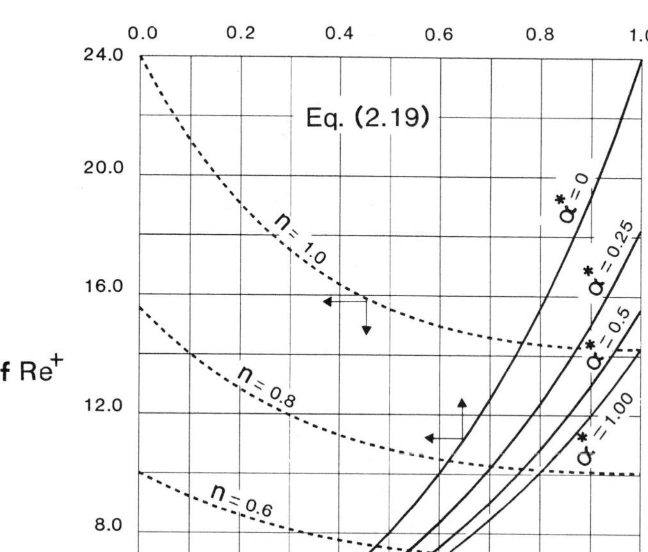

FIG. 3. Dimensionless friction factor, $f\,Re^+$, as a function of power law index n and aspect ratio α^*.

TABLE VIII

THE RATIO OF $f\,Re^*$ CALCULATED WITH NEW PARAMETERS ε_0 and ε_1 [$f\,Re^*$ (new)] TO $f\,Re^*$ CALCULATED WITH ORIGINAL VALUES OF a^* AND b^* [$f\,Re^*$ (original)]

	$f\,Re^*$ (new)/$f\,Re^*$ (original)			
n	$\alpha^* = 0.25$	$\alpha^* = 0.50$	$\alpha^* = 0.75$	$\alpha^* = 1.00$
0.00	1.0000	1.000	1.000	1.000
0.25	0.980	0.981	0.989	1.000
0.50	0.968	0.971	0.984	1.000
0.75	0.972	0.975	0.985	1.000
1.00	0.971	0.974	0.985	0.999

(Carbopol 934 and 960) solutions were measured in a square duct by Hartnett *et al.* [28]. The results are shown in Fig. 4 as a function of the Reynolds number Re* proposed by Kozicki *et al.* The Carbopol experimental results are in excellent agreement with Kozicki's prediction [Eq. (2.16)]. Wheeler and Wissler [22] also conducted experimental measurements of the laminar friction factor with aqueous solutions of carboxymethyl cellulose, CMC, (0.28–1.0% by weight) in a square duct. Even though aqueous CMC solutions are known to be weakly viscoelastic, the experimental results agree well with the power law predictions. Other viscoelastic fluids, such as aqueous polyacrylamide solutions (Separan AP-273), were used in rectangular ducts having aspect ratios of 0.5 and 1 by Hartnett *et al.* [28] and Hartnett and

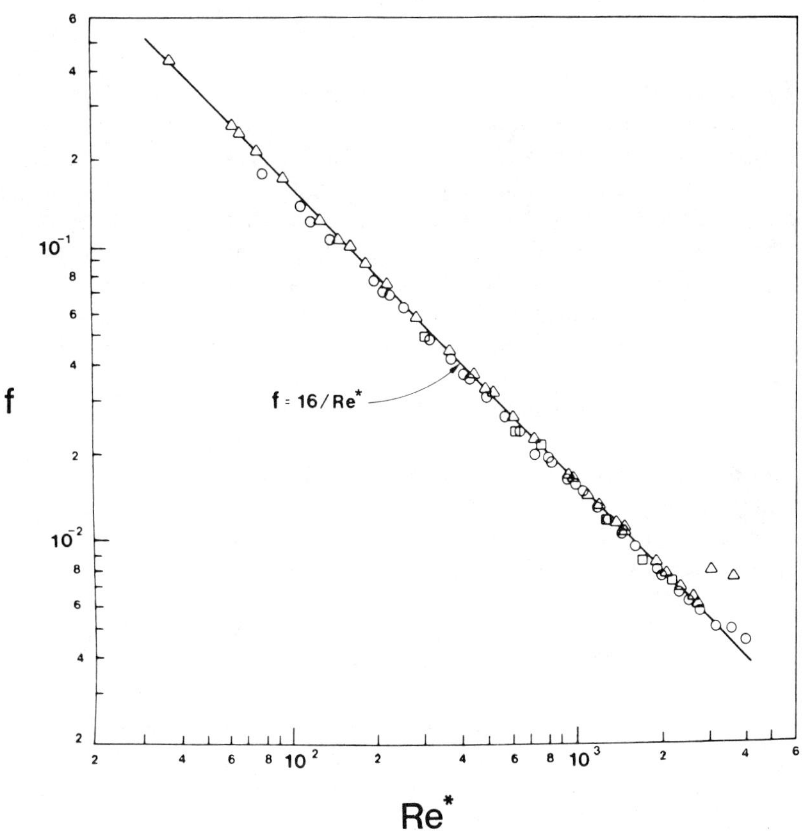

FIG. 4. Experimental friction factor measurements for non-Newtonian fluids in fully established laminar flow through rectangular channels. Results of Wheeler and Wissler [22] (○), Hartnett *et al.* [28] (△), and Hartnett and Kostic [29] (□).

Kostic [29], and again their results also show good agreement with analytical and numerical studies for a power law fluid.

Analytical and experimental studies [30–32] suggest that the presence of normal forces in the case of viscoelastic fluids may induce secondary flows in noncircular ducts even in steady laminar flow. However, measurements to date suggest that if such secondary flows do exist they have negligible influence on the laminar friction factor. In brief, insofar as the fully established pressure drop is concerned, there is no difference in the behavior of viscoelastic and purely viscous non-Newtonian fluids in laminar flow through rectangular channels.

B. HYDRODYNAMICALLY DEVELOPING FLOW

1. *Introduction*

The velocity distribution of a fluid flowing through a duct will undergo a development from some initial profile at the entrance of the duct to a fully developed profile at locations far downstream. Correspondingly, the pressure gradient in the region of flow development will differ from that of a fully developed flow. The length of the duct in which such a velocity development occurs is called the hydrodynamic entrance length or entrance region. In view of the engineering importance of duct flows, a number of analytical and experimental studies have been carried out to determine the velocity and pressure distribution in the entrance region of parallel plates and rectangular ducts.

Most of the analytical studies involve approximations, such as neglecting the cross-stream velocities, in order to solve the continuity and momentum equations simultaneously. In the case of plane parallel plates, the integral method [33–40] and the patching (or matching) method [41–44] have been widely used. Linearization of the inertia terms in the momentum equation has led to solutions for the plane parallel plate case and for rectangular channels [45–51]. In addition, numerical finite-difference methods have been applied to the rectangular channel geometry [22, 23, 52–58]. A summary is presented in Table IX. The The reader is referred to Shah and London [1] for more details.

Experimental velocity and pressure measurements have been carried out in the entrance region of rectangular ducts using such techniques as flow visualization [13, 16], pitot tube measurements [15, 59], and laser doppler anemometry [14]. In general, these measurements reveal that the approximate solutions using the linearization method predict developments in velocity and pressure in the entrance region of rectangular ducts that are too rapid. Numerical methods that take account of the cross-stream velocities show better agreement with experimental measurements.

2. *Developing Velocity Profiles*

Yau and Tien [40] used an integral method to solve for the velocity distribution of a power law fluid in a duct formed by parallel plates, while Collins and Schowalter [42] used a patching procedure to solve the same problem. In the case of power law fluids in rectangular channels of finite aspect ratio, Curr *et al.* [57] and Chandrupatla [23] used a finite-difference procedure to solve for the velocity profile and pressure in the entrance region of a square duct. Representative developing velocity distributions for a Newtonian fluid and for a power law fluid with a power law index of 0.2 and 1/3 in laminar flow through parallel plates and a square duct are presented in Fig. 5a and b respectively, under the assumption that a uniform velocity profile exists at the entrance.

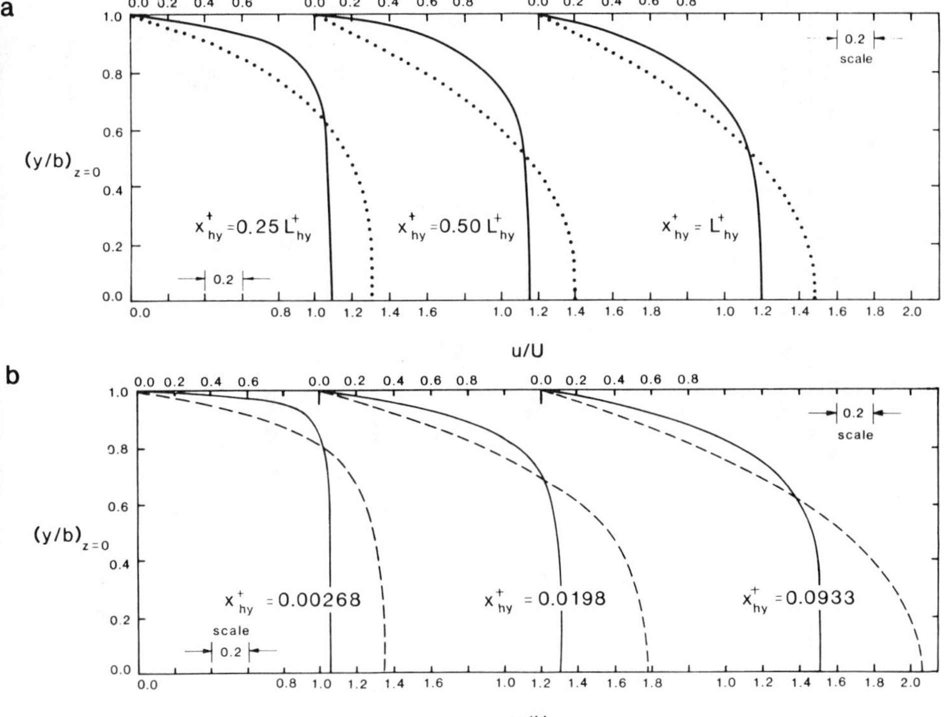

FIG. 5. Developing centerline velocity distribution for laminar flows of a power law fluid in plane parallel plates [42] (a) and a square duct [57] (b). In (a), $\alpha^* = 0$ and $n = 1.0$ (···) and 0.2 (——). In b, $\alpha^* = 1.0$ and $n = 1.0$ (– –) and $\frac{1}{3}$(——).

TABLE IX

SOLUTIONS FOR HYDRODYNAMICALLY DEVELOPING RECTANGULAR DUCT FLOW

Method	Investigator(s)	Aspect ratio	Comments
Integral	Schiller [33]	0	Parabolic velocity distribution in boundary layer and Bernoulli's equation in inviscid core
	Gupta [34]; Williamson [35]	0	Extended the method of Campbell and Slattery [36] to plane parallel plates
	Bhatti and Savery [37]	0	Refined approach of Ref. 33, employing parabolic velocity distribution in boundary layer, an integral form of the continuity equation, and the mechanical energy equation
	Naito and Hishida [38]; Naito [39]	0	Refined Ref. 33 by using a polynomial of fourth degree for the velocity profile in boundary layer
	Yau and Tien [40]	0	Extended the integral method to power law fluid in parallel plate duct
Patching (matching)	Schlichting [41]	0	Boundary layer theory upstream and perturbed fully developed downstream. Solutions are patched together at appropriate axial location
	Collins and Schowalter [42]; Roidt and Cess [43]	0 0	Refined Ref. 41 using more terms in the series describing the upstream and downstream velocity
	Collins and Schowalter [44]	0	Extended the method of Ref. 41 to power law fluid in parallel plate channel
Linearization	Han [45]	0	Extended Langhaar's method [46] to plane parallel plates
	Han [47]	0, 0.125, 0.25, 0.5, 0.75, 1.0	Extended Langhaar's method [46] to rectangular geometry

	Sparrow *et al.* [48]	0	Stretched coordinate in flow direction used for linearization
	Miller and Han [49]	0, 0.2, 0.5, 1.0	Modifies Han [45] by evaluating the pressure drop from mechanical energy equation
	Fleming and Sparrow [50]	0.2, 0.5	Applied stretched coordinate method [35] to rectangular duct flow
	Wiginton and Dalton [51]	0.2, 0.25, 0.5, 1.0	Slight modification of previous method [50]
Numerical finite difference	Bodoia and Osterle [52]	0	Linearized momentum equation; velocity and pressure drop calculated by finite-difference method
	Shah and Farnia [53]	0	Used finite-difference method of Patankar and Spalding [54]
	Patankar and Spalding [55]	1.0	A general numerical marching procedure for calculation of three-dimensional duct flow
	Caretto *et al.* [56]	0.2, 0.5, 1.0	Two marching procedures. One is similar to Ref. 55 but solved differently. Another procedure uses vorticity instead of the pressure as the main variable
	Curr *et al.* [57]	0.2, 0.5, 1.0	Above method applied to Newtonian fluid flow in rectangular ducts, and to power law fluid in square duct
	Carlson and Hornbeck [58]	1.0	Assume transverse velocities given by $(v/w) = (y/z)$. Pressure distribution evaluated from the integral continuity equation. Demonstrated accuracy of the method by solving Navier–Stokes equations and continuity equation for Re = 2000
	Chandrupatla [23]	1.0	Adopted the previous method [58] for power law fluid in square duct

3. *Incremental Pressure Drop*

a. *Newtonian fluids*

In the hydrodynamic entrance region, the pressure decreases as a result of the wall shear and the change in momentum. The pressure drop in this region may be treated as consisting of two components: (1) the pressure drop based on the fully developed flow and (2) an additional pressure drop due to the momentum change and accumulated increment in wall shear between the developing flow and the developed flow, designated as the incremental pressure drop number $K^*(x)$:

$$\Delta p^* = \frac{\Delta p(x)}{\rho U^2/2} = f_{\mathrm{fd}} \frac{x}{(D_{\mathrm{h}}/4)} + K^*(x) \tag{2.20}$$

Introducing Eqs. (2.15)–(2.19), we may rewrite Eq. (2.20):

$$\Delta p^* = (f_{\mathrm{fd}}\,\mathrm{Re}^+)4x^+_{\mathrm{hy}} + K^*(x^+_{\mathrm{hy}}) = 64x^*_{\mathrm{hy}} + K^*(x^*_{\mathrm{hy}}) \tag{2.21}$$

where $x^+_{\mathrm{hy}} = x/(D_{\mathrm{h}}\,\mathrm{Re}^+)$ and $x^*_{\mathrm{hy}} = x/(D_{\mathrm{h}}\,\mathrm{Re}^*)$. Equation (2.21) implies that the dimensionless pressure drop in the fully developed region has a constant slope independent of the shape of the duct with the value of K^* dependent on duct geometry, if the coordinate x^*_{hy} is used.

In the case of Newtonian fluids flowing through rectangular channels, the experimental measurements of Beavers *et al.* [59] constitute the most comprehensive data set available. Figure 6 recasts these results in the form of K^* versus x^*_{hy} with the eight solid curves representing experimental data for the eight aspect ratio rectangular ducts studied. For each aspect ratio, the K^* values are compared with the results for the square duct. In general, as the aspect ratio decreases, the K^* value decreases at a fixed x^*_{hy} position. The limiting value of K^*, designated as $K^*(\infty)$ is presented on Fig. 7, where the results of Beavers *et al.* are compared with the measurements of Sparrow *et al.* [15] and with several analytical predictions [49–51, 57, 60]. In general, the analytical predictions are higher than the experimental measurements. This may be due to the neglect of the transverse velocities in the linearization method used in most analyses.

b. *Non-Newtonian fluids*

The incremental pressure drop parameter K^* has been reported for power law fluids for the two limiting rectangular duct geometries, the parallel plates channel, and the square duct. Figure 8 reveals K^* as a function of x^*_{hy} as reported by Yau and Tien [40], for plane parallel plates with n values of 0.25, 0.50, and 0.75. Also shown on the figure is the prediction of Bodoia for $n = 1.0$

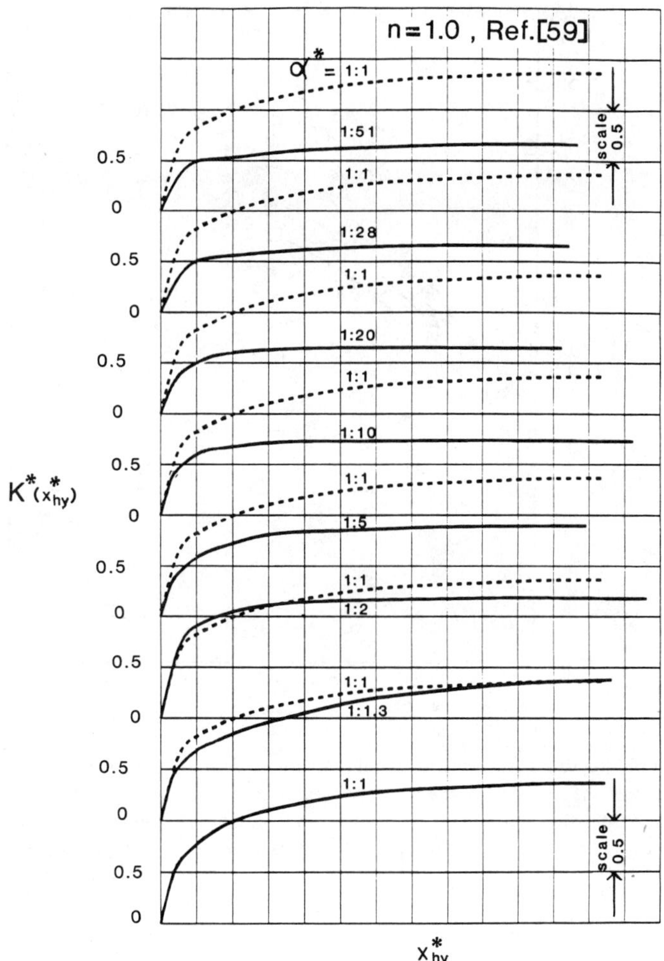

FIG. 6. Experimental measurements of the incremental pressure drop factor, K^*, for laminar flows of a Newtonian fluid in rectangular channels of various aspect ratios.

[52]. As expected, the value of K^* decreases with decreasing values of n, with the limiting value of K^* being zero at $n = 0$.

A similar trend occurs in the square duct geometry as shown in Fig. 9. The results of Chandrupatla [23] and Curr *et al.* [57], which cover the range of n from 0.333 to 1.0, are seen to be in good agreement.

The limiting values of the incremental pressure drop number, $K^*(\infty)$, are provided on Fig. 10 as a function of the power law index n, for $\alpha^* = 0$ and 1.

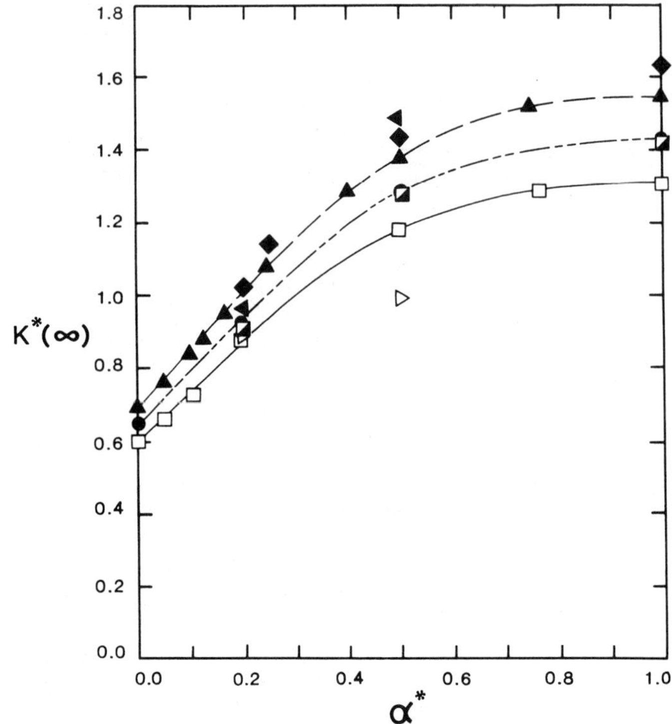

FIG. 7. Limiting values of the incremental pressure drop factor, $K^*(\infty)$, for Newtonian fluids as a function of the rectangular channel aspect ratio. Experimental values from Beavers *et al.* [59] (□) and Sparrow *et al.* [15] (▷). Analytical values from Wiginton and Dalton [51] (◆), Miller and Han [49] (●), Fleming and Sparrow [50] (◀), Lundgren *et al.* [60] (▲), and Curr *et al.* [57] (◪).

The trends are clear, with $K^*(\infty)$ increasing monotonically from a value of zero at $n = 0$ to the Newtonian value at $n = 1$. A reasonable estimate of $K^*(\infty)$ for any aspect ratio and any value of n can be accomplished by combining Fig. 7, which reveals $K^*(\infty)$ for $n = 1$ for any value of α^*, and Fig. 10.

4. *Hydrodynamic Entrance Length*

The hydrodynamic entrance length L_{hy} is defined as the duct length required to achieve velocity or pressure development within a given range (for example, 1 or 2%) of the corresponding fully developed values. It is well known that the development of the velocity field is substantially slower than

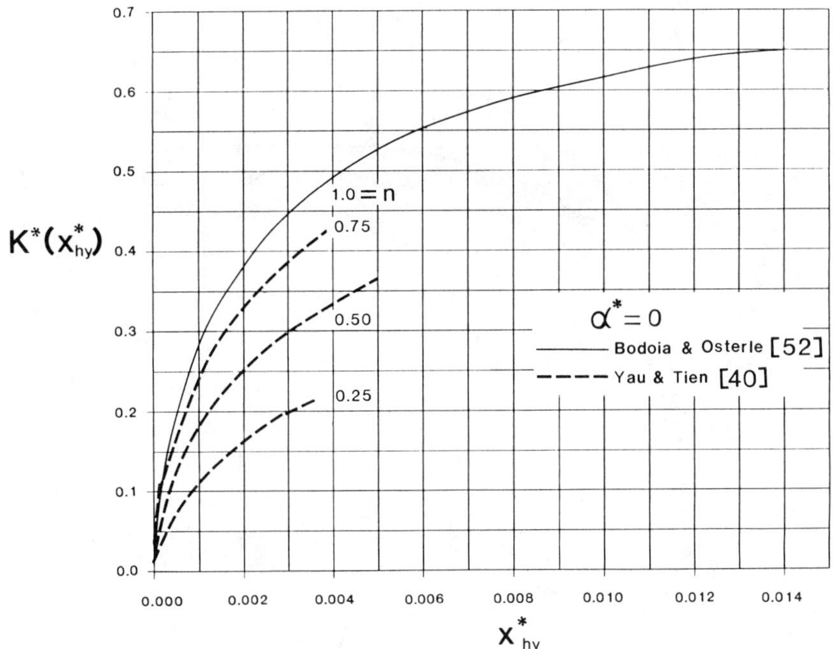

FIG. 8. Incremental pressure drop factor K^* for laminar flow of a power law fluid in parallel plates channel.

that of the pressure field. Furthermore, the central portion of the velocity profile generally develops more slowly than the portion of the velocity profile adjacent to the wall. Therefore, the hydrodynamic entrance length is defined here as the length necessary to achieve a centerline velocity equal to 98 or 99% of the fully developed value.

a. *Newtonian fluids*

The hydrodynamic entrance lengths for rectangular ducts reported by various investigators are compared in Fig. 11, which shows L_{hy}^+ as a function of α^*. Here the 99% criterion was used.

b. *Non-Newtonian fluids*

Collins and Schowalter [44], using an integral method, reported L_{hy}^+ for a power law fluid in a parallel plate. They determined L_{hy}^+ based on 98% development of the central velocity, and the results are shown in Fig. 12. Their

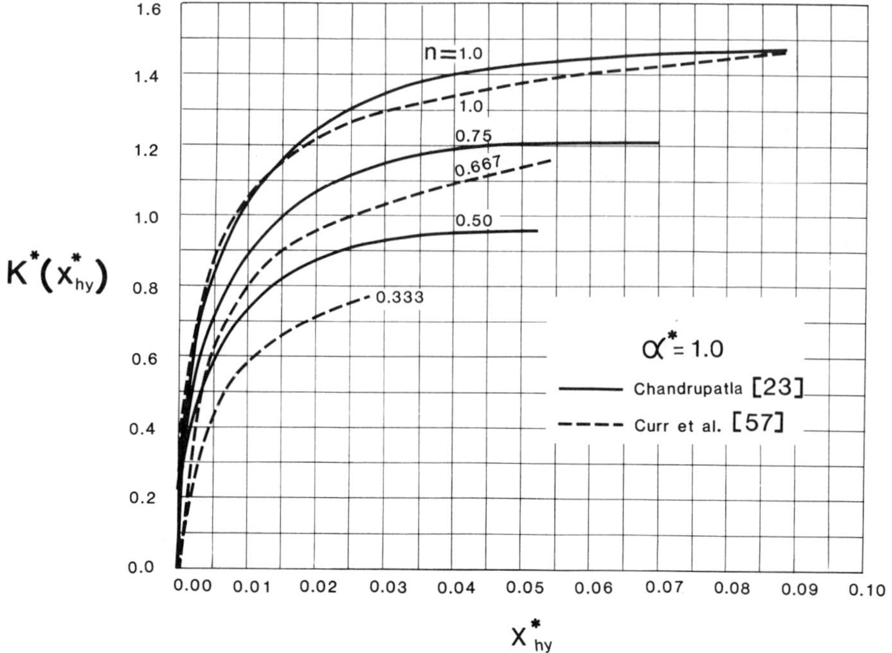

FIG. 9. Incremental pressure drop factor K^* for laminar flow of a power law fluid in a square duct.

value of $L_{hy}^+ = 0.00875$ for a Newtonian fluid is shorter than Bodoia's numerical result 0.011 based on 99% development of the central velocity. The hydrodynamic entrance length L_{hy}^+ is seen to increase with a decrease in the power law index n down to a certain value of about 0.3. Below this value, L_{hy}^+ decreases with an decrease in the power law index, approaching zero as the power law index becomes zero, that is, slug flow. The entrance length results of Yau and Tien [40], who also used an integral method, are shown in Fig. 12, where it may be seen that the Yau and Tien predictions are considerably lower than those of Collins–Schowalter, especially at the intermediate values of n.

Chandrupatla [23] solved the hydrodynamic entrance problem of a power law fluid in a square duct numerically and determined L_{hy}^+ based on 99% development of the central velocity. The results shown in Fig. 12 reveal that L_{hy}^+ increases with a decrease in the power law index down to 0.5. However, his prediction of 0.07 for a Newtonian fluid is somewhat shorter than the value of 0.09 shown on Fig. 11.

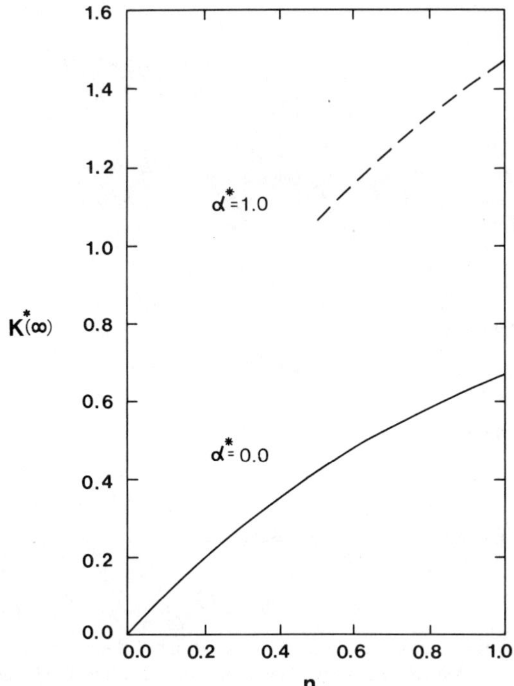

FIG. 10. Limiting values of the incremental pressure drop factor, $K^*(\infty)$, for power law fluids in laminar flow through rectangular channels. Values are from Collins and Schowalter [42] (——) and Chandrupatla [23] (– –).

C. THERMALLY DEVELOPED FLOW

1. *Newtonian Fluids*

An extensive review of heat transfer to Newtonian fluids in laminar flow through rectangular channels is presented by Shah and London [1] and Shah and Bhatti [2]. A very brief summary of their results is given here since Newtonian fluids represent a special limiting case of power law fluids.

1. Specified constant wall temperature on all walls: T boundary condition.

Shah and London [1] propose the following formula, which correlates their more exact results within $\pm 0.1\%$.

$$\mathrm{Nu}_T = 7.541(1 - 2.610\alpha^* + 4.970\alpha^{*2} - 5.119\alpha^{*3} + 2.702\alpha^{*4} - 0.548\alpha^{*5}) \tag{2.22}$$

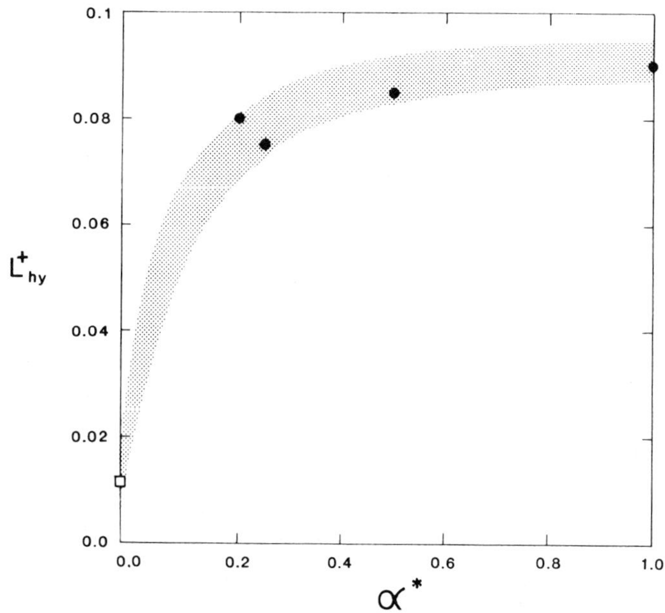

FIG. 11. Hydrodynamic entrance length of Newtonian fluids in laminar flow through rectangular channels. Values are from Wiginton and Dalton [51] (●) and Shah and London [1] (□).

2. Specified heat flux per unit length on all walls, peripheral wall temperature constant at any axial position: *H*1 boundary condition.

Shah and London [1] propose the following approximation to correlate their more exact results within $\pm 0.03\%$.

$$\mathrm{Nu}_{H1} = 8.235[1 - 2.0421\alpha^* + 3.0853\alpha^{*^2} - 2.4765\alpha^{*^3} + 1.0578\alpha^{*^4} - 0.1861\alpha^{*^5})\tag{2.23}$$

3. Specified heat flux on all walls: *H*2 boundary condition.

This case was studied by Cheng [61], Sparrow and Siegel [62], Iqbal *et al.* [63], and Shah [64]. Results are available for aspect ratios from 0.1 to 1.0 for Newtonian fluids [1].

Figure 13 summarizes the fully developed Nusselt numbers for the three boundary conditions, *T*, *H*1, and *H*2, as a function of the aspect ratio when all boundary walls are heated. For completeness, the values of $f\,\mathrm{Re}$ are also presented. Generally Nu_{H1} and Nu_T reach their maximum values for the plane parallel plate case, (8.235 and 7.541, respectively) decreasing monotonically

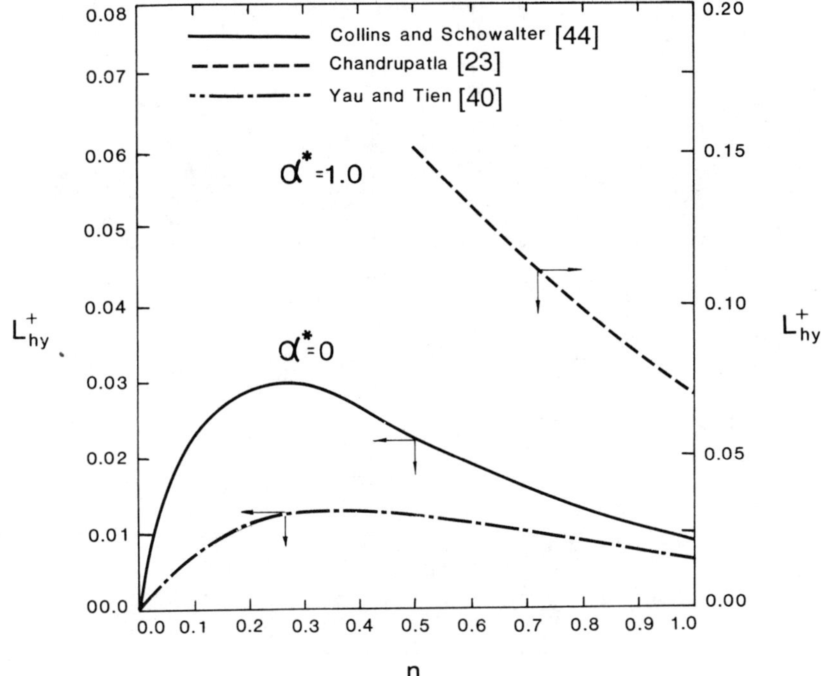

FIG. 12. Hydrodynamic entrance length of power law fluids in laminar flow through rectangular channels.

with the aspect ratio, with the minimum value corresponding to the square duct geometry (3.608 and 2.976, respectively). The $H1$ boundary condition yields the highest Nusselt values of the three cases considered for any aspect ratio rectangular duct. The behavior of Nu_{H2} is completely different than Nu_T or Nu_{H1}, reflecting the fact that the wall temperatures in the neighborhood of the corners reach high values for the $H2$ condition [62]. Accordingly, the value of Nu_{H2} decreases slightly with an increasing aspect ratio from a value of 2.95 at $\alpha^* = 0.1$ to 2.93 at $\alpha^* = 0.2$, then increasing to a value of 3.091 at $\alpha^* = 1.0$.

In many practical applications involving a rectangular duct, other combinations of wall heating may be encountered. In the case of a Newtonian fluid, the recomputed values of Schmidt and Newell [65], as reported by Shah and London [1], are presented in Fig. 14 for the case where one or more walls are held at constant temperature while the remaining walls are adiabatic.

A comparable curve for the case where the $H1$ condition prevails on one or more walls with the other wall adiabatic is shown as Fig. 15. From Figs. 14 and 15 it is interesting to note for a constant finite value of the aspect ratio that the

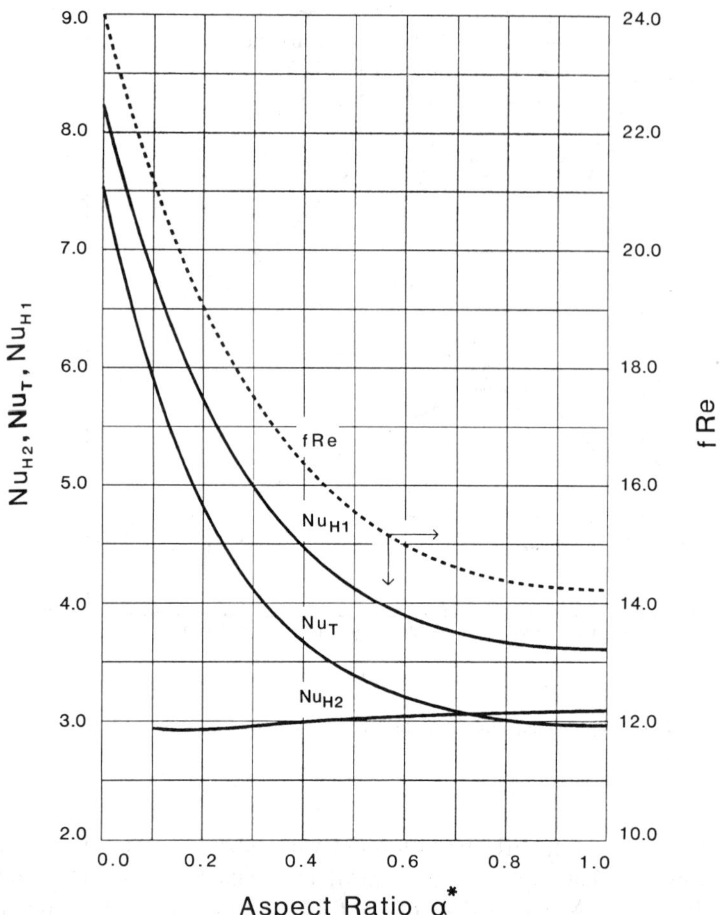

FIG. 13. Fully established laminar Nusselt values (Nu_T, Nu_{H1}, Nu_{H2}) as a function of the aspect ratio. Values are from Shah and London [1]; $n = 1$.

highest Nusselt number for both the T and $H1$ boundary conditions occurs for the case when the two long walls are heated (2L), and that the one long wall heated Nusselt number (1L) falls below the Nusselt value for four heated walls (4).

2. *Non-Newtonian Fluids*

A number of analytical results are available for fully developed Nusselt values for the laminar flow of power law fluids in rectangular channels. A

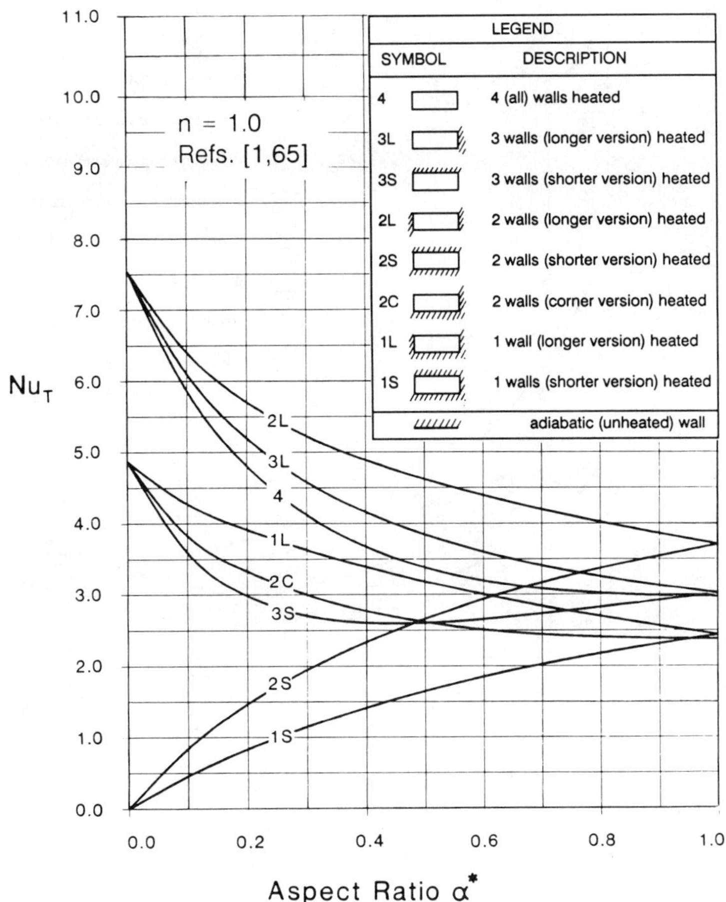

FIG. 14. Fully established laminar Nusselt values for T boundary condition as a function of the aspect ratio for different combinations of heated and adiabatic walls.

listing of these solutions is presented in Table X. As already mentioned, the Newtonian results ($n = 1$) are available for the T, $H1$, and $H2$ boundary conditions for the complete range of aspect ratios. Another limiting case for which many results are available is the slug or plug flow condition, which corresponds to $n = 0$ [66–69]. At other values of n, results are available for plane parallel plates [40, 70–80] and for the square duct [23, 81–83].

In the case of plane parallel plates, the fully developed laminar heat transfer results for a power law fluid for the constant wall temperature boundary

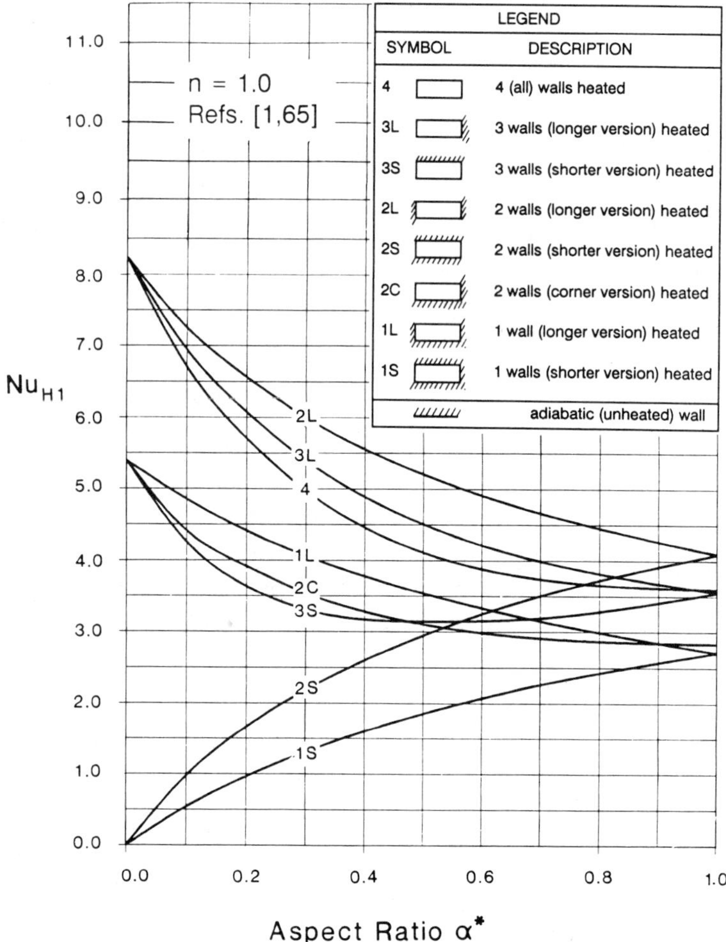

FIG. 15. Fully established laminar Nusselt values for $H1$ boundary condition as a function of the aspect ratio for different combinations of heated and adiabatic walls.

condition (T condition) were obtained by extending the Graetz method. The Nusselt number for the constant wall temperature in the thermally developed region was given by the following equation [75]:

$$\mathrm{Nu}_T = \frac{4(n+1)}{2n+1}\zeta_1^2 \tag{2.24}$$

where ζ_1 is the first eigenvalue for the following boundary value problem

$$\frac{d^2\Psi_i}{d\xi^2} + \zeta_i^2(1 - \xi^{(n+1)/n})\Psi_i = 0 \tag{2.25}$$

with

$$\Psi_i(1) = 0 \quad \text{and} \quad \Psi_i'(0) = 0 \tag{2.26}$$

where $\xi = y/b$. The values of the Nusselt number for Newtonian fluids and slug flow are 7.541 and 9.870, respectively. An approximate solution for the constant wall temperature was given by Skelland [19]:

$$\mathrm{Nu}_T = \frac{12}{\phi}\frac{n+1}{2n+1} \tag{2.27}$$

where

$$\phi = \frac{5}{4} - \frac{2n}{2n+1} + \frac{3n}{4n+1} - \frac{n}{5n+1} \tag{2.28}$$

However, this expression does not reduce to the Newtonian value and predicts higher values than other analytical solutions. Other methods have been used to analyze the heat transfer of a power law fluid in the thermal entrance region of a parallel plate duct, as shown on Table X.

The fully developed Nusselt number for a power law fluid in the parallel plates geometry is easily determined for the H boundary condition (note that $H1$ and $H2$ are identical for the parallel plates case). The solution of the energy equation with the fully developed velocity distribution [Eq. (2.9)] leads to the following expression for the fully established Nusselt number [19].

$$\mathrm{Nu}_H = (4/\psi)[(n+1)/(2n+1)] \tag{2.29}$$

where

$$\psi = (1/3)\left(\frac{5n+1}{4n+1}\right) - \left(\frac{n}{2n+1}\right)\left(\frac{n}{3n+1}\right) + \left(\frac{n}{3n+1}\right)\left(\frac{n}{4n+1}\right)\left(\frac{n}{5n+2}\right) \tag{2.30}$$

This expression for the Nusselt number for a power law fluid with constant heat flux reduced to the Newtonian value of 8.235 with n equal to unity and the slug flow value of 12 when n is equal to zero.

In the case of the square duct, the most extensive analysis has been carried out by Chandrupatla [23], who treated the T, $H1$, and $H2$ boundary conditions. Chandrupatla's predictions cover the range of n from 0.0 to 1.0, Javeri [66, 69] presents results for the limiting case of $n = 0$.

Figure 16 presents the fully established Nusselt values for the T boundary condition as a function of the power law index n with the aspect ratio α^* as a parameter. As noted, many predictions are shown for the plane parallel plates

TABLE X

SUMMARY OF HEAT TRANSFER SOLUTIONS FOR POWER LAW FLUIDS IN LAMINAR FLOW THROUGH RECTANGULAR CHANNELS

Investigator	Aspect ratio α^*	Power law index n	Thermal boundary condition	Comments
Tien [70]	0	0.25, 0.50, 0.75	T	Velocity profile fully developed at start of heating; Schechter's approximate velocity profile used
Suckow *et al.* [71]	0	Any value	T	Velocity profile fully developed at start of heating; separation of variables solution
Vlachopoulos and Keung [72]	0	0.25, 0.50, 1.0, 2.0	T	Velocity profile fully established at start of heating; viscous dissipation also considered
Kwant and Van Ravenstein [73]	0	0.333, 0.50, 0.75, 1.0	T	Velocity profile fully developed at start of heating; considers influence of variable consistency index
Richardson [74]	0	0.5, 1.0, 2.0	T	Velocity profile fully established; considers effects of viscous dissipation
Cotta and Osizik [75]	0	0.333, 1.0, 3.0	T	Velocity profile fully established; considers effects of viscous dissipation
Javeri [67]	0	0, 1.0	T	Velocity profile fully established; considered MHD effects

Javeri [68]	0	0, 1.0	T	Velocity profile fully established; also considered case where local wall heat flux is linear function of local wall temperature
Gottifredi and Flores [76]	0	0.5, 1.0, 2.0	T	Velocity profile fully established; considers effects on viscous dissipation
Lin and Hsu [77]	0	0.5, 1.0, 2.0	T	Assumes consistency index to be temperature dependent; velocity profile is temperature sensitive; thermally developing flow with viscous dissipation
Yau and Tien [40, 78]	0	0.25, 0.50, 0.75	T	Simultaneous development of velocity and temperature
Lin [79]; Lin and Shah [80]	0	0.5, 0.75, 1.0, 1.5	T, H	Simultaneous development of velocity and temperature fields for $\mathrm{Pr} = 0.1, 1, 10$
Javeri [66, 69]	0, 0.125, 0.25, 0.50 1.0	0, 1.0	T	Velocity profile fully established; also considered other thermal boundary conditions
Chandrupatla [23]; Chandrupatla and Sastri [81]	1.0	0.5–1.0	$T, H1\ H2$	Velocity profile fully developed at start of heating
Chandrupatla [23]; Chandrupatla and Sastri [82]	1.0	0.5, 1.0	T	Simultaneous development of velocity and temperature fields for $\mathrm{Pr} = 0, 0.1, 1, 10, \infty$
Lawal and Majumdar [83]	1	0.5, 1.0, 1.25	T	Simultaneous development of velocity and temperature

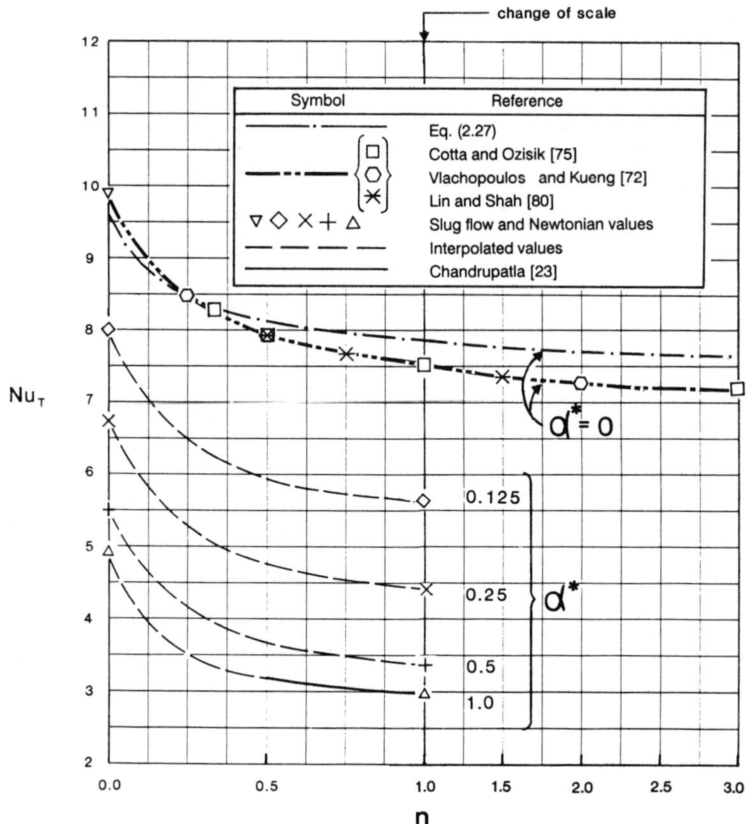

FIG. 16. Fully established laminar Nusselt values for T boundary condition as a function of the power law index for various values of the aspect ratio.

case covering the range of n values from 0 to 3. The corresponding Nusselt number decreases rather rapidly from a value of 9.87 at $n = 0$ to 7.94 at $n = 0.5$, then decreases more slowly to a value of 7.54 at $n = 1.0$

In the case of the square duct geometry, the Nusselt number also undergoes a large decrease from $n = 0$ to $n = 0.5$ (from 4.918 to 3.184), with the change from $n = 0.5$ to $n = 1.0$ being much more modest (from 3.184 to 2.975). Against this background, with the Newtonian and slug flow limits available for all aspect ratios, it is a simple exercise to estimate the fully established Nusselt numbers for the T boundary condition for any aspect ratio for any power law index, n. The dashed lines in Fig. 16 represent such estimates.

Turning next to the $H1$ boundary condition, Fig. 17 presents the fully developed Nusselt number predictions for the plane parallel plates case covering the power law index range from 0 to 3. The available predictions for the square duct, with n varying from 0.5 to 1 are also shown. As in the case of the T boundary condition, the slug flow and Newtonian flow limits are also available for the $H1$ condition for all aspect ratios. As in the constant temperature case, a large decrease in the Nusselt number occurs for any aspect ratio when n increases from 0 to 0.5, and the subsequent decrease from 0.5 to 1.0 is more gentle. The dashed lines represent estimates of the fully established Nusselt values for intermediate values of the aspect ratio and power law index.

In the case in which the heat flux is constant on all boundary walls, the $H2$ condition, only results for the square duct and the parallel plates channel are available as shown in Fig. 18. As noted earlier in Fig. 13, the fully established Nusselt number for the $H2$ condition for Newtonian flow, $n = 1$, is

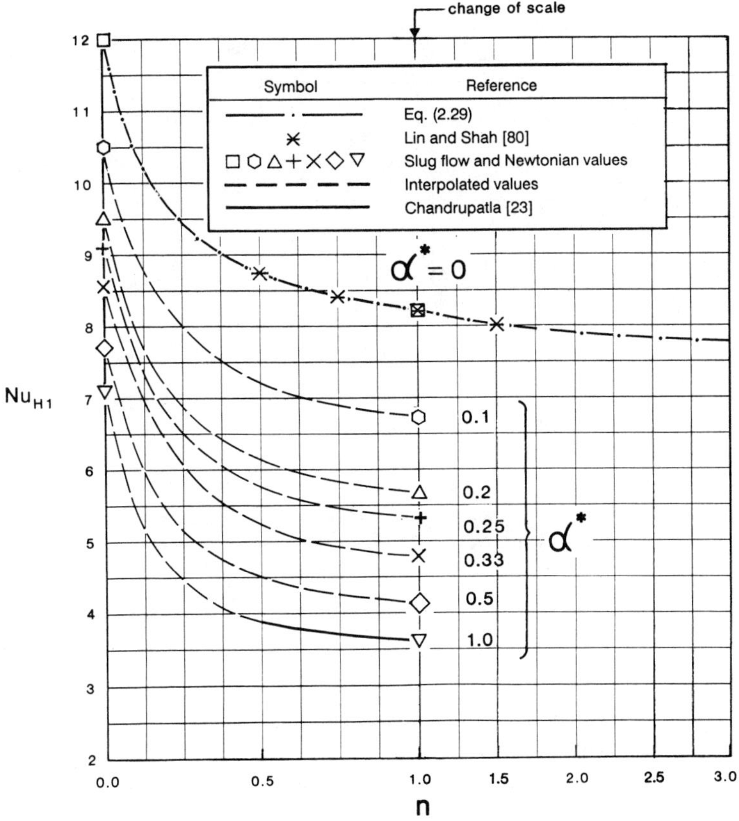

FIG. 17. Fully established laminar Nusselt values for $H1$ boundary condition as a function of the power law index for various values of the aspect ratio.

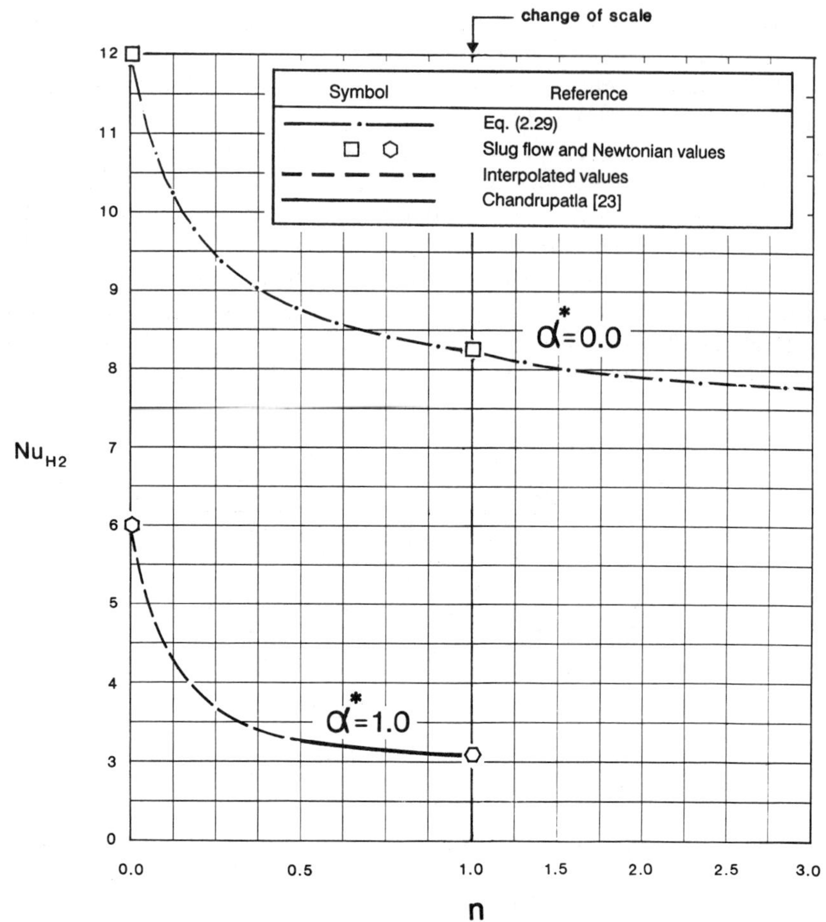

FIG. 18. Fully established laminar Nusselt numbers for $H2$ boundary condition as a function of the power law index for plane parallel plates and for a square channel.

approximately 3 over the range of the aspect ratio from 0.1 to 1.0. It is suggested that the results given in Fig. 18 for $\alpha^* = 1.0$ be used as an estimate of the Nusselt number for other aspect ratios down to a value of $\alpha^* = 0.1$.

The fully established Nusselt numbers for the limiting case of slug flow, $n = 0$, are available for a wide range of conditions involving the T or $H1$ boundary condition on one or more of the boundary walls. Relatively few solutions are available for the $H2$ boundary condition. Given the fact that the velocity is uniform over the duct cross section for $n = 0$, the analytic solutions are equivalent to the corresponding solutions of the heat conduction equation. Accordingly, it is possible to take advantage of symmetry conditions to generate new solutions from the baseline T, $H1$, and $H2$ conditions. For example, the T solution for the $\alpha^* = 0.5$ duct leads directly to two solutions: (1) $\alpha^* = 1$, with three walls at constant temperature T while the fourth wall is insulated $T(3)$; and (2) $\alpha^* = 0.25$, with one long wall insulated while the other three walls are at constant temperature $T(3S)$. In turn, these new solutions can be used to generate results for the T condition on two adjacent walls [i.e., $T(2C)$] whereas the opposite walls are insulated. Table XI tabulates the resulting fully established Nusselt number for a wide range of thermal conditions for the rectangular duct geometry. Taken together with Fig. 14 and 15, the upper and lower limits on fully established heat transfer to a pseudoplastic fluid flow through rectangular channels are established.

D. THERMALLY DEVELOPING FLOWS

1. *Newtonian Fluids*

Figures 19 and 20 show the thermally developing Nusselt number values for the T and $H1$ boundary conditions, respectively, for Newtonian fluids flowing through rectangular channels under conditions where the velocity profiles are

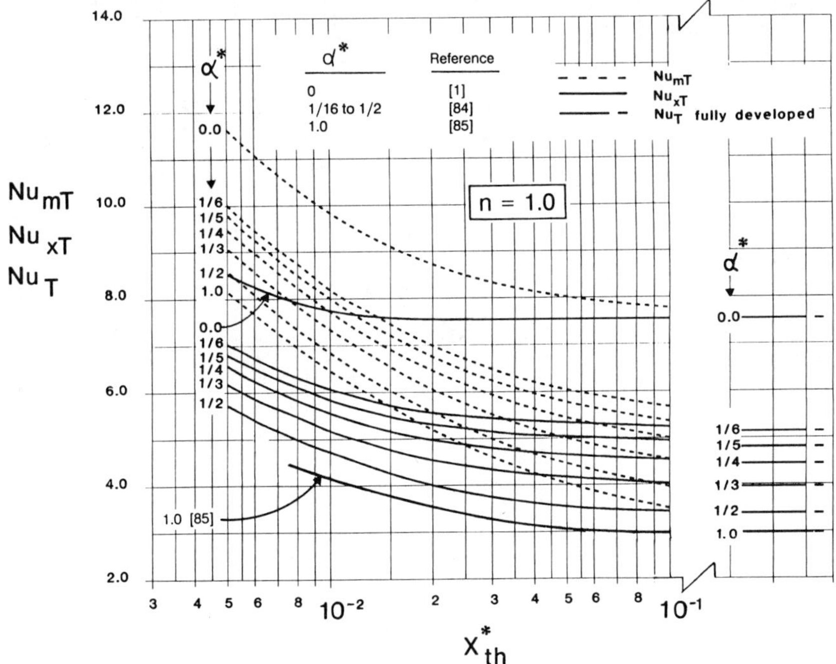

FIG. 19. Thermally developing Newtonian Nusselt numbers as a function of x^*_{th} and the aspect ratio for the T boundary condition.

TABLE XI

FULLY DEVELOPED NUSSELT VALUES FOR SLUG FLOW, $n = 0$, IN RECTANGULAR CHANNELS[a]

α^*	Nu_T				Nu_{H1}				Nu_{H2}			
	2b / 2a		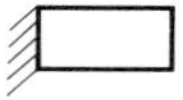	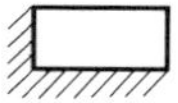			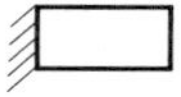	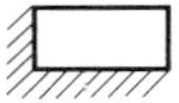			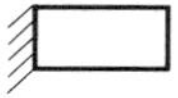	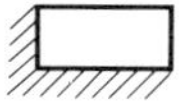
1.0	4.94	4.11	4.11	2.47	7.11	5.82	5.82	3.56	5.99			3.00
0.6666					7.36	5.07	6.84	3.68				
0.5	5.48	3.29	5.62	2.74	7.77	4.74	7.60	3.88		3.99		
0.4					8.18	4.62	8.17	4.09				
0.3333					8.55	4.60	8.63	4.27				
0.25	6.74	3.29	7.19	3.37	9.12	4.66	9.27	4.56				
0.20					9.54	4.77	9.70	4.77				
0.1666					9.86	4.88	10.03	4.93				
0.125	7.99	3.74		4.00	10.30	5.06	10.46	5.15				
0.10					10.58	5.20	10.728	5.29				
0.0625		4.23			11.07	5.46	11.18	5.54				
0.05					11.24	5.54	11.34	5.62				
0.0	9.87	4.94	9.87	4.94	12.00	6.00	12.00	6.00	12.00	6.00	12.00	6.00

[a] ////////, adiabatic wall.

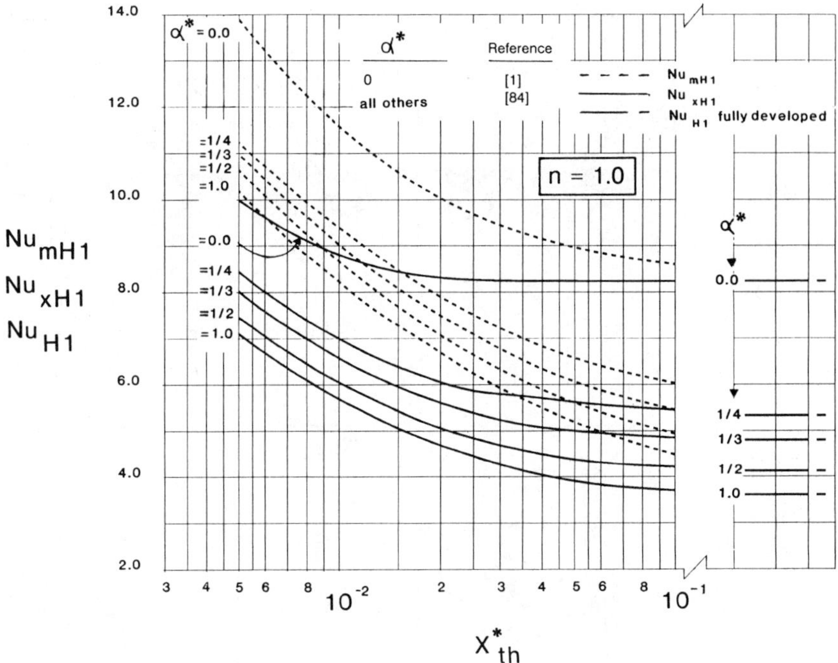

FIG. 20. Thermally developing Newtonian Nusselt numbers as a function of x^*_{th} and the aspect ratio for the $H1$ boundary condition.

fully established at the start of heating. The corresponding fully established Nusselt values are also given in the figures. The majority of these results are from Wibulswas [84]. The exceptions are the values for $\alpha^* = 0$, where the result tabulated in Shah and London [1] are given, and the values of Nu_{xT} for $\alpha^* = 1$, where the results of Lyczkowski *et al.* [85] are shown, inasmuch as Wilbulswas appears to be in error in this particular case. Approximate equations, which are sufficiently accurate for engineering purposes, are also available in Shah and London [1] for the plane parallel plates duct and the square channel.

2. *Non-Newtonian Fluids*

The thermally developing Nusselt numbers for the limiting case of slug flow, $n = 0$, have been calculated for the plane parallel plates case [2, 67, 68] and for the square duct [23, 66, 69]. Figure 21 presents Nusselt values for the T, H, $T(1)$, and $H(1)$ boundary conditions for the plane parallel plates case while Fig. 22 presents the T, $H1$, and $H2$ Nusselt values for the square duct geometry.

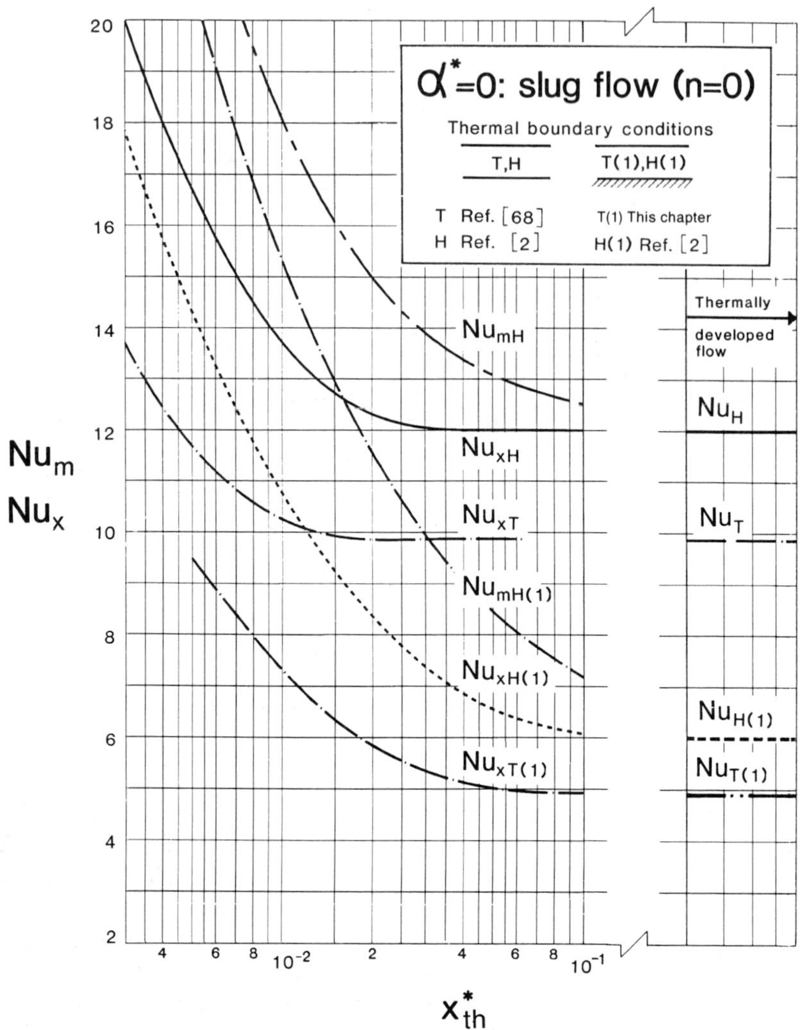

FIG. 21. Thermally developing Nusselt numbers as function of x^*_{th} for plane parallel plates, $n = 0$.

Also, in Table XII, calculated values of thermally developing Nusselt numbers for slug flow and H and $H(1)$ boundary conditions are presented. Against this background of establishing the limiting Nusselt values for $n = 0$ and $n = 1$ for the plane parallel plates case and the square duct, there are a number of analytical solutions for intermediate values of n, as listed in Table X.

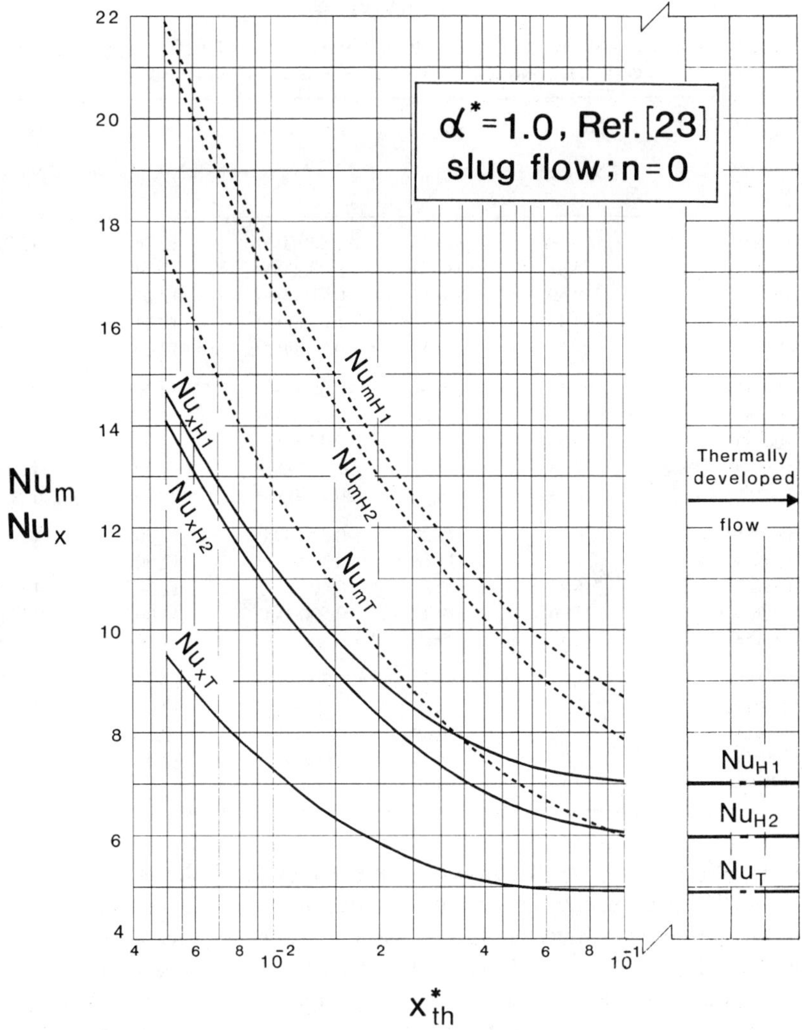

FIG. 22. Thermally developing Nusselt numbers as a function of x^*_{th} for a square duct, $n = 0$.

Tien [70] analyzed the thermal entry length problem of a power law fluid in a parallel plate duct having constant wall temperature boundary condition by applying the Graetz method. In his study, the fully developed velocity profile was obtained using a variational method developed by Schechter [20] instead of employing the exact velocity profile. By using this approximation, the restriction due to the power law index in the velocity profile is removed. Tien

TABLE XII

THERMALLY DEVELOPING SLUG FLOW ($n = 0$) IN PARALLEL PLATE DUCT WITH BOTH OR ONE WALL HEATING FOR H AND $H(1)$ BOUNDARY CONDITIONS

x^*_{th} (both walls heated) $x^*_{th}/4$ (one wall heated)	Nu_{xH} $2Nu_{xH(1)}$	Nu_{mH} $2Nu_{mH(1)}$	x^*_{th} $x^*_{th}/4$	Nu_{xH} $2Nu_{xH(1)}$	Nu_{mH} $2Nu_{mH(1)}$
0.0000005	1168.40	1460.35	0.0300000	12.06	13.91
0.0000010	875.61	1184.54	0.0350000	12.03	13.61
0.0000050	399.50	590.83	0.0400000	12.01	13.39
0.0000100	283.43	422.11	0.0450000	12.01	13.22
0.0000500	128.55	191.52	0.0500000	12.00	13.09
0.0001000	91.88	136.54	0.0600000	12.00	12.89
0.0005000	43.05	63.20	0.0700000	12.00	12.75
0.0010000	31.56	45.89	0.0800000	12.00	12.65
0.0020000	23.55	33.73	0.0900000	12.00	12.58
0.0030000	20.08	28.40	0.1000000	12.00	12.51
0.0040000	18.06	25.26	0.2000000	12.00	12.25
0.0050000	16.73	23.15	0.3000000	12.00	12.16
0.0060000	15.77	21.61	0.4000000	12.00	12.12
0.0070000	15.06	20.43	0.5000000	12.00	12.09
0.0080000	14.51	19.49	0.6000000	12.00	12.08
0.0090000	14.07	18.73	0.7000000	12.00	12.07
0.0100000	13.72	18.10	0.8000000	12.00	12.06
0.0150000	12.72	16.08	0.9000000	12.00	12.00
0.0200000	12.32	15.00	1.0000000	12.00	12.00
0.0250000	12.14	14.34			

reported the local Nusselt number for the power law index n equal to 0.25, 0.5, and 0.75 graphically. This extended solution of the Graetz–Nusselt problem for a power law fluid has been refined by using an exact velocity profile, Eq. (29) [71–76]. Recently, Cotta and Ozisik [75] determined the eigenvalues of the extended Graetz solution by the sign–count numerical method. Their local Nusselt number for Newtonian fluids are in excellent agreement with the results calculated by Shah [86] using the Leveque solution even at x^+_{th} equal to 10^{-6}. Cotta and Ozisik reported the local Nusselt values numerically and graphically for the power law index n of $\frac{1}{3}$, 1, and 3.

Richardson [74] and Gottifredi and Flores [76] extended the Leveque approach for a power law fluid in the thermal entrance region of a parallel plate duct having a constant wall temperature boundary condition. These extended Leveque solutions can be applied at the region very near to the entrance of duct. As pointed out by Cotta and Ozisik [75], even the second-order Leveque solution by Richardson [74] becomes smaller than the exact solution at x^*_{th} larger than 2×10^{-4} and the magnitude of the error increases with increasing x_{th}^* and power law index n. The detailed comparison between these two different types of solutions can be found in their papers.

Vlachopoulos and Keung [72] and Kwant and Van Ravenstein [73] analyzed numerically the thermal entrance length problem of a power law fluid in a parallel plate duct having constant wall temperature. They reported the numerical results of the local Nusselt number for several values of the power law index n.

Representative values of the thermally developing Nusselt numbers for the T boundary condition are given on Fig. 23 as a function of the power law index for the plane parallel plates case, $\alpha^* = 0$. Also shown are local Nusselt numbers for the square duct geometry for the T boundary condition as reported by Chandrupatla [23]. As expected, the higher Nusselt values are associated with the parallel plates geometry. Also, for a given aspect ratio, the local heat transfer coefficient increases with a decreasing n value.

In comparison with the T boundary condition, there are relatively few solutions available for the H boundary condition. Shah and Bhatti [2] report thermally developing Nusselt numbers for the slug flow case, $n = 0$, for the parallel plates geometry; while Shah and London [1] give comparable values for the Newtonian case, $n = 1$. Kwack [87] reports values of Nu_H for values of n equal to 0.5 and 2.0. For the square duct, Chandrupatla [23] presents values of Nu_{xT}, Nu_{xH1}, and Nu_{xH2} for thermally developing flow. These results for thermally developing flows subject to the H boundary condition are presented on Fig. 24.

The laminar heat transfer results in the thermally developing region in circular duct for power law fluids may be accurately expressed by the following asymptotic relationships [10]:

$$\mathrm{Nu}_{xH} = 1.30[(3n + 1)/(4n)]^{1/3}\,\mathrm{Gz}^{1/3} \qquad \text{for} \qquad \mathrm{Gz} > 100 \tag{2.31}$$

$$\mathrm{Nu}_{xT} = 1.08[(3n + 1)/(4n)]^{1/3}\,\mathrm{Gz}^{1/3} \qquad \text{for} \qquad \mathrm{Gz} > 150 \tag{2.32}$$

The term $[(3n + 1)/(4n)]^{1/3}$ in Eqs. (2.31) and (2.32) comes from the fact that the Nusselt number is proportional to the one-third power of the velocity gradient at the duct wall. The average velocity gradient at rectangular (or circular or parallel plates) duct walls may be expressed as

$$\dot{\gamma}_{\mathrm{w,av}} = [(a^* + b^*n)/n](8U/D_{\mathrm{h}}) = \begin{cases} [(1 + 3n)/(4n)](8U/D_{\mathrm{h}}) & \text{for circle} \\ [(1 + 2n)/(2n)](8U/D_{\mathrm{h}}) & \text{for parallel plates} \end{cases} \tag{2.33}$$

Therefore, the entrance region Nusselt numbers for power law non-Newtonian fluids in a rectangular duct may be expressed as

$$\mathrm{Nu}_x(\alpha^*, n) = [(a^* + b^*n)/n]^{1/3}\,\mathrm{Nu}_x(\text{circle}, n = 1) \tag{2.34}$$

where $\mathrm{Nu}_x(\text{circle}, n = 1)$ is the corresponding Newtonian circular duct correlation, that is, Eqs. (2.31) or (2.32).

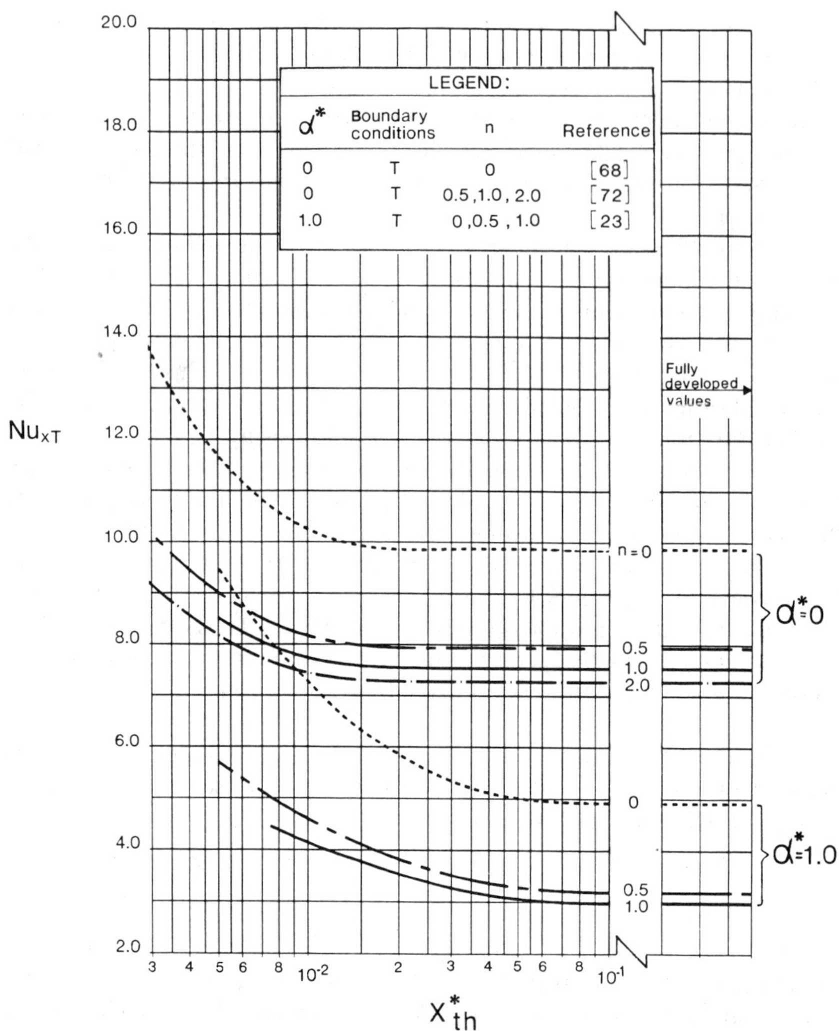

FIG. 23. Thermally developing Nusselt numbers as a function of x_{th}^* and the aspect ratio for T boundary condition, $n = 0$, 0.5, 1.0, and 2.0.

For cases when it is impossible to match the thermal boundary condition of a rectangular duct and a circular one (i.e., not all walls are heated, etc.), it is necessary to recast Eq. (2.34) in the following form:

$$\text{Nu}_x(\alpha^*, n) = \{[(a^* + b^*n)]/[(a^* + b^*)n]\}^{1/3}\text{Nu}_x(\alpha^*, n = 1) \quad (2.35)$$

where $\text{Nu}_x(\alpha^*, n = 1)$ is the corresponding Newtonian Nusselt number with the same α^* and the same thermal boundary condition. Equation (2.35) is in

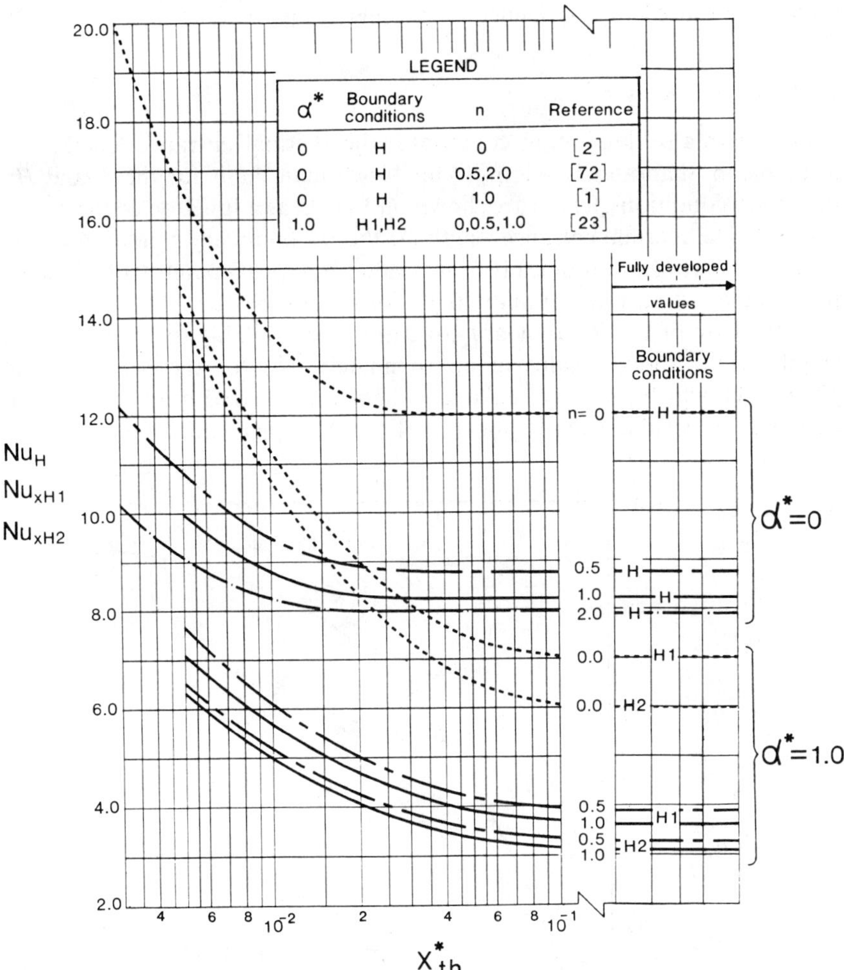

FIG. 24. Thermally developing Nusselt numbers as a function of x^*_{th} and the aspect ratio for H boundary condition, $n = 0$, 0.5, 1.0, and 2.0.

very good agreement with other numerical results [23]. Even for the case of thermally developed flow, Eq. (2.35) gives a reasonable estimate for $0.5 < \alpha^* < 2$, as mentioned by Irvine and Karni [4].

3. *Thermal Entrance Length*

The thermal entrance length L_{th} is the duct length required for the local Nusselt number to come within 5% of the fully developed value; that is, the thermal entrance length is defined as the length required for Nu_x to reach 1.05 $(Nu)_{fd}$.

a. *Newtonian fluids*

For thermally developing conditions, the thermal entrance lengths are available in Shah and London [1] for Newtonian fluids for the T and $H1$ boundary conditions. These are shown on Fig. 25 as a function of the aspect ratio α^*. The thermal entrance length for the $H1$ boundary condition is well behaved, varying monotonically from a value of approximately 0.01 for plane parallel plates to a value of about 0.066 for the square duct.

In the case of the T boundary condition, the available thermal entrance length results show a value of approximately 0.01 for the plane parallel plates duct, which rises to a maximum slightly above 0.05 at $\alpha^* = 0.25$ then decreases to a value slightly above 0.04 for the square duct. Since there is

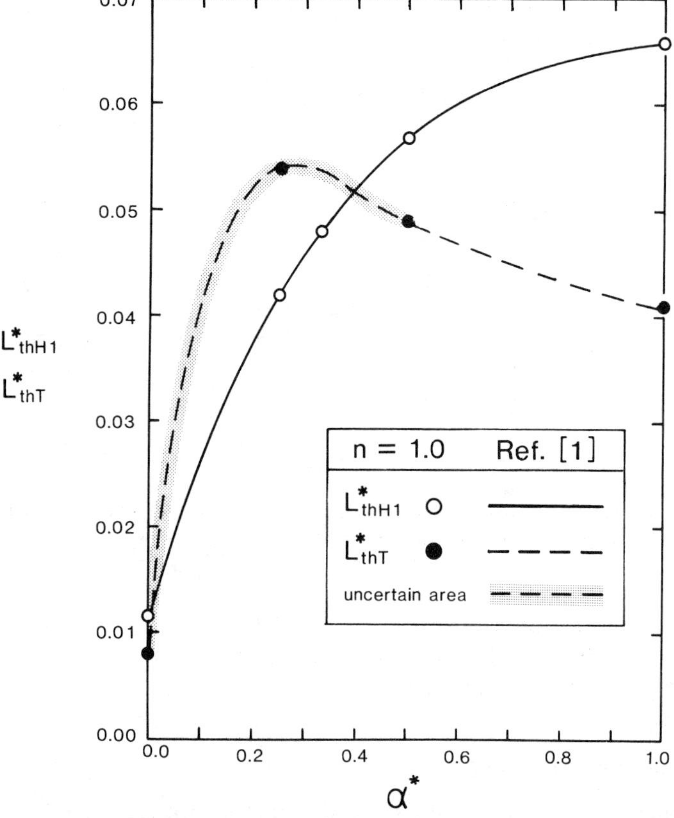

FIG. 25. Thermal entrance lengths for T and H boundary conditions for Newtonian fluid.

some uncertainty connected with these values, the curve is shown dashed and shaded. For engineering estimates, it is suggested that a value of the dimensionless thermal entry length of the order of 0.055 be used for a rectangular duct of aspect ratios equal to 0.2 or greater for both the T and $H1$ boundary conditions.

b. *Non-Newtonian fluids*

The influence of the power law exponent on the thermal entrance length for the case where the velocity distribution is fully established at the start of heating is shown in Fig. 26 for the plane parallel plates channel and for the

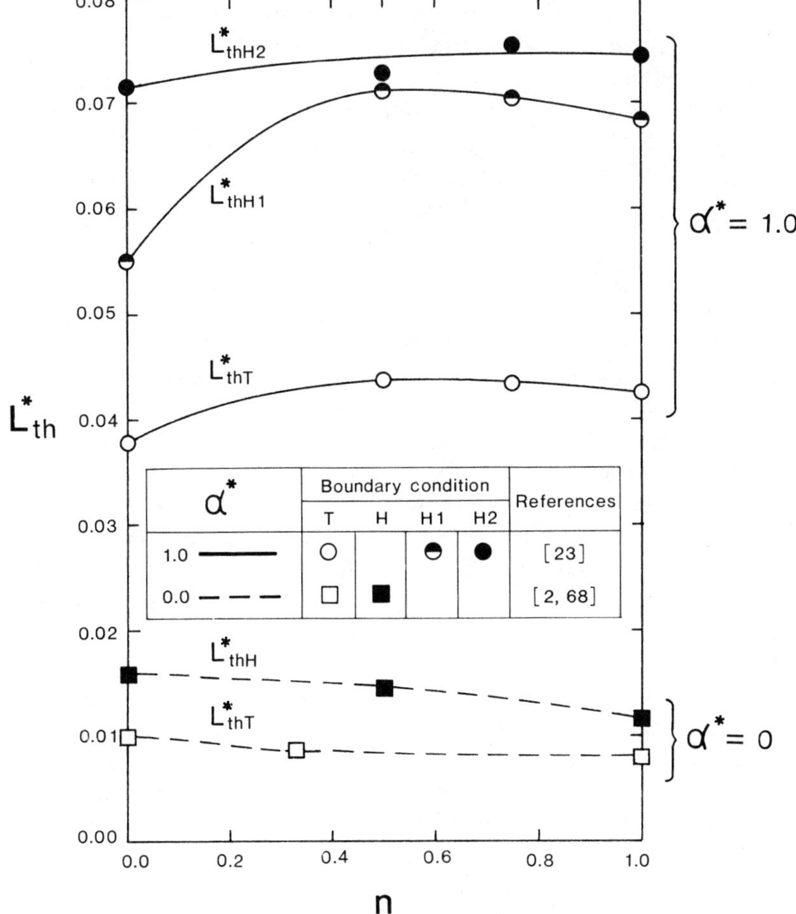

FIG. 26. Thermal entrance lengths for T and H boundary conditions for power law fluids.

square duct. For these two limiting cases for the T, $H1$, and $H2$ boundary conditions, there is no major shift in the thermal entrance length and the Newtonian results are recommended. That is, for all values of n for laminar flow in a rectangular channel of an aspect ratio equal to 0.2 or greater, it is recommended that the dimensionless thermal entrance length be taken as 0.055 for the T and $H1$ boundary conditions.

E. Simultaneously Developing Flow

In the case of simultaneous development of the velocity and temperature, the departure from the thermally developing Nusselt number at a given axial position, x^*_{th}, depends on the Prandtl number. Generally speaking, if the Prandtl number is of the order of 50 there is essentially no difference, and the thermally developing Nusselt values may be used. For many liquids, particularly pseudoplastics, the Prandtl number satisfies this requirement and consequently the simultaneously developing flow case is primarily of academic interest. Since there are a few analytical results available in the literature, they will be presented for the sake of completeness.

1. *Newtonian Fluids*

Shah and London [1] presented the simultaneous developing Nusselt values for the plane parallel plates channel for both T and H boundary conditions. Wibulswas [84] treated simultaneous flow in rectangular ducts of various aspect ratios ranging from $\frac{1}{6}$ to 1, for both the T and $H1$ boundary conditions. Most of Wibulswas' results are for a Prandtl number of 0.72. However, he does present local Nusselt values for the $H1$ boundary condition; simultaneous development; for Prandtl numbers of 0, 0.10, 0.72, 10, and ∞; and $\alpha^* = 0.5$. These are presented in Shah and London [1]. In the case of the square duct, Chandrupatla [23] reports results for the T, $H1$, and $H2$ boundary conditions for simultaneous flow and temperature development.

Incropera and Mahaney [88] have solved the problem of the simultaneous development of a Newtonian fluid of Prandtl number 6.5 in a rectangular duct with an aspect ratio of 0.5 under the following thermal boundary conditions: $H2$, $H2$(2L), and $H2$(1L). These results are shown on Fig. 27. It should be noted that the $H2$ condition yields the lowest Nusselt values of the three cases shown. This is in contrast with the $H1$ boundary condition, which yields Nusselt numbers that lie between those for the $H1$(2L) and the $H1$(1L) condition. This brings out the fact that the $H2$(2L) heating configuration is preferred to the $H2$ condition in that it yields higher Nusselt values and avoids the extremely high corner temperatures associated with the $H2$ boundary condition.

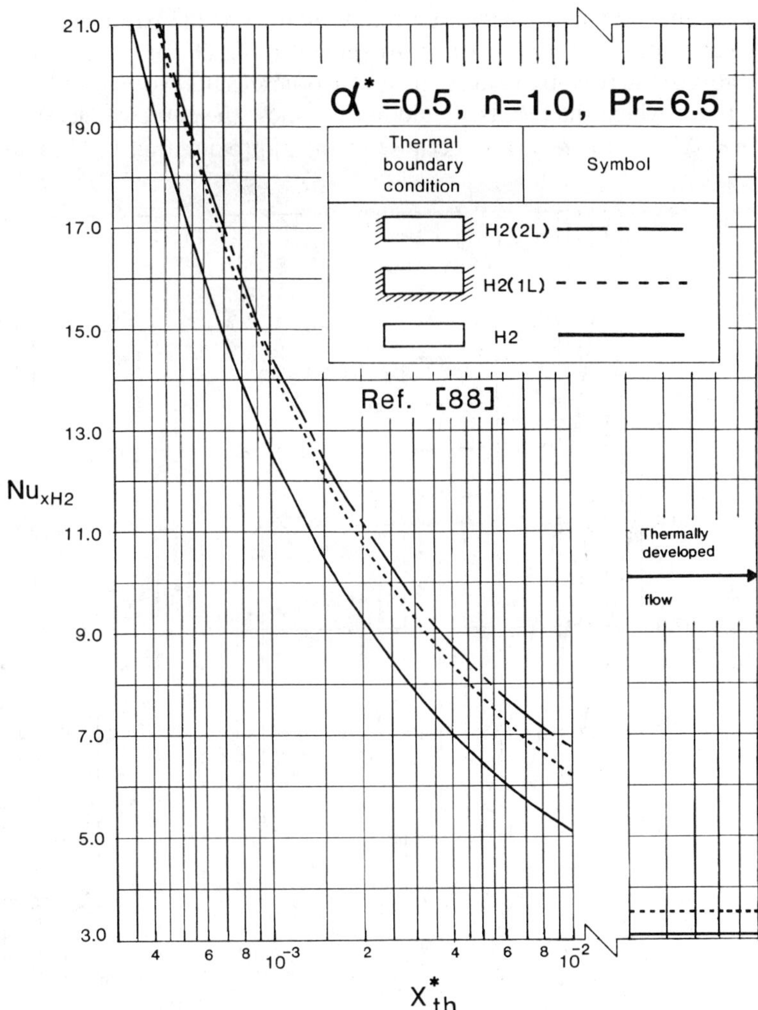

FIG. 27. Simultaneously developing Nusselt number as a function of x^*_{th} for Prandtl number = 6.5; $H2$, $H2(1L)$, and $H2(2L)$ boundary condition; and $\alpha^* = 0.5$.

2. *Non-Newtonian Fluids*

Yau and Tien [40, 78] and Lin [79] reported the Nusselt values associated with the simultaneous development of the temperature and velocity fields for power law fluids in a parallel plates channel covering a range of n values for the T boundary condition. The Yau and Tien analysis used an integral method and their Nusselt predictions appeared to be in error when recalculated on the basis of $Nu_{x,T}$ versus x^*_{th} with the Prandtl number as the parameter. On the other hand, the results of Lin, which are shown on Fig. 28, for a power law fluid of $n = 0.5$, are in reasonable agreement with other results except for the case of $Pr = 10$ for which the results should not be below the curve labeled $Pr = \infty$ and $n = 0.5$. This is probably because of the approximation in his analysis.

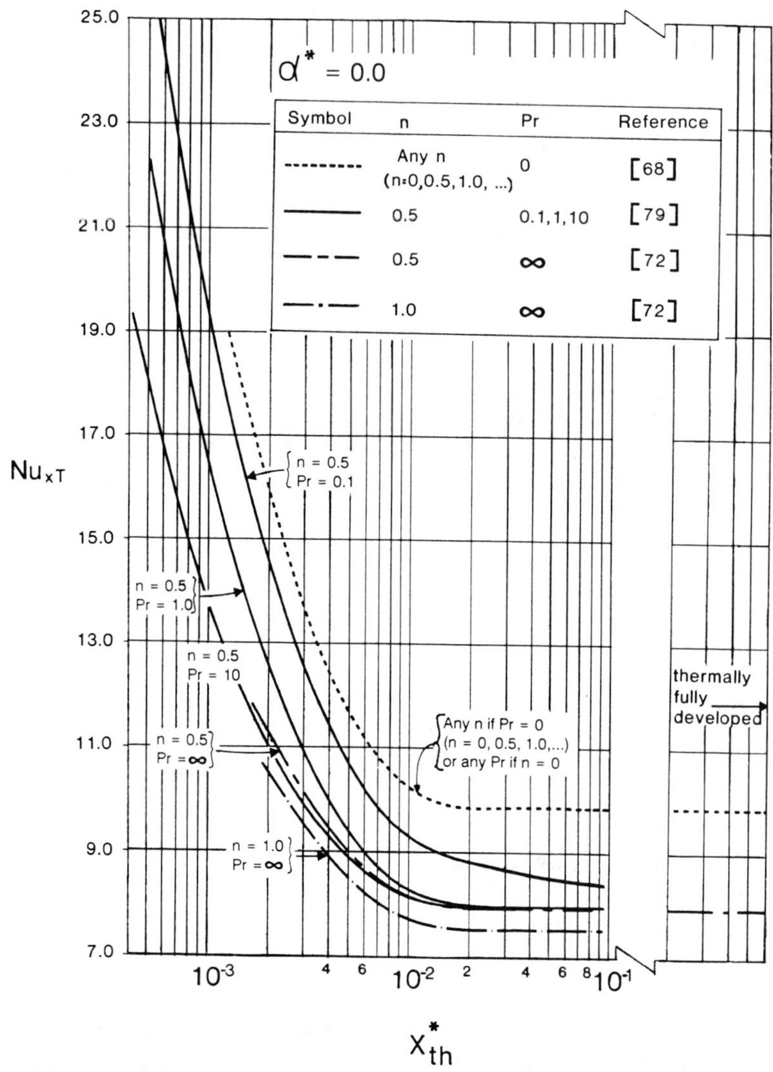

FIG. 28. Simultaneously developing Nusselt number as a function of x^*_{th} and Prandtl number, plane parallel plates. T boundary condition; $n = 0, 0.5$, and 1.0.

The results of Chandrupatla [23] for the square duct are shown on Figs. 29–31 for the three boundary conditions of T, $H1$, and $H2$ for values of n equal to 0, 0.5, and 1.0. These figures bring out the fact that there is little difference between the thermally developing Nusselt number and the Nusselt number for simultaneous development if the Prandtl number is larger than 50, a situation that usually prevails for pseudoplastics.

The final two figures, Figs. 32 and 33, summarize the limiting values for the local and for the mean Nusselt numbers, respectively. In each case, the Nusselt values corresponding to the T and H boundary conditions are given for the plane parallel plates duct and the square duct. The Nusselt values corresponding to the two limiting values of the power law index for pseudoplastics, $n = 0$ and $n = 1$, are presented. Finally, the values corresponding to a Prandtl number of zero and a Prandtl number of infinity are also shown. These two figures should serve as a check on solutions for different aspect ratios, the power law index, and Prandtl numbers, which should fall between the limiting values shown on these figures.

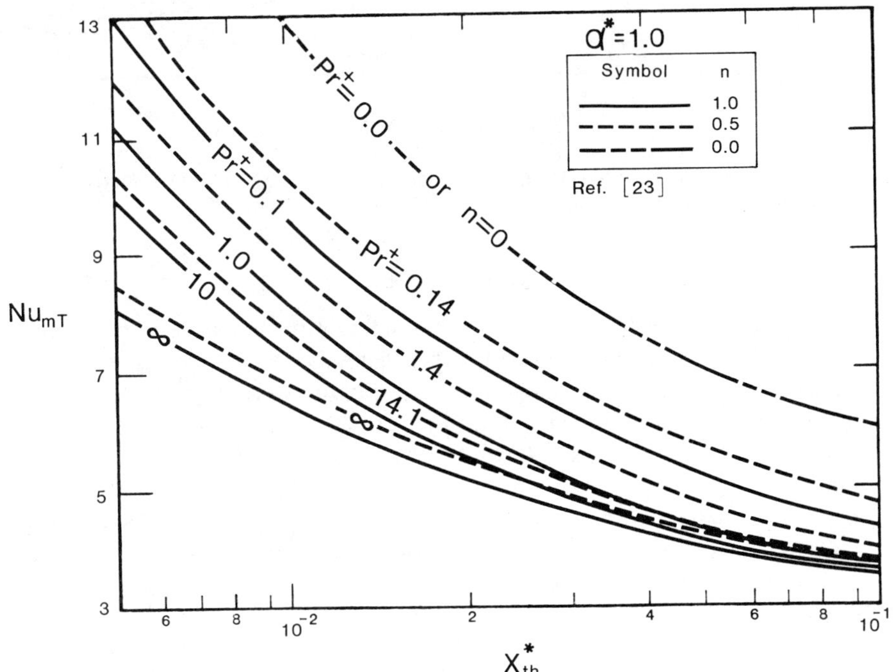

FIG. 29. Simultaneously developing Nusselt number as a function of x^*_{th} and Prandtl number, square duct. T boundary condition; $n = 0, 0.5$, and 1.0.

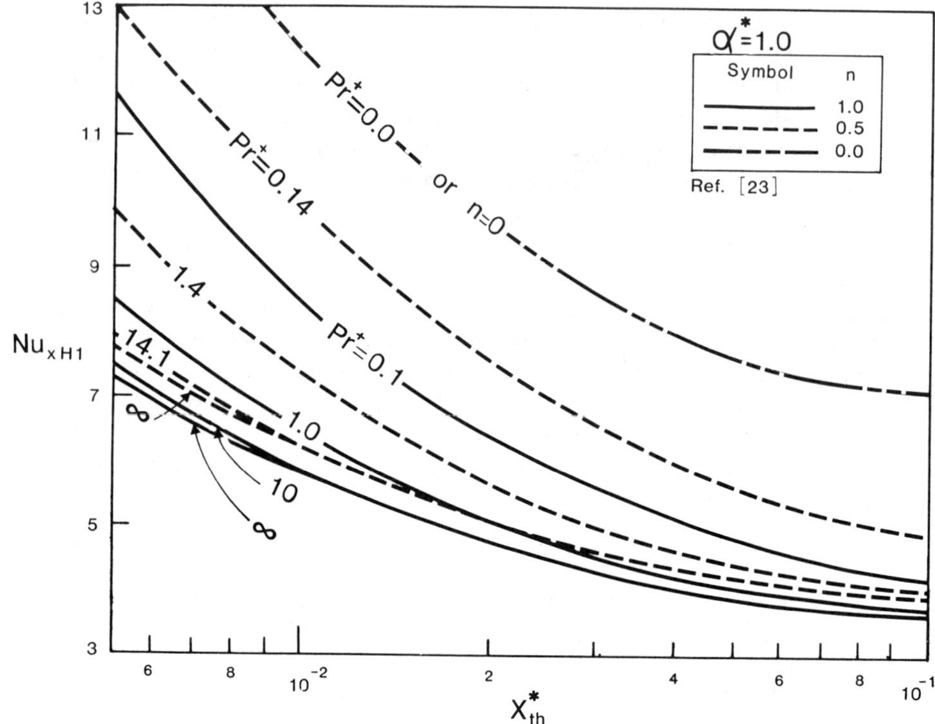

FIG. 30. Simultaneously developing Nusselt number as a function of x^*_{th} and Prandtl number, square duct, $H1$ boundary condition; $n = 0$, 0.5, and 1.0.

F. EXPERIMENTAL HEAT TRANSFER STUDIES

The heat transfer behavior of Newtonian fluids in laminar motion through circular tubes has been studied experimentally for many years and a considerable body of knowledge has been accumulated [89]. The analytical forced convection solutions have been verified for small temperature differences between the wall and the fluid. At a larger temperature difference, the influence of natural convection comes into play, and in this case the tube orientation is important as are the flow Reynolds number and the Rayleigh number. A number of empirical equations have been developed for superposed free and forced convection for laminar flow of Newtonian fluids in the circular tube geometry [89].

In the case of non-Newtonian fluids in laminar flow through circular tubes, there are experimental results to validate the power law analytical forced convection predictions [90]. Furthermore, it turns out that viscoelastic fluids in laminar pipe flow, for small temperature differences between the wall and

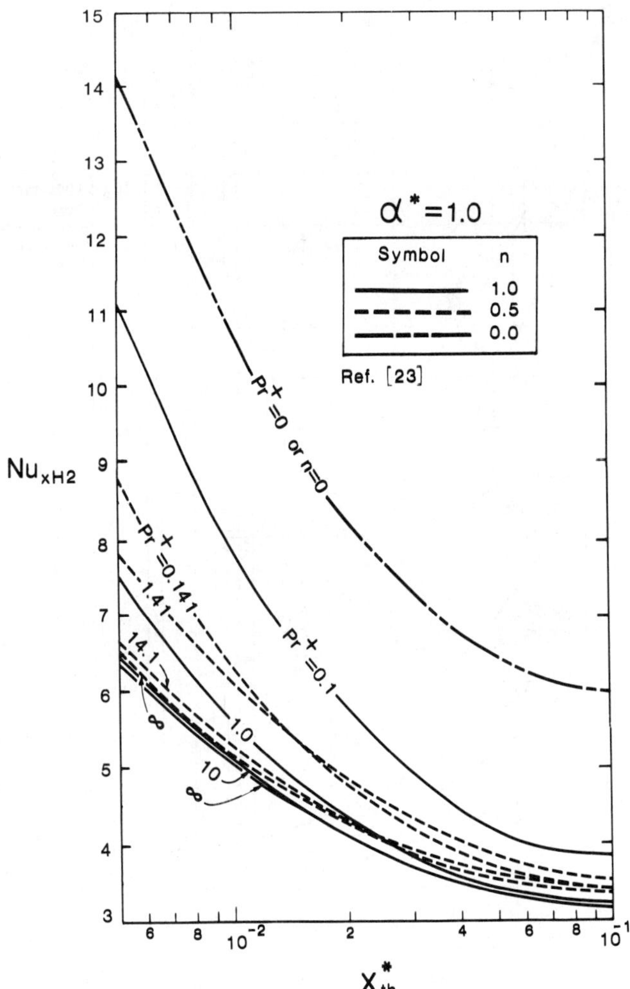

FIG. 31. Simultaneously developing Nusselt number as a function of x^*_{th} and Prandtl number, square duct, $H2$ boundary condition; $n = 0, 0.5$, and 1.0.

the fluid, behave as purely viscous power law fluids. At higher temperature differences, where free convection effects become important, there are a number of empirical equations available [90].

Experimental heat transfer studies of laminar flow in rectangular geometries are not as common as in the case of the circular tube. Nevertheless there is a growing body of literature on the subject. In this section, these studies are reviewed for both Newtonian and non-Newtonian fluids.

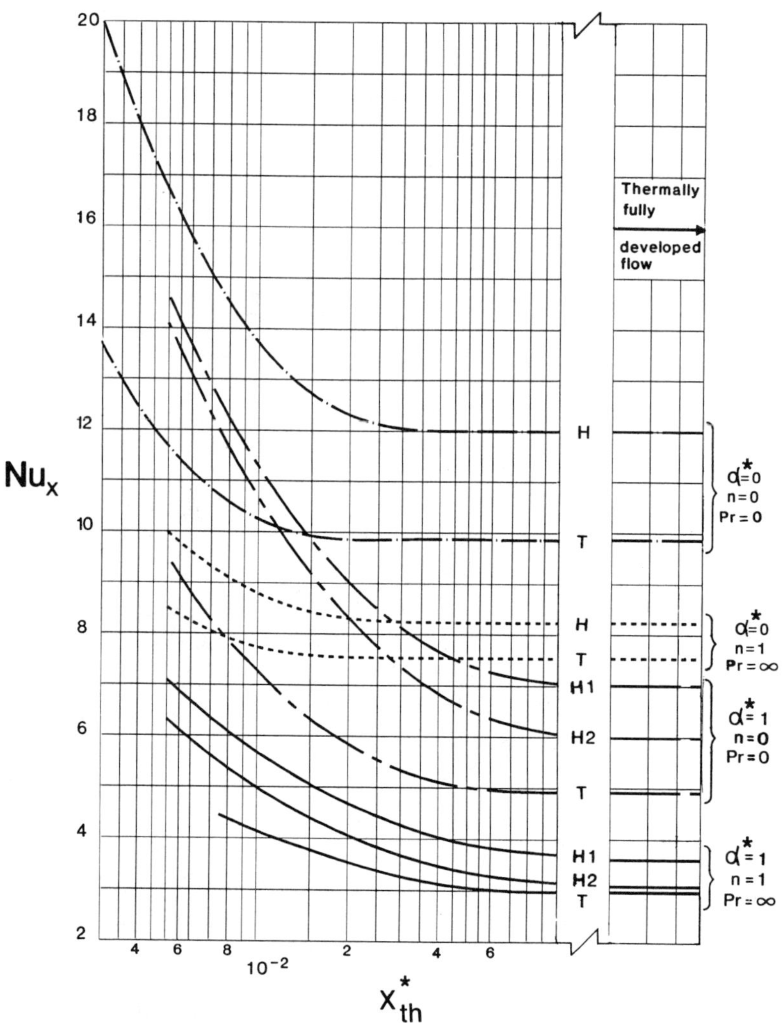

FIG. 32. Summary of axially local Nusselt numbers for limiting values of the aspect ratio, power law index, and Prandtl number. Results are from Shah and London [1], Shah and Bhatti [2], Chandrupatla [23], and Javeri [68].

1. *Newtonian Fluids*

During the past quarter of a century, there has been increased interest in studying the experimental heat transfer behavior of Newtonian fluids in laminar flow through rectangular ducts. Some representative examples of these experimental studies are listed in Table XIII. The early work of Clark

TABLE XIII

Experimental Studies of Newtonian Laminar Heat Transfer in a Rectangular Channel

Investigator(s)	Aspect ratio	Hydraulic diameter (cm)	Thermal boundary condition	Fluid studied	Flow condition	Comments
Clark and Kays [91]	1.0 0.38	0.45 0.55	T, $H1$ T	Air	Simultaneous development	Heating accomplished by counter flow of water. Experiments verified predicted values of fully established Nusselt numbers. Limited data presented on entrance length effects.
Wibulswas [84]	0.5	3.38	T, $H1$	Air	Velocity distribution fully established at start of heating	Good agreement of experimental thermally developing Nusselt numbers with predictions for $H1$ boundary condition. Measured Nusselt values higher than predicted for T boundary condition.
Mori and Uchida [96]	0.1 or less		Lower plate heated electrically; upper plate cooled by flowing water	Air	Simultaneous development	When temperature difference between top and bottom walls exceeds a critical value, vortex rolls appear. Influence of these vortex rolls on flow and heat transfer studied.

(*Continued*)

TABLE XIII (*Continued*)

Investigator(s)	Aspect ratio	Hydraulic diameter (cm)	Thermal boundary condition	Fluid studied	Flow condition	Comments
Mercer *et al.* [92]	0.041	2.44	1. Top and bottom plate at constant temperature 2. Bottom plate at constant T, upper plate insulated 3. Upper plate at constant temperature, bottom plate insulation	Air	Simultaneous development	An interferometric study giving experimental Nusselt values within 10% of prediction.
Akiyama *et al.* [97]	0.045 0.091	2.43 4.66	1. Upper wall heated, lower wall cooled. 2. Upper wall cooled, lower wall heated 3. Upper and lower walls at same temperature	Air	Simultaneous development	Experimental results confirmed a critical Rayleigh number of 1708 for the onset of longitudinal columnar vortices due to buoyant forces. Concludes that longitudinal vortices must be taken into account for channels with $\alpha^* \leq 0.1$
Kwant and Van Ravenstein [73]	0.102 0.275	0.74 1.14	T	Oil	Velocity distribution fully developed at beginning of heating	Experimental results for $\alpha^* = 0.102$ in good agreement with theory. For $\alpha^* = 0.275$, measured values are higher than predicted, possibly because of free convection.

Lombardi and Sparrow [93]	0.01	0.294	Mass transfer analogy corresponds to constant wall temperature on top and bottom plates; side walls are adiabatic	Naphthalene diffusing into air	Simultaneous development	Experimental results were in agreement within 5% with theoretical predictions.
Ostrach and Kamotani [98]	0.028–0.056	1.46–2.85	Lower plate heated electrically; upper plate cooled by water	Air	Velocity fully developed at start of heating	Sharp increase in heat transfer when longitudinal vortex rolls appear. With increasing Ra number, a second type of vortex rolls appears and distorts existing periodic temperature distribution.
Kamotani and Ostrach [99]	0.048–0.071	2.42–3.56	Lower plate heated electrically; upper plate cooled by water	Air	Velocity profile fully established at start of heating	Extended range of Rayleigh and Reynolds numbers, $10^3 < \mathrm{Ra} < 3 \times 10^4$, $30 < \mathrm{Re} < 1100$. Flow reported to be more stable in thermal entrance region than in fully developed region. Critical Rayleigh numbers higher than predicted.

(*Continued*)

TABLE XIII (*Continued*)

Investigator(s)	Aspect ratio	Hydraulic diameter (cm)	Thermal boundary condition	Fluid studied	Flow condition	Comments
Hwang and Liu [100]	0.067 0.10 0.15	3.70 5.45 7.83	Lower plate heated electrically; upper plate cooled by free convection to surrounding air	Air	Velocity fully developed at start of heating	Critical Rayleigh numbers determined by flow visualization were 1.4–10 times the value predicted by linear theory.
Kamotani *et al.* [101]	0.067	7.12	Lower plate heated electrically; upper plate cooled by water flow	Air	Velocity fully established at start of heating	Rayleigh range from 2×10^4 to 2×10^5 studied. The Nusselt number is increased by a factor of 1.4–4.4 because of thermal instability. Report some influence of Re^2/Gr on Nu.
Nakamura *et al.* [94]	1.	0.75	T boundary condition; steam heated test section	Ethylene glycol	Velocity fully developed at start of heating	Reported experimental fully developed friction factors and developing Nusselt numbers, taking into account the influence of temperature-dependent viscosity. Reasonable agreement with theoretical predictions.

Kostic [95]	0.5	1.20	Top and bottom plates heated by passing electricity through them; side walls adiabatic	Water	Simultaneous development	Measured local Nusselt number of upper wall in good agreement with prediction.
Osborne and Incropera [102]	0.20 0.066	10.02 3.75	Top and bottom plates heated using electrical heaters, approaching *H*2 boundary conditions on these surfaces; side walls adiabatic; various combinations of heating upper and lower walls were studied	Water	Velocity fully developed at start of heating	Experimental results are correlated by equation: $Nu_x = [C_1 Gz + 0.012\, Ra_q^{3/4}]^{1/3}$ where $C_1 = C_1(\alpha^*)$
Rao [103]	0.2	3.75	Bottom plate heated by passing electric current through it; other walls are insulated	Water, oil	Simultaneous development	Experimental results are correlated by $Nu_x = [2.676\, Gz + 0.0163\, Ra_q^{3/4}]^{1/3}$
Hartnett *et al.* [104]	0.5	1.20	Top and bottom plates heated by passing electric current through them	Water	Simultaneous development	Correlation equation provided for two plates heated and for one plate heated.

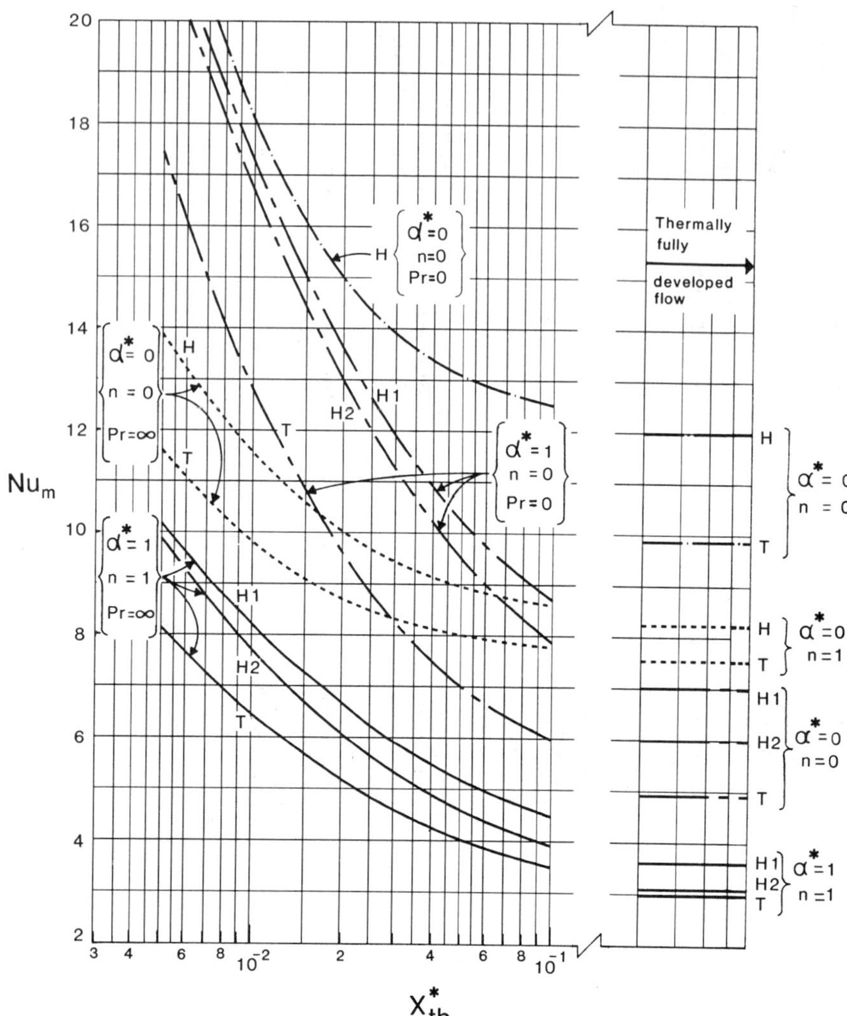

FIG. 33. Summary of mean Nusselt numbers for limiting values of the aspect ratio, power law index, and Prandtl number. Results are from Shah and London [1], Shah and Bhatti [2], Chandrupatla [23], and Javeri [68].

and Kays [91] provided experimental confirmation of the analytical predictions for the T and $H1$ thermal boundary conditions. Other investigators, including Wibulswas [84], Mercer *et al.* [92], Kwant and Van Ravenstein [73], Lombardi and Sparrow [93], and Nakamura *et al.* [94], report experimental verification of the laminar flow predictions.

An example of the agreement of experiment and theory may be seen in Fig. 34, where the local Nusselt numbers along the top wall of a rectangular duct of aspect ratio 0.5 are compared with available predictions [95]. In contrast, the local values on the bottom wall are considerably higher than those on the top wall, showing the influence of free convection.

A number of the studies shown in Table XIII are concerned with thermal instabilities, which give rise to vortex rolls with axes parallel to the flow direction [96–101]. These results are of particular importance in understanding the heat transfer behavior of small aspect ratio channels ($\alpha^* < 0.1$).

The recent work of Osborne and Incropera [102] for laminar flow of water in rectangular channels with asymmetric heating of the upper and lower walls (approximating $H2$ conditions on each of the two walls) is of special interest inasmuch as these authors present a relatively simple correlation equation for symmetric and asymmetric heating.

$$\mathrm{Nu}_x = [\mathrm{C}_1\mathrm{Gz} + 0.012\ \mathrm{Ra}_q^{3/4}]^{1/3} \tag{2.36}$$

for $(q)_{\text{top}}/(q)_{\text{bottom}} < 2$. Here C_1 is a function of α^*.

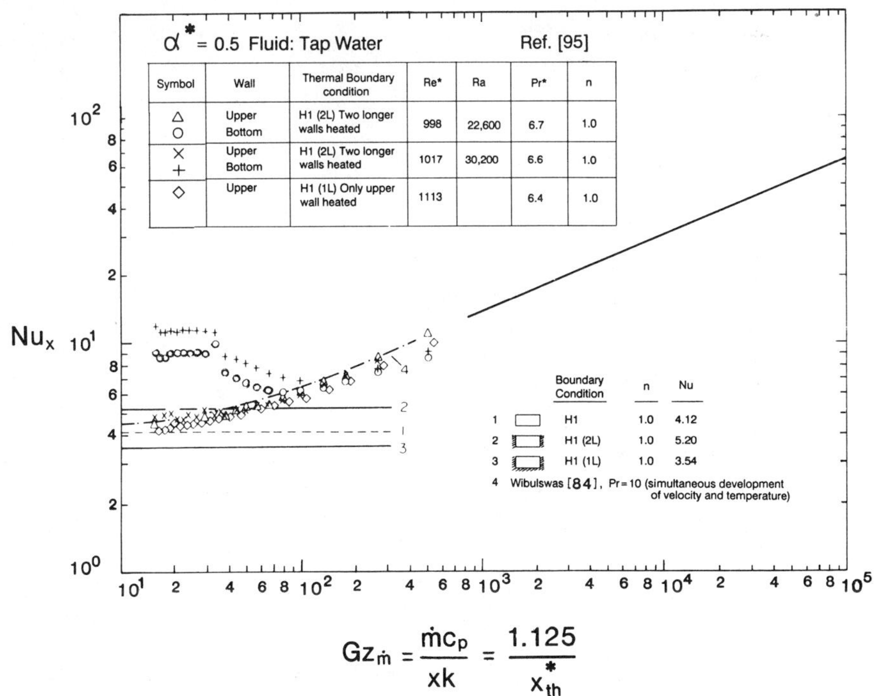

FIG. 34. Experimental laminar Nusselt numbers for Newtonian flow in rectangular duct, $\alpha^* = 0.5$, $H1$(1L), $H1$(2L) boundary condition.

Rao [103] reported results for laminar mixed convection in a duct of aspect ratio 0.2, with only the bottom wall heated electrically and the other three walls adiabatic. His experimental Nusselt numbers are in reasonable agreement with Eq. (2.36)[1], as shown on Fig. 35.

Hartnett *et al.* [104] analyzed some 20 runs carried out in a rectangular duct of aspect ratio 0.5 with the bottom wall heated [$H1(1L)$] and recommended the following correlation equation:

$$
\begin{aligned}
\mathrm{Nu}_x &= [2.14\,\mathrm{Gz} + 0.018\,\mathrm{Re}^{0.037}\,\mathrm{Ra}_q^{3/4}]^{1/3} \\
&= [1.9\,\mathrm{Gz}_{\dot{m}} + 0.018\,\mathrm{Re}^{0.037}\,\mathrm{Ra}_q^{0.75}]^{1/3} \\
&\qquad 100 < \mathrm{Re} < 1900 \\
&\qquad 2 \times 10^5 < \mathrm{Ra}_q < 4 \times 10^6
\end{aligned}
\tag{2.37}
$$

Figure 36 shows data for two typical runs along with the Osborn–Incropera

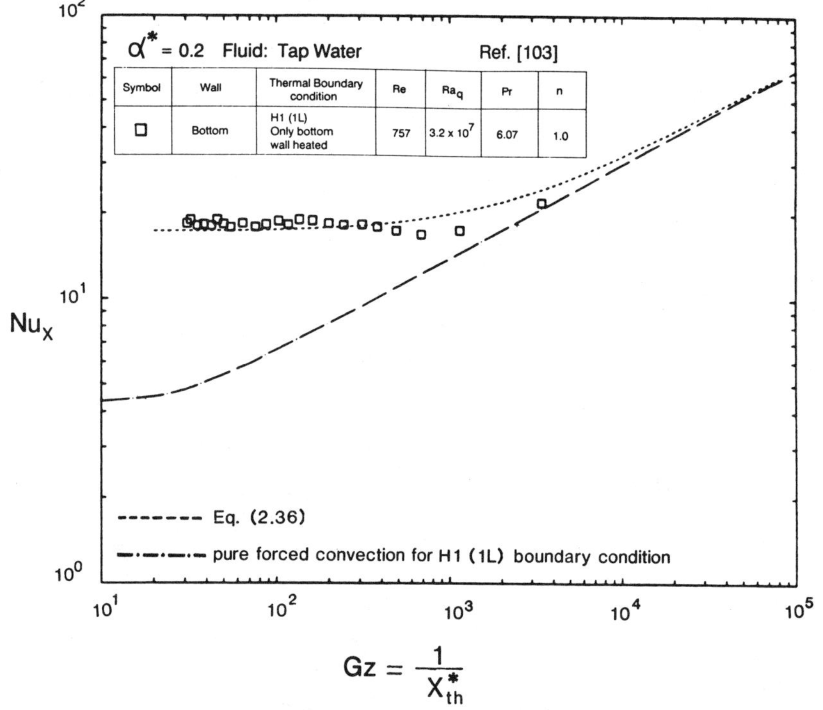

FIG. 35. Experimental laminar Nusselt numbers for Newtonian flow in rectangular duct, $\alpha^* = 0.2$, $H1(1L)$ boundary condition.

[1] The value of C_1 is 2.76 for $\alpha^* = 0.2$ and 2.21 for $\alpha^* = 0.5$.

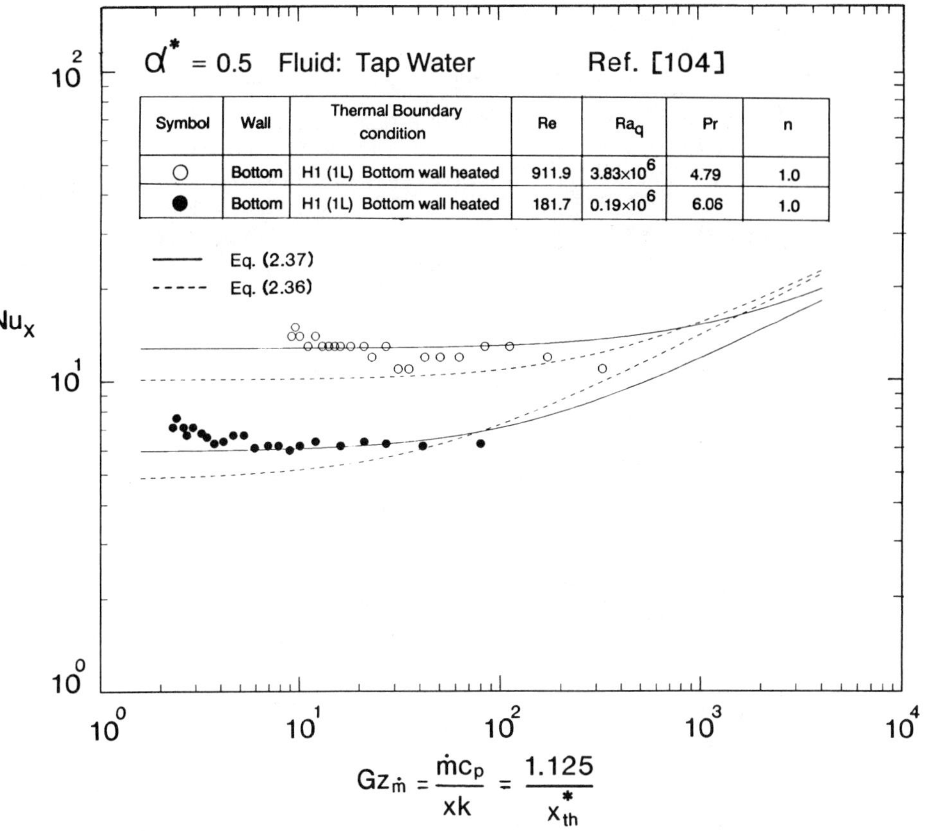

FIG. 36. Experimental laminar Nusselt numbers for Newtonian flow in rectangular duct, $\alpha^* = 0.5$, $H1$(1L) boundary condition.

equation [Eq. (2.36)][1] and the predictions of Eq. (2.37). The difference in the two correlations may reflect the difference in the thermal boundary conditions in the two studies.

In the case where the top and bottom walls are heated symmetrically Hartnett *et al.* propose the following equation for the average local Nusselt numbers on the basis of statistical analysis of 19 runs:

$$\begin{aligned} \mathrm{Nu} &= [2.14\ \mathrm{Gz} + 0.005\ \mathrm{Re}^{0.13}\ \mathrm{Ra}_q^{3/4}]^{1/3} \\ &= [1.9\ \mathrm{Gz}_{\dot{m}} + 0.005\ \mathrm{Re}^{0.13}\ \mathrm{Ra}_q^{3/4}]^{1/3} \qquad (2.38) \\ &350 < \mathrm{Re} < 1600 \\ &2 \times 10^5 < \mathrm{Ra}_q < 10^6 \end{aligned}$$

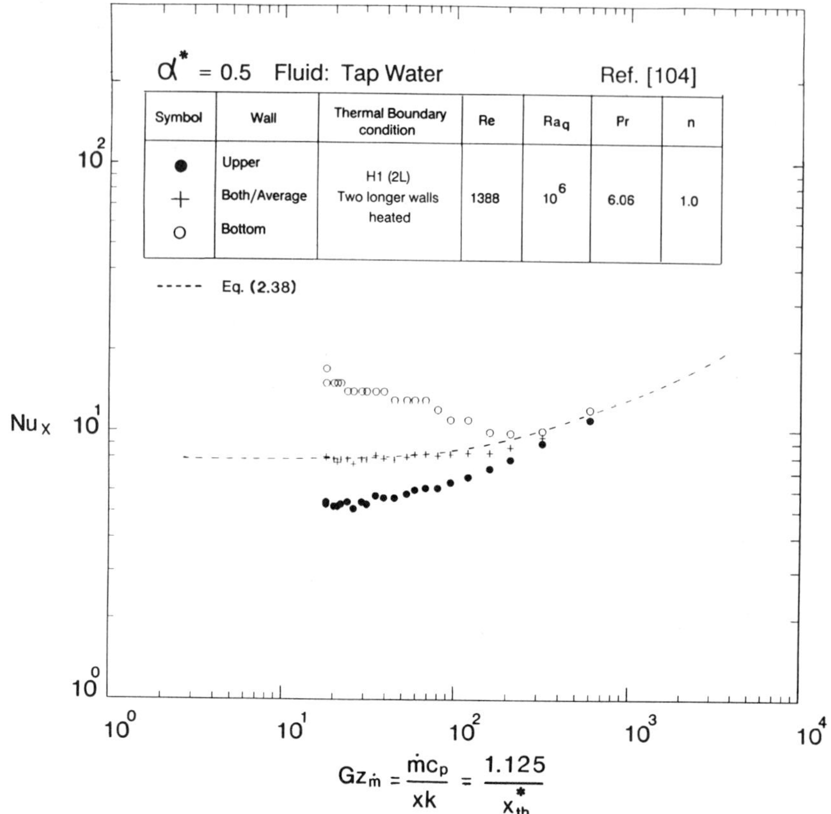

FIG. 37. Experimental laminar Nusselt numbers for Newtonian flow in rectangular duct, $\alpha^* = 0.5$, $H1(2L)$ boundary condition.

Figure 37 shows the results of a typical run, where the influence of free convection on the lower wall Nusselt number behavior is apparent. It may be seen that the lower wall Nusselt number is of the order of three times the upper wall value. The mean value is also shown along with the prediction given by Eq. (2.38).

2. *Non-Newtonian Fluids*

It was pointed out in the previous section that considerable experimental evidence exists to validate the heat transfer predictions in the case of laminar flow of Newtonian fluids in rectangular channels. This does not appear to be true in the case of non-Newtonian flow in rectangular channels. As brought out in Table XIV, all of the available experimental data lie above the predicted values. Many of the cited investigators [29, 105–109] suggested that the high experimental values were due to secondary flows resulting from the visco-elastic behavior of the fluids studied. Such viscoelastic fluids exhibit normal force differences along the boundary surface that could give rise to increased heat transfer. Such secondary flows have been predicted by Green and Rivlin [30], and there is a considerable body of evidence [31, 32] to support their existence. Figure 38 [31] reveals the general character of these secondary motions. It is interesting to note that the direction of this flow is counter to the direction of secondary flows that arise in the case of turbulent flow through rectangular passages [110]. The available analyses indicate that the secondary flow effects are greatest in the square duct geometry, and their influence decreases as the aspect ratio decreases. In addition, for any aspect ratio, the secondary flow increases as the main flow Reynolds number increases.

Hartnett and Kostic [29, 108] report local Nusselt numbers for a duct of aspect ratio 0.5 with the upper and lower walls heated symmetrically [$H1(2L)$]. Their results, shown in Fig. 39, reveal little difference between the local Nusselt numbers on the upper and lower plates, and both are considerably higher than the prediction. Ordinary natural convection, resulting

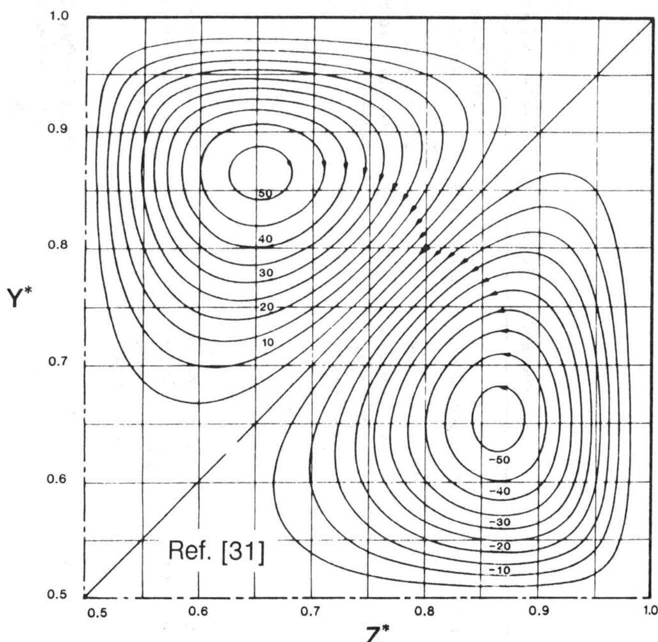

FIG. 38. Secondary flow patterns in laminar flow of aqueous CMC solution in square duct.

TABLE XIV

EXPERIMENTAL STUDIES OF NON-NEWTONIAN LAMINAR HEAT TRANSFER IN RECTANGULAR CHANNELS

Investigator(s)	Aspect ratio	Hydraulic diameter (cm)	Thermal boundary condition	Fluid studied	Flow condition	Comments
Oliver [105]	Flattened copper tube originally of diameter equal to 0.635 cm		Heated by hot water flowing in surrounding jacket	Aqueous solutions of polyacrylamide and polyethylene oxide		Original heat transfer measurements with Newtonian fluid, 85% glycerol in water in good agreement with theory. Viscoelastic fluids yielded Nusselt values 40% higher than the Newtonian results.
Oliver and Karim [106]	Flattened tubes of aspect ratios of 0.71, 0.63, 0.50, 0.38, and 0.18		Heated by hot water flowing in surrounding jacket	250 and 500 wppm polyacrylamide in water		Finds higher heat transfer than predicted, up to 45% increase. Maximum effect at $\alpha^* \cong 0.667$
Mena *et al.* [107]	1	1	Heated by fluid flowing through a surrounding jacket	Aqueous solutions of polyacrylamide	Simultaneous development	Secondary flows are assumed to be responsible for the observed large increase in heat transfer in square duct as compared to results found in circular tube.

Hartnett and Kostic [29] Kostic and Hartnett [108]	0.5	1.2	Upper and lower walls heated electrically; Symmetric heating	Aqueous solutions of polyacrylamide (1000 wppm Separan AP 273)	Simultaneous development	Measured mean and local Nusselt values much higher than predicted and much higher than found for water under comparable conditions.
Lawal [109] Lawal and Majumdar [5]	1.0	1.575	Test section electrically heated $H1$	Aqueous solution of Carbopol 934; 1% unneutralized solution; 0.3% partially neutralized solution	Velocity fully developed at start of heating	Square duct heat transfer data are 50% higher than predicted.
Rao [103]	0.2	3.75	Lower wall heated electrically $H1(1L)$	Aqueous solution of polyethylene oxide, hydroxyethyl cellulose	Simultaneous development	Experimental Nusselt numbers only slightly higher than water when compared at same Gz, Re, and Ra.
Hartnett *et al.* [104]	0.5	1.2	Upper and lower wall heated electrically $H1(1L)$ and $H1(2L)$	Aqueous solutions of Carbopol (1000 wppm)	Simultaneous development	Measured mean and local values much higher than predicted and much higher than found for water under comparable conditions.

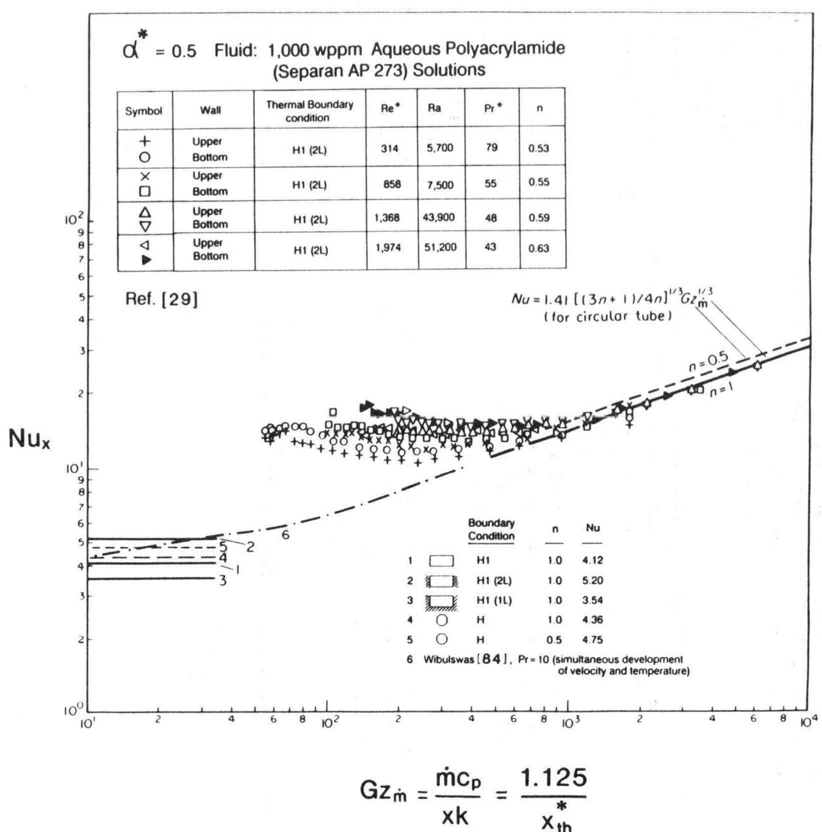

FIG. 39. Experimental laminar Nusselt number for aqueous polyacrylamide solutions in rectangular duct, $\alpha^* = 0.5$, $H1$(2L) boundary condition.

from density differences in the flow field, gives rise to a substantial difference in the local Nusselt numbers on the upper plate and on the lower plate in the case of a Newtonian fluid, as clearly seen in Fig. 34. It is obvious that such density-driven natural convection is not the major influence in the flow of a viscoelastic fluid in a noncircular channel.

Another interesting feature of viscoelastic flow in a rectangular channel is that the pressure drop is not greatly influenced by these secondary motions, and the experimental friction factor measured in the same experiment [29] was in agreement with the value predicted for a power law purely viscous fluid (i.e., $f = 16/\mathrm{Re}^*$).

Recent experiments have been carried out in a rectangular duct of aspect ratio 0.5 with the upper and lower walls electrically heated, with an aqueous solution of Carbopol as the working fluid. It is widely reported in the literature that aqueous Carbopol solutions are purely viscous inasmuch as they do not show drag reduction under turbulent flow conditions and their turbulent heat transfer behavior on a dimensionless basis is the same as aqueous clay suspensions. As it turns out, 1000 wppm Carbopol aqueous solution is elastic, as determined in an oscillatory viscometer. The heat transfer behavior, as presented in Fig. 40, also suggests that aqueous Carbopol solutions are viscoelastic. From Fig. 40, it may be seen that there is little difference between the upper and lower plate Nusselt values. The lower wall shows slightly higher Nusselt values probably reflecting the influence of ordinary natural convection. Furthermore, there is a substantial increase in the Nusselt number as the Reynolds number is increased, an expected result as described previously.

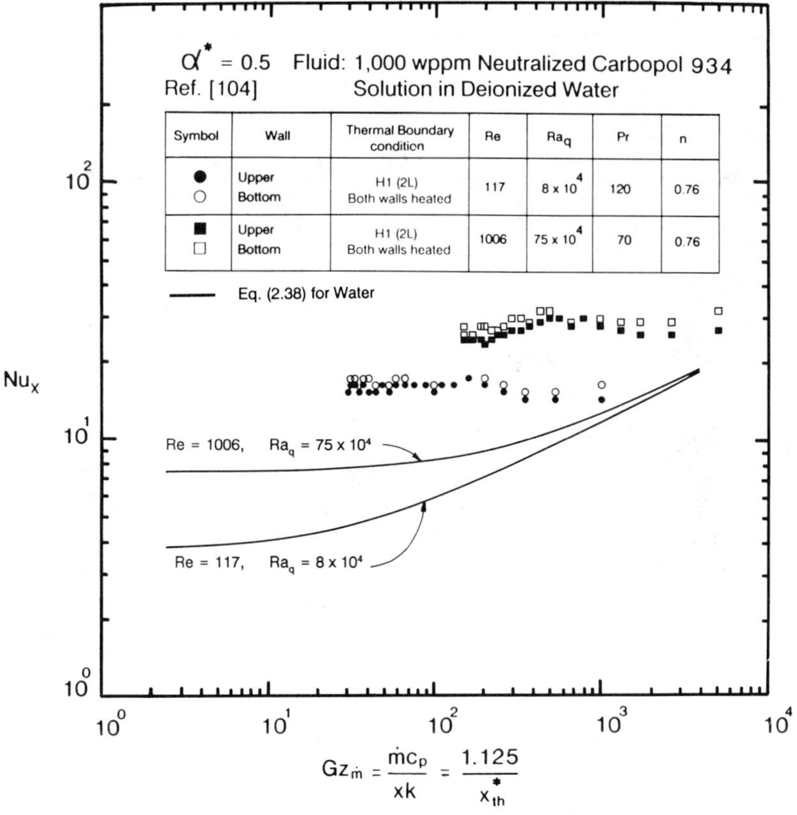

FIG. 40. Experimental laminar Nusselt number for aqueous Carbopol solutions in rectangular duct, $\alpha^* = 0.5$, $H1(2L)$ boundary condition.

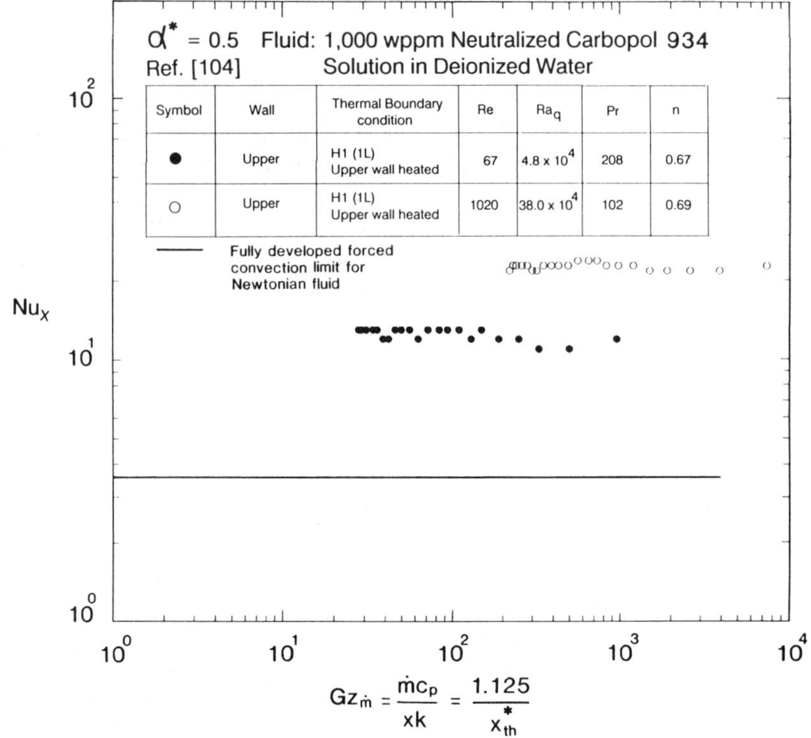

FIG. 41. Experimental laminar Nusselt number for aqueous Carbopol solutions in rectangular duct, $\alpha^* = 0.5$, $H1(1L)$ boundary condition.

Further evidence to support the conclusion that secondary flow plays a major role in increasing the heat transfer may be seen in Fig. 41, which reveals local Nusselt values for two runs where only the top wall is heated [104]. Here, again, the local values are much higher than predicted for a Newtonian or for a purely viscous non-Newtonian fluid. Therefore, it is concluded that viscoelastic fluids generate secondary flow in noncircular channels, which give substantial increase in heat transfer while not noticeably increasing friction. This behavior may make such fluids candidates for heat exchanger applications.

III. Turbulent Flow

The subject of turbulent convective heat transfer behavior of Newtonian fluids in channel flow has recently received considerable attention by Bhatti and Shah [3]. Unlike laminar flow, where most of the available literature is of an analytical nature, our knowledge of turbulent flow rests on a broad base of experimental studies. In this section, we will provide a brief coverage of the fluid mechanical and heat transfer performance of Newtonian fluids in turbulent flow through rectangular channels. Against this background, the behavior of purely viscous non-Newtonian fluids and viscoelastic fluids will be reviewed.

A. FLUID MECHANICS

1. *Newtonian Fluids*

In the case of turbulent channel flow of Newtonian fluids, most of the resistance to the transfer of heat and momentum is concentrated in the laminar sublayer region near the wall. Accordingly, the temperature and velocity profiles are relatively flat over most of the cross section. Hence, the influence of duct shape in turbulent flow is generally not as great as in laminar flow. Therefore, it has been a common practice to estimate the friction factors and dimensionless heat transfer coefficients for noncircular ducts by using corresponding circular pipe results, replacing the diameter by the hydraulic diameter. However, in recent years, more accurate procedures were introduced especially for predicting the fully developed friction factors in noncircular ducts, including rectangular ducts [111].

The hydrodynamic entrance length of Newtonian fluids in turbulent duct flow is relatively short, generally not more than 10 to 20 hydraulic diameters. Therefore, the fully developed velocity profile and friction factor are of major practical interest. Most analytical studies of turbulent flow through noncircular ducts have involved an extension of theories developed for circular pipe flow. An early example is the use of the well-known Karman–Nikuradse–Martinelli three-layer model developed for fully established turbulent pipe flow of a Newtonian fluid [112] in which the flow region is divided into a laminar sublayer, a buffer layer, and a turbulent core region described by the following relation:

$$u^+ = \begin{cases} y^+ & \text{for} \quad y^+ \leq 5 \\ 5 \ln y^+ - 3.05 & \text{for} \quad 5 < y^+ \leq 30 \\ 2.5 \ln y^+ + 5.5 & \text{for} \quad 30 < y^+ \end{cases} \tag{3.1}$$

Integration of this velocity profile across the tube cross section leads to the well-known Karman–Nikuradse equation:

$$1/\sqrt{f} = 4.0 \log(\mathrm{Re}\sqrt{f}) - 0.4 \tag{3.2}$$

The coefficients of this equation have been slightly modified to fit the experimental results.

Deissler and Taylor [113] employed the same velocity model to predict the fully developed friction factor and heat transfer in a square duct, assuming that the universal turbulent velocity profile for the circular pipe flow is also valid for a noncircular passage. Hartnett *et al.* [114] applied the Deissler–Taylor analysis to rectangular ducts having aspect ratios of 0.1, 0.2, and 1. They also performed experimental measurements of the fully developed friction factor in the same geometries for turbulent flow conditions. They found that the experimental turbulent results for all three rectangular ducts lie above the predicted values. The measured friction factors of the square duct show the greatest departure from their prediction, being 12% higher at Reynolds numbers larger than 10,000. The experimental results show a more modest aspect ratio dependency than the Deissler–Taylor analysis.

This discrepancy between the analytical and experimental results may be due to the neglect in the analysis of the secondary flow that exists in turbulent flow through noncircular ducts. The existence of these secondary flows has been known for over 50 years, from the time of the pioneering experiments of Nikuradse [115]. In more recent times, Hoagland [116] and other investigators (see Ref. 3) have provided more detailed information on these secondary flows. As shown in the studies by Emery *et al.* [110, 117], the secondary flow continuously transports momentum from the center to the corners and generates higher velocities there. Hence, the influence of secondary flow on pressure drop and heat transfer is the greatest for a square duct and decreases with a decreasing aspect ratio. This is consistent with the results of the Deissler–Taylor analysis, which showed the largest deviations from the experimental value to occur in the case of the square duct.

Jones [111] proposed a simple procedure to predict the fully developed turbulent friction factor for Newtonian fluids flowing in rectangular ducts and annuli. In the case of rectangular ducts, he compiled the available friction factor data for ducts having aspect ratios from 0.025 to 1.0. After eliminating some of the data that appeared questionable, Jones reported that the remaining experimental data fell within a 25% band relative to the Karman–Nikuradse equation based on the hydraulic diameter (−5 to +20%). The friction factor monotonically increased with the aspect ratio. Based on these observations, he rejected the Reynolds number based on the hydraulic diameter and proposed instead a different Reynolds number to correlate the fully developed friction factor in noncircular ducts. In particular, the Reynolds number proposed by Jones, Re_N^*, is defined such that all fully established laminar friction factors are given by $16/\mathrm{Re}_\mathrm{N}^*$. This Reynolds number proposed by Jones is consistent with the Kozicki Reynolds number, Re^*, introduced earlier:

$$\mathrm{Re}_\mathrm{N}^* = (\mathrm{Re}^*)_{n=1} = \Phi(\alpha^*) \cdot \mathrm{Re} \tag{3.3}$$

$$\Phi(\alpha^*) = \left(\frac{2}{3}\right)(1 + \alpha^*)^2\left[1 - \frac{192}{\pi^5}\alpha^* \sum_{i=0}^{\infty}(2i + 1)^{-5}\tanh\frac{(2i + 1)\pi}{2\alpha^*}\right] = \frac{1}{a^* + b^*} \tag{3.4}$$

and a^* and b^* are tabulated on Table VI.

An approximate equation for $\Phi(\alpha^*)$, accurate to 2%, was provided by Jones.

$$\Phi(\alpha^*) = \tfrac{2}{3} + \tfrac{11}{24}\alpha^*(2 - \alpha^*) \tag{3.5}$$

Jones reported that the experimental friction factor data for rectangular ducts lie within ±5% of the Karman–Nikuradse expression if Re_N^* is used:

$$1/\sqrt{f} = 4.0\log(\mathrm{Re}_N^*\sqrt{f}) - 0.4 \tag{3.6}$$

A similar conclusion was obtained for concentric annular ducts.

2. *Non-Newtonian Fluids*

While rather extensive studies have been conducted for non-Newtonian fluids in turbulent circular pipe flow, less attention has been paid to these fluids in noncircular flow. Since the relationship between the shear rate and shear stress is nonlinear in non-Newtonian fluids, the turbulent flow behavior is generally more complicated than that of Newtonian fluids. If the non-Newtonian fluid is viscoelastic, the turbulent flow behavior may be even more complicated since the fluid structure and elasticity may play a role in damping turbulence (partially laminarizing flow) [118] and storing some of the flow energy, which would otherwise be dissipated by the eddy motion. As in the case of Newtonian fluids, the hydrodynamic entrance lengths for purely viscous fluids are relatively short, being of the order of 20 hydraulic diameters. Again, the fully developed friction factor results are of major practical interest for this class of non-Newtonian fluids.

In the case of elastic fluids, such as aqueous solutions of such high molecular weight polymers as polyacrylamide and polyethylene oxide, the hydrodynamic entrance lengths may be of the order of 100 diameters. This reflects the laminar-like nature of such flows.

a. *Purely viscous fluids*

A major advance in the studies of purely viscous fluids was made by Dodge and Metzner [119]. They derived a general equation for the fully developed friction factor of purely viscous fluids in turbulent circular pipe flow by extending Millikan's approach to Newtonian fluids [120].

$$1/\sqrt{f} = M\log(\mathrm{Re}'f^{(2-n)/2}) - N \tag{3.7}$$

They determined M and N as functions of the power law index, n, for power law fluids using their pressure drop measurements of aqueous Carbopol solutions and slurries of Attagel. The experimental results covered a range of the power law index from 0.3 to 1. They proposed the following equation for predicting the fully developed friction factor for these so-called purely viscous fluids in turbulent flow through a circular pipe:

$$1/\sqrt{f} = \frac{4.0}{n^{0.75}} \log(\mathrm{Re}' f^{(2-n)/2}) - \frac{0.4}{n^{1.2}} \tag{3.8}$$

This equation is in agreement with the Karman–Nikuradse equation for the case of $n = 1$. They also suggested a universal velocity profile for a power law fluid in turbulent circular pipe flow based on Eq. (3.7):

$$u^+ = \begin{cases} (y^+)^{1/n} & \text{for} \quad y^+ < 5^n \\ \dfrac{5.0}{n} \ln y^+ - 3.05 & \text{for} \quad 5^n < y^+ < y_2^+ \\ \dfrac{2.78}{n} \ln y^+ + \dfrac{3.8}{n} & \text{for} \quad y_2^+ < y^+ \end{cases} \tag{3.9}$$

where y_2^+ is the value of y^+ at the interface between the buffer layer and turbulent core.

Kozicki *et al.* [25] suggested that the Dodge–Metzner equation with the hydraulic diameter replacing the pipe diameter be used to predict the fully developed friction factor for turbulent flow of power law fluids in noncircular ducts. Subsequently, Kostic and Hartnett [121] modified this suggestion and proposed that the Dodge–Metzner equation be used with Re′ replaced by Re*, a recommendation that is consistent with Jones' approach. The proposed equation for predicting the fully established friction factor of a purely viscous non-Newtonian fluid in turbulent flow through a rectangular channel then becomes:

$$1/\sqrt{f} = \frac{4.0}{n^{0.75}} \log(\mathrm{Re}^* f^{(2-n)/n}) - \frac{0.4}{n^{1.2}} \tag{3.10}$$

The validity of this approach was confirmed by Hartnett *et al.* [28] and Hartnett and Rao [122] for a wide range of power law index values. Pressure drop measurements of a number of aqueous Carbopol 934, Carbopol 960, and Attagel solutions were carried out in a square duct and in a 2:1 rectangular channel. The pressure drops were measured at x/D_h equal to 200 and represent the fully developed friction factors. The resulting fully developed Fanning friction factors are given in Fig. 42 as function of Re*. For comparison, the predicted values are also shown as solid lines for n values ranging from 0.4 to

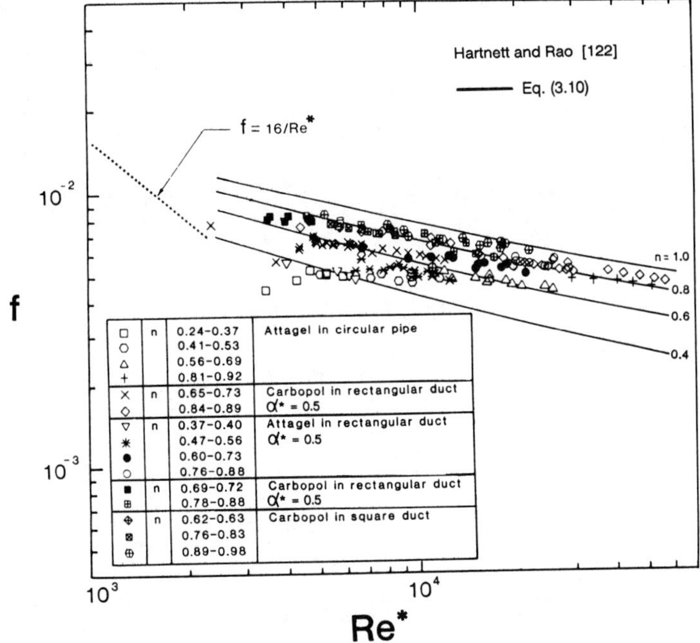

FIG. 42. Measured friction factors of aqueous solutions of Attagel and Carbopol in turbulent flow through circular pipes and rectangular ducts.

1.0. The experimental results show generally good agreement with the values predicted by Eq. (3.10). This is brought out in Fig. 43, which shows the measured friction factors plotted against the predicted values. With the exception of a few measurements, which fall in the transition region, the majority of the predictions were within $\pm 5\%$ of the measured values. It is concluded that the modified Dodge–Metzner equation, Eq. (3.10), can be used with confidence to predict the fully developed turbulent friction factor of a so-called purely viscous non-Newtonian fluid in a rectangular duct.

A minor disadvantage of the Dodge–Metzner formulation is its implicit form. As a consequence, a number of explicit formulations for the non-Newtonian friction factor have been introduced. These have been examined [123], and it was concluded that the modified Yoo equation [124] yields results that are within $\pm 10\%$ of Eq. (3.11) over the range of the power law index from 0.4 to 1.0.

$$f = 0.079n^{0.675}(\mathrm{Re}^*)^{-0.25} \quad \text{for} \quad 0.4 \leq n \leq 1.0, \quad 5000 < \mathrm{Re}^* < 50{,}000 \tag{3.11}$$

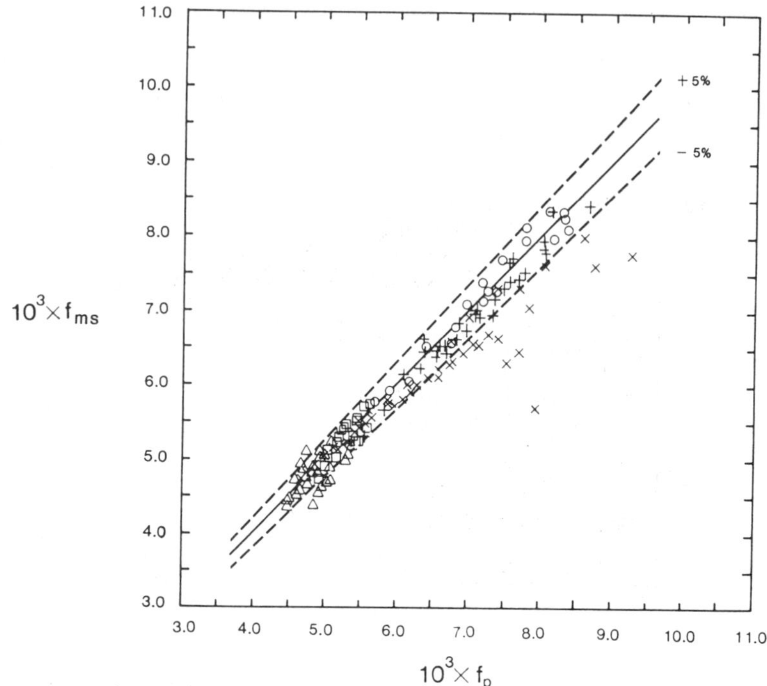

FIG. 43. A comparison of the measured turbulent friction factors, f_{ms} from Hartnett and Rao [122], with values predicted by the modified Dodge–Metzner equation, Eq. (3.10) (f_p). (△), Attagel in circular pipe; (×), Carbopol in circular pipe; (□), Attagel in rectangular duct ($\alpha^* = 0.5$); (○), Carbopol in rectangular duct ($\alpha^* = 0.5$); and (+), Carbopol in square duct.

The reliability of Eq. (3.11) is demonstrated in Fig. 44, which compares the predicted friction factors with the measured values previously shown on Fig. 42.

In summary, the modified Dodge–Metzner equation, Eq. (3.10), is recommended for predicting the fully established turbulent friction factor of a purely viscous fluid over the complete range of aspect ratios and power law index values. For values of n between 0.4 and 1.0, the explicit formulation of Yoo, Eq. (311), may be used.

b. *Drag-reducing viscoelastic fluids*

Viscoelastic fluids in turbulent channel flow may exhibit more complex friction factor and heat transfer behavior than Newtonian and so-called purely viscous fluids. Viscoelastic non-Newtonian fluids are usually shear thinning and also exhibit normal force differences. Even in a simple shear flow vis-

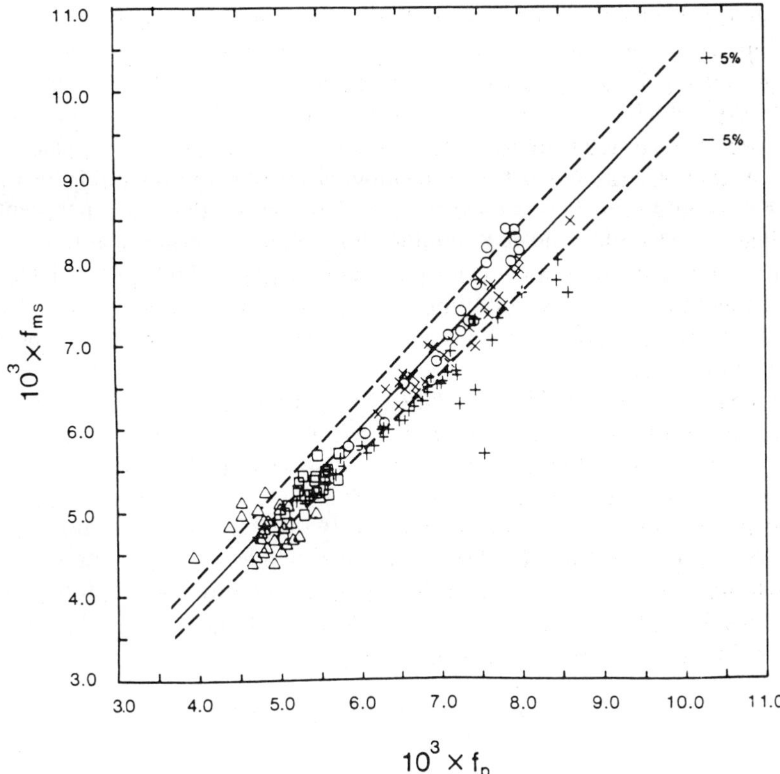

FIG. 44. A comparison of the measured turbulent friction factors, f_{ms} from Hartnett and Rao [122], with values predicted by the modified Yoo equation, Eq. (3.11) (f_p). (△), Attagel circular pipe; (+), Carbopol circular pipe; (□), Attagel rectangular duct ($\alpha^* = 0.5$); (○), Carbopol rectangular duct ($\alpha^* = 0.5$); and (×), Carbopol square duct.

coelastic fluids develop normal stress differences as well as shear stresses, a factor that is believed to play a major role in the behavior of some viscoelastic fluids. Such fluids may be obtained by the addition to a solvent of small amounts of certain long chain polymers: polyethylene oxide (Polyox), polyacrylamide—PAM (Separan), carboxymethylcellulose (CMC), and hydroxyethylcellulose (Natrosol). The resulting frictional drag and heat transfer in turbulent pipe flow are substantially reduced below the values associated with the pure solvent or with those fluids designated as purely viscous non-Newtonian fluids. Furthermore, for these drag-reducing viscoelastic fluids, the hydrodynamic entrance length [125] increases with increasing fluid elasticity reaching an asymptotic value of approximately 100 hydraulic diameters for highly elastic fluids.

Against this background, it should be noted that not all viscoelastic fluids exhibit drag reduction. Furthermore, there is some question as to whether all drag-reducing fluids are elastic. For example, a solution of 1000 wppm of carboxypolymethylene (Carbopol 934, B. F. Goodrich Co.) exhibits elastic behavior when tested in an oscillatory viscometer; that is, there is a phase shift smaller than $\pi/2$ radians between the input shear rate and the fluid stress response. In addition, as discussed in an earlier section, there is experimental evidence of secondary flows in laminar flow of such Carbopol solutions in noncircular geometries, a behavior predicted for elastic fluids [30, 31]. However, under turbulent flow conditions the aqueous Carbopol solutions behave as purely viscous power law fluids, giving the same pressure drop as aqueous Attagel solutions.

In contrast, the turbulent behavior of drag-reducing viscoelastic fluids is quite different from that of aqueous Carbopol solutions. The measured friction factors of these fluids decreases with increasing polymer concentration at a given Reynolds number. However, there is evidence that the extent of drag reduction is limited by a unique asymptote. In other words, the friction factor values of drag-reducing viscoelastic fluids are bounded by two limits, with the Newtonian value giving the upper limit and the minimum asymptote giving the lower limit. In a circular pipe flow, the lower limiting value is called Virk's minimum drag asymptote [126] and is given by

$$1/\sqrt{f} = 19.0 \log(\mathrm{Re}\sqrt{f}) - 32.4 \tag{3.12}$$

or

$$f = 0.59\,\mathrm{Re}^{-0.58} \tag{3.13}$$

Cho and Hartnett [90] proposed a minimum asymptote that is slightly lower than Virk's:

$$f = 0.2\,\mathrm{Re}^{-0.48} \tag{3.14}$$

Tung *et al.* [127] proposed a minimum drag asymptote based on Re′:

$$f = 0.332\,\mathrm{Re}'^{-0.55} \tag{3.15}$$

For these two limiting cases, the fully developed friction factor is a function only of the Reynolds number.

The upper limit of the fully developed friction factor in turbulent flow through rectangular ducts is the same as that for circular pipe flow if the Reynolds number proposed by Jones [111] or Kozicki [25] is used. The applicability of the lower limit pipe flow equations [Eqs. (3.14) and (3.15)] to the rectangular duct geometry has been studied by Hartnett *et al.* [121], Kostic [95], and Rao [103]. The fully developed turbulent friction factors of aqueous polyacrylamide solutions measured at x/D_{h} equal to 200 in a square duct are shown in Fig. 45 as a function of Re*. As in turbulent pipe flow, the friction factor decreases with increasing concentration (i.e., increasing fluid elasticity) up to 100 wppm. Any further increase in concentration beyond 100 wppm has no effect on the friction factor, and the friction factor remains at a minimum asymptote. However, the measured minimum square duct asymptotic friction factors are greater than the minimum circular pipe asymptotic values by 15 to 20% when compared at a fixed value of Re*. This difference may be due to the fact that the Kozicki Reynolds number, Re*, takes into account geometry and pseudoplasticity, but it does not account for the influence of viscoelasticity.

Alternatively, the same experimental results are replotted in Fig. 46 as a function of Re, the Reynolds number based on the apparent viscosity at the wall. In this presentation, the measured minimum friction factors for turbulent flow in a square duct are in good agreement with the minimum asymptotic values found for a circular pipe (i.e., $f = 0.2\,\mathrm{Re}^{-0.48}$). Further support for this conclusion is given in Fig. 47, which presents data for a rectangular duct of

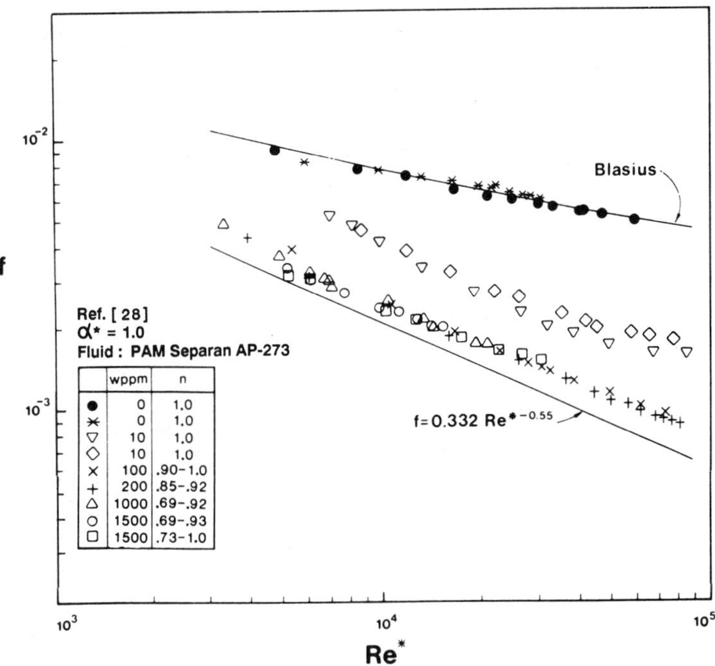

FIG. 45. Measured friction factors of aqueous polyacrylamide solutions in a square duct as a function of the Kozicki Reynolds number, Re*.

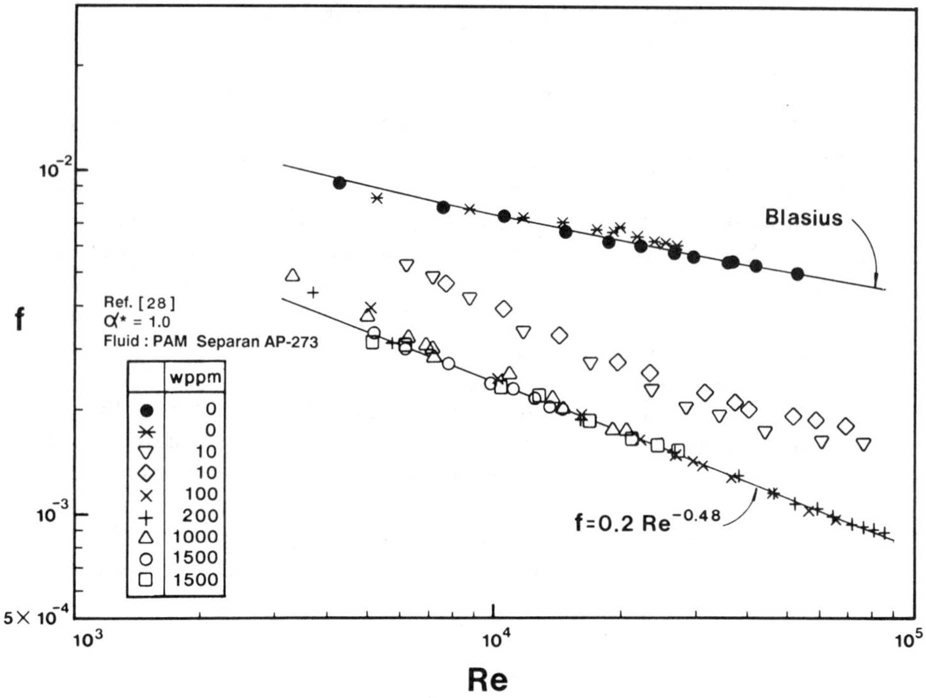

FIG. 46. Measured friction factors of aqueous polyacrylamide solutions in a square duct as a function of the Reynolds number based on the apparent viscosity, Re.

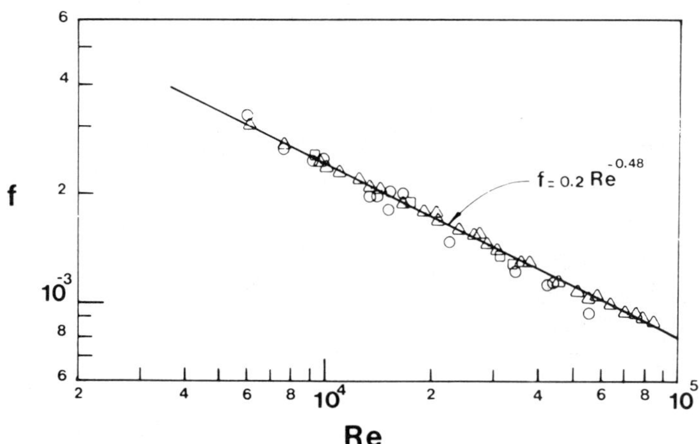

FIG. 47. Measured friction factors of aqueous polyacrylamide solutions in a rectangular duct ($\alpha^* = 0.5$ and 1.0) as a function of the Reynolds number based on the apparent viscosity, Re. Values are from Kostic and Hartnett [128] (○), Kwack *et al.* [129] (□), and Hartnett *et al.* [28] (△).

aspect ratios 0.5 and 1.0. [28, 128, 129]. Recent findings in a 5:1 rectangular duct also support this conclusion [103].

In the intermediate region, between the Newtonian and the minimum asymptote limits, the friction factor decreases with increasing fluid elasticity. Since the elasticity of the fluid is very sensitive to polymer concentration, degradation, and solvent and solute chemistry, these factors influence the friction factor behavior in this intermediate region. The detailed influence of these factors on circular pipe flow is discussed in the review by Cho and Hartnett [90], who conclude that another dimensionless parameter, the Weissenberg number, which takes account of the elasticity of the fluid, is required to correlate the fully developed friction factor of viscoelastic fluids between these two limits. It was found by Hartnett and Kwack [130] and by Ghajar and Azar [131] that the fully developed turbulent friction factor is a function only of the Weissenberg and Reynolds numbers.

Figure 48 shows typical values of the fully developed friction factor of aqueous polyacrylamide solutions (Separan AP-273) in a circular pipe as a function of the Weissenberg number for various Reynolds numbers from 10,000 to 70,000 [130]. The Weissenberg number was calculated based on the Powell–Eyring characteristic time, λ, which was determined by fitting the measured values of steady shear viscosity to the Powell–Eyring model:

$$\frac{\eta - \eta_\infty}{\eta_0 - \eta_\infty} = \frac{\sinh^{-1}(\lambda\dot{\gamma})}{(\lambda\dot{\gamma})} \tag{3.16}$$

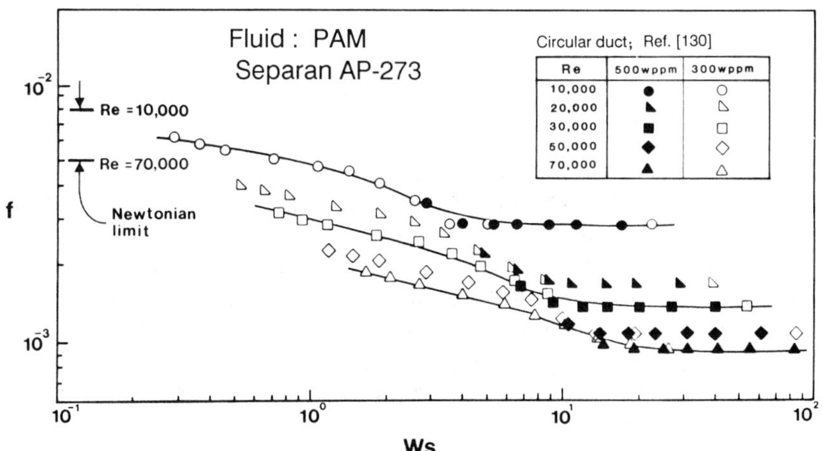

FIG. 48. Fanning friction factors for fully established turbulent pipe flow of aqueous polyacrylamide solutions as a function of the Weissenberg number for various Reynolds numbers.

The fully established friction factor is a function only of the Reynolds number in the high Weissenberg number region. However, as the Weissenberg number decreases for a fixed Reynolds number, a critical Weissenberg number is reached; below this critical value, the friction factor increases with decreasing Weissenberg number, approaching the Newtonian value as the Weissenberg number approaches zero.

Ghajar and Azar [131], on the basis of experiments with two polyacrylamides, using Separan AP-273 and Separan AP 30, extended the work of Kwack and proposed the following correlations for the friction factor, which are in agreement with Kwack's earlier results.

$$f = 0.20\,\mathrm{Re}^{-0.48}\exp[1.4(1\text{-}\mathrm{Ws}/\mathrm{Ws}_{cf})^{1.1}] \qquad \text{for AP-273} \qquad (3.17)$$

$$f = 0.20\,\mathrm{Re}^{-0.48}\exp[1.0(1\text{-}\mathrm{Ws}/\mathrm{Ws}_{cf})^{1.1}] \qquad \text{for AP-30} \qquad (3.18)$$

Here the critical Weissenberg numbers for friction, Ws_{cf}, are given by the following expressions:

$$\mathrm{Ws}_{cf} = 1.5 \times 10^{-3}\,\mathrm{Re}^{0.85} \qquad \text{for AP-273} \qquad (3.19)$$

$$\mathrm{Ws}_{cf} = 2.1 \times 10^{-3}\,\mathrm{Re}^{0.85} \qquad \text{for AP-30} \qquad (3.20)$$

Since these correlations are based on experimental pipe flow results for aqueous polyacrylamide solutions only, the general applicability for other polymer solutions and other geometries needs to be examined. For preliminary designs, these correlations may be used to estimate the fully developed friction factor of aqueous polyacrylamide solutions in a rectangular duct, if the hydraulic diameter is used as a characteristic length.

B. HEAT TRANSFER

1. *Newtonian Fluids*

The thermal entrance region for turbulent duct flow of Newtonian fluids is relatively short, not more than 15 to 20 hydraulic diameters for various thermal boundary conditions. Therefore, the fully developed heat transfer results are of major practical interest. Although the heat transfer behavior of Newtonian fluids in turbulent flow through rectangular ducts had not been investigated as widely as the circular pipe geometry, nevertheless there is a substantial body of experimental information on this topic.

Table XV presents a summary of typical experimental studies of turbulent heat transfer of Newtonian fluids flowing through rectangular ducts having various thermal boundary conditions [103, 132–148]. Most of the studies involved the thermal boundary condition of constant heat flux on one or more of the bounding walls. Several investigators imposed constant heat flux on

TABLE XV

TURBULENT HEAT TRANSFER MEASUREMENTS—NEWTONIAN FLUIDS IN RECTANGULAR DUCTS

Investigator	Aspect ratio	Hydraulic diameter (cm)	L/D_h	Fluid studied	Boundary conditions[a]	Temperature	Remarks and results
Washington and Marks [132]	0.025 0.050 0.1124	0.051 0.102 0.213	197 101 47	Air	T_w (top), T_w (bottom)	$T_w = 366$ K	Steam heated copper plate. Critical Re = 3400 for nonisothermal flow. Heat transfer data agree with $\mathrm{Nu} = 0.0203\ \mathrm{Re}^{0.8}$.
Lowdermilk *et al.* [133]	1.0 0.2	1.14 1.07	53 57	Air	q	$T_w = 300–900$ K $T_w/T_b = 1.2–2.3$	Good agreement with Dittus–Boelter equation if properties evaluated at film temperature. $\frac{hD_h}{k} = 0.023\left[\frac{\rho V D_h}{\eta}\right]^{0.8}(\mathrm{Pr})^{0.4}$ for $L/D_h > 50$
Lancet [134]	0.2	0.099	128	Air	q	$T_w = 340–425$ K	Reasonable agreement with Colburn equation, Re > 10,000: $\frac{hD_h}{k} = 0.023\left[\frac{\rho V D_h}{\eta}\right]^{0.8}(\mathrm{Pr})^{1/3}$
Levy *et al.* [135]	0.04	0.55	95–190	Water	q (top), q (bottom)	$T_b = 310$ K $T_w/T_b = 1.2–2.6$	About 30% below all circular tube correlations.
Heineman [136]	0.038	0.230	133	Superheated steam	q	$T_w = 530–800$ K $T_w - T_b = 280–420$ K	Approximately 8% below the Colburn equation: $\frac{hD_h}{k} = 0.023\left[\frac{\rho V D_h}{\eta}\right]^{0.8}(\mathrm{Pr})^{1/3}$ for $L/D_h > 60$

(Continued)

TABLE XV (*Continued*)

Investigator	Aspect ratio	Hydraulic diameter (cm)	L/D_h	Fluid studied	Boundary conditions[a]	Temperature	Remarks and results
Gambill and Bundy [137]	0.1	0.246	103	Water	q (all walls)	$T_w - T_b \geq 15$ K	Fair agreement ($\pm 25\%$) with Sieder–Tate correlation: $\mathrm{Nu} = 0.027(\mathrm{Re}_b)^{0.8}(\mathrm{Pr}_b)^{1/3}\left(\frac{\eta_w}{\eta_b}\right)^{0.14}$
Battista and Perkins [138]	1.0	0.241	155	Air	q (all walls)	$T_w - T_b = 195$–250 K	Measured friction factors 20% higher and heat transfer coefficients 10% lower than circular tube results.
Novotny *et al.* [139, 140]	1.0 0.2 0.1	2.49 2.54 1.85	132–206	Air	q (top), q (bottom), insulated sides	$T_w = 310$–365 K $T_w - T_b = 10$–20 K	Experimental data in good agreement with analytical prediction for parallel plate duct. Data in agreement with Colburn equation at $\mathrm{Re} = 10^4$ but 15% higher than Colburn equation at $\mathrm{Re} = 10^5$.
Barrow [141]	0.034	1.57	97	Air	q_1 (top), q_2 (bottom), insulated sides	$T_w = 300$ K	For symmetrical heating, measurements were below the Dittus–Boelter equation. For asymmetrical heating, scatter of data was too great to yield definitive conclusions.
Sparrow *et al.* [142]	0.2	2.54	140	Air	q_1 (top), q_2 (bottom), insulated sides	$T_w = 310$–350 K $T_w - T_b = 5$–20 K	Measured heat transfer coefficients for higher heat flux wall are lower and for lower heat flux wall are higher than symmetic heating results.

James *et al.* [143]	1, 0.5, 0.4 0.333 0.25	0.254–0.85	>67 hydro 51–67 thermal	Distilled water and mixture with glycerine, Pr = 6.5 to 100	T_w	$\Delta T_b = 20\text{–}80$ K	The results are correlated by: $\mathrm{Nu} = \dfrac{0.104\ \mathrm{Re}^{0.016\,\mathrm{Pr}+0.75}\ \mathrm{Pr}^{0.4}\ R^{213\,\mathrm{Re}^{-0.9}}}{e^{0.0134\,\mathrm{Pr}}(2.05 + 1.62e^{-A})}$ $A = \begin{cases} 1/\alpha^* & \text{if shorter side heated} \\ \alpha^* & \text{if longer side heated} \end{cases}$ $R = \eta_b/\eta_w$
Brundrett and Burroughs [144]	1.0	9.57	96	Air	q		Measured friction factors and Nusselt numbers of the order of 10% lower than in circular tube.
Tan and Charters [145]	0.333	7.11	103	Air	q	$T_w = 310$ K $T_w - T_b = 40$ K	Measured fully developed heat transfer coefficients given by $\mathrm{Nu} = 0.018\ \mathrm{Re}^{0.8}\ \mathrm{Pr}^{0.4}$
Haynes and Ashton [146]	0.1	4.60	796	Water	q	$T_w = 275\text{–}280$ K	Bottom wall cooled by flowing refrigerant. Measured heat transfer coefficient approximately 50% higher than predicted by Dittus–Boelter equation.
Sparrow *et al.* [147]	0.2	4.445	40 hydro 13 thermo	Air	T_w	Mass transfer: naphthalene to air was measured	Measured fully developed Nusselt numbers approximately 20% below Dittus–Boelter equation. Thermal entrance length of the order of 40 hydraulic diameters.
Kostic and Hartnett [148]	0.5	1.2	532	Water	q_1 q_2	$T_w = 300$ K $T_w - T_b = 2$ K	Measured Nusselt number in agreement with Dittus–Boelter equation for symmetric heating. Asymmetric heating influence on Nusselt number also reported.
Rao [103]	0.2	3.75	150	Water	q	$T_w = 300$ K $T_w - T_b = 3$ to 5 K	Measured Nusselt numbers are approximated 10% below Dittus–Boelter predictions.

[a] T_w, uniform surface temperature; q, uniform heat flux; ▨, insulated wall.

the upper and lower walls with the side walls being insulated. Heat transfer results for the case where the upper and lower walls are heated at constant temperature are also available. The most common test fluid was air with a few studies using water, while superheated steam was used by one investigator. With the exception of Levy *et al.* [135] and Haynes and Ashton [146], all of the investigators using symmetrical thermal boundary conditions reported reasonable agreement with one of the commonly used circular tube correlations when the characteristic length is taken as the hydraulic diameter.

Shah and Johnson [149] reviewed a number of correlations for heat transfer and for the friction factor of Newtonian fluids in circular pipes and noncircular ducts. In most correlations, a value of 0.4 is used for the power of the Prandtl number. Hence, the experimental results of Novotny *et al.* [139, 140] and Kostic and Hartnett [148] are presented on Fig. 49 as $Nu/Pr^{0.4}$ as a function of Re. The experimental results are in good agreement with the Dittus–Boelter prediction in the Reynolds number range of 10,000–50,000. At lower and higher Reynolds numbers, the Dittus–Boelter equation predicts

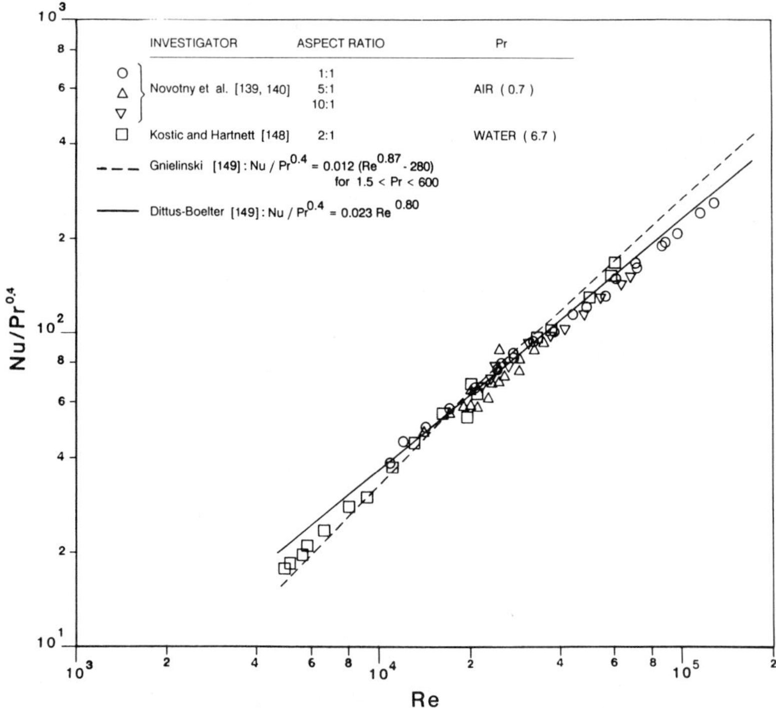

FIG. 49. Comparison of experimental heat transfer results for turbulent flow of Newtonian fluids in rectangular channels.

Nusselt values that are somewhat higher than the experimental values. Emery *et al.* [110], using the k-ε model, predicted velocity and temperature profiles for the square duct geometry and reported heat transfer coefficient which are approximately 10% below the Dittus–Boelter prediction. Thus, it is proposed that the fully established turbulent heat transfer behavior of Newtonian fluids in rectangular ducts in the range of Reynolds numbers from 10^4–10^6 be predicted by the following equation:

$$\mathrm{Nu} = 0.90 \times 0.023\, \mathrm{Re}^{0.8}\, \mathrm{Pr}^{0.4} \tag{3.21}$$

In the case of asymmetric heating of a Newtonian fluid in turbulent flow through a rectangular duct, there are several experimental studies available, including the work of Sparrow *et al.* [142], Kostic and Hartnett [148], and Rao [103]. In addition, there are analytical studies available for the limiting case of a parallel plate duct with asymmetric thermal boundary conditions: Barrow [150], Hatton and Quarmby [151], and Hatton *et al.* [152]. They suggest that the Prandtl number has a strong effect on the Nusselt number for a fixed Reynolds number and heat flux ratio. In particular, the effect of asymmetric heating decreases with an increasing Prandtl number.

On the basis of the experimental and analytical studies, it is clear that the Nusselt numbers for the lower heat flux wall of an asymmetrically heated rectangular duct are higher than the Nusselt numbers in a symmetrically heated case. Conversely, the higher heat flux wall Nusselt number is lower than the symmetric case value. This is brought out on Fig. 50. In the case where only one long side is heated, the Nusselt numbers for air ($\mathrm{Pr} = 0.7$) are approximately 20% lower than the symmetric heating case and 10% lower for water ($\mathrm{Pr} = 7.0$) for $\alpha^* = 0.2$ and 0.5.

2. *Non-Newtonian Fluids*

Relatively few experimental studies involving non-Newtonian fluids in turbulent flow through rectangular ducts have been carried out. These will be discussed in this section against the background of circular pipe flow studies of non-Newtonian fluids.

a. *Purely viscous fluids (nondrag-reducing fluids)*

Metzner and Friend [153], using the approach introduced by Reichardt, determined the Stanton number of a purely viscous power law fluid in turbulent pipe flow as a function of the friction factor and Prandtl number:

$$\mathrm{St} = \frac{f/2}{1.2 + 11.8(f/2)^{1/2}(\mathrm{Pr} - 1)\,\mathrm{Pr}^{-1/3}} \tag{3.22}$$

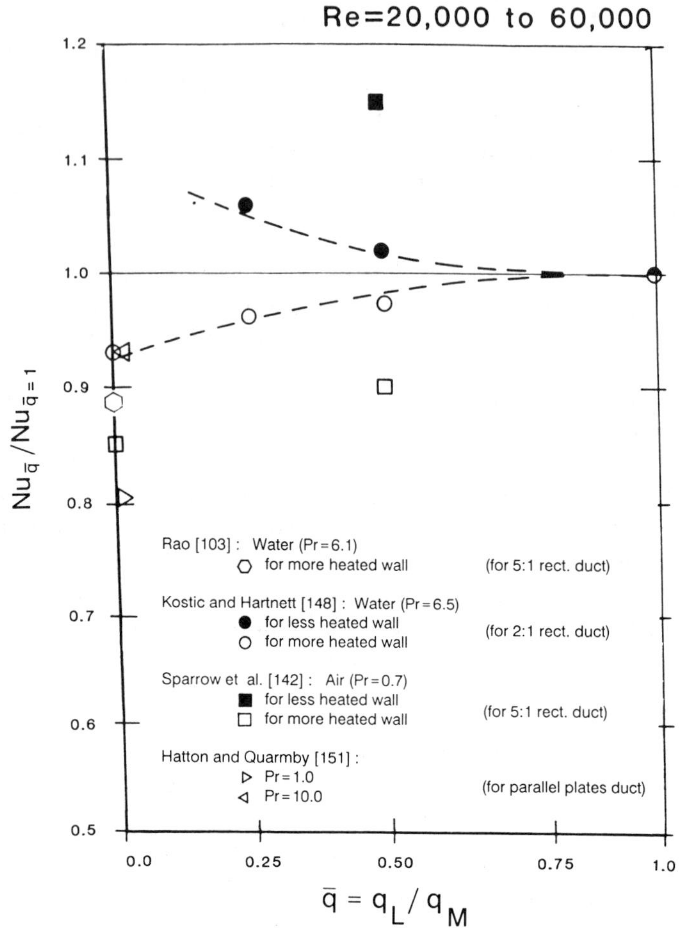

FIG. 50. Nusselt number ratio, $Nu_{\bar{q}}/Nu_{\bar{q}=1}$ as a function of $\bar{q}$ for asymmetric heating of a Newtonian fluid in rectangular channels.

The friction factor is determined by the Dodge–Metzner equation, Eq. (3.8). Metzner and Friend reported that the prediction was in good agreement with turbulent flow heat transfer data obtained with aqueous solutions of Carbopol, corn syrup, and slurries of Attagel.

Yoo [124] reported better agreement of his circular tube data with an empirical correlation:

$$St = 0.0152\,Re^{-0.155}\,Pr^{-2/3} \tag{3.23}$$

He reported that his experimental pipe flow data taken with aqueous solutions of Carbopol and Attagel agreed with Eq. (3.23) with a standard deviation of 24%. Hartnett and Rao reported that Eq. (3.23) could be used for rectangular channels as well as for circular tubes, yielding predictions that are within ±20% of experimental data [122]. They also proposed a new equation to bring the prediction for circular pipes as well as rectangular channels into better agreement with generally accepted Newtonian heat transfer results:

$$\mathrm{Nu} = (0.0081 + 0.0149n)\,\mathrm{Re}^{0.8}\,\mathrm{Pr}^{0.4} \tag{3.24}$$

The agreement of Eq. (3.24) with experiment is brought out in Fig. 51, where it may be seen that the equation is generally with ±20% of the experimental data.

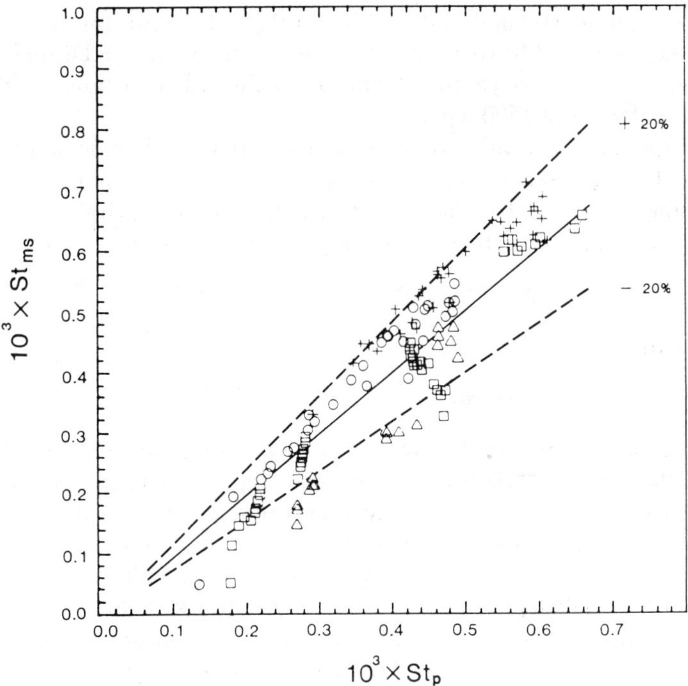

FIG. 51. Comparison of measured fully developed Stanton numbers, St_{ms} from Hartnett Rao [122], with predicted Stanton number, St_p from Eq. (3.24), for turbulent flow of purely viscous non-Newtonian fluids in circular tubes and rectangular ducts, $\alpha^* = 0.5$. (+), Attagel in rectangular duct; (○), Attagel in circular pipe; (△), Carbopol in rectangular duct; and (□), Carbopol in circular pipe.

b. *Drag-reducing viscoelastic fluids*

Cho and Hartnett [90] reviewed the available turbulent pipe flow heat transfer information for drag-reducing viscoelastic fluids. They reported that a lower asymptotic limit is reached for the dimensionless heat transfer coefficient, j_H, and that the thermal entrance length in this asymptotic case is of the order of 400 to 500 pipe diameters. These low heat transfer values and the long thermal entrance region reflect the laminar-like character of the flow. The reduction in heat transfer from the Newtonian value is found to be greater than the reduction in the friction factor.

Heat transfer results for drag-reducing viscoelastic fluids in turbulent flow through rectangular channels are relatively rare. Kostic and Hartnett [128] reported results for aqueous polyacrylamide solutions in an aspect ratio of 0.5 with the upper and lower walls heated by passing an electric current through them while the side walls were insulated [i.e., the $H1(2L)$ boundary condition]. The length of the test section was 532 hydraulic diameters. As mentioned in an earlier section, Kostic and Hartnett found the upper limit for zero polymer concentration ($n = 1.0$) to be in good agreement with the Dittus–Boelter equation. The lower asymptotic limit was achieved with polymer concentrations of 1000 and 1500 wppm.

The experimental results for these aqueous polyacrylamide solutions are shown in Fig. 52, where the asymptotic values are compared with the empirical correlations reported by Cho and Hartnett [90] for turbulent pipe flow asymptotic conditions. The pipe flow asymptotic results are correlated by

$$j_H^* = \mathrm{Nu}/(\mathrm{Re}^* \, \mathrm{Pr}^{*1/3}) = 0.02\,\mathrm{Re}^{*-0.4} \tag{3.25}$$

or alternatively by

$$j_H = \mathrm{Nu}/(\mathrm{Re}\,\mathrm{Pr}^{1/3}) = 0.03\,\mathrm{Re}^{-0.45} \tag{3.26}$$

Although the rectangular channel data are somewhat higher (5 to 10%) than the circular tube correlation equations, it appears that the circular tube predictions may be used for engineering estimates of the asymptotic heat transfer for rectangular ducts having an aspect ratio of approximately 0.5.

In this same spirit, it is proposed that the intermediate values of the heat transfer coefficient lying between the Newtonian value and the lower asymptotic limit be estimated from the pipe flow correlation shown on Fig. 53 [130]. This approach should give reasonable estimates, at least for aqueous solutions of Separan AP-273.

Given the fact that the influence of this class of polymer additives laminarizes the flow, it is important that the influence of the aspect ratio and asymmetric heating be studied. These effects are major for laminar flow as brought out earlier in Section II.

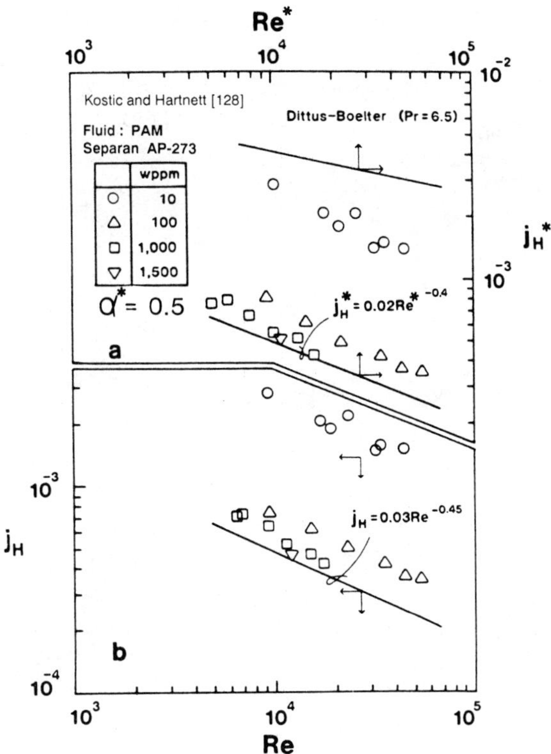

FIG. 52. Heat transfer factor, j_H, versus Reynolds number for turbulent flow of aqueous polyacrylamide solution, $\alpha^* = 0.5$.

Limited experimental data have been reported by Rao [103] for aqueous polyacrylamide solutions flowing turbulently in a one-side heated duct [i.e., the $H1$(1L) boundary condition] of aspect ratio 0.2. However, Rao's test section was only 150 hydraulic diameters and many of his results are in the thermal entrance region. Furthermore, it is unclear whether Rao reached the asymptotic heat transfer limit. A continuation of this work in a longer duct should provide answers as to the influence of the aspect ratio and asymmetric heating.

IV. Suggestions for Future Research

An overview of the analytical and experimental hydrodynamics and heat transfer studies of Newtonian and non-Newtonian fluids in laminar and turbulent flow through rectangular tubes has been presented. In most cases,

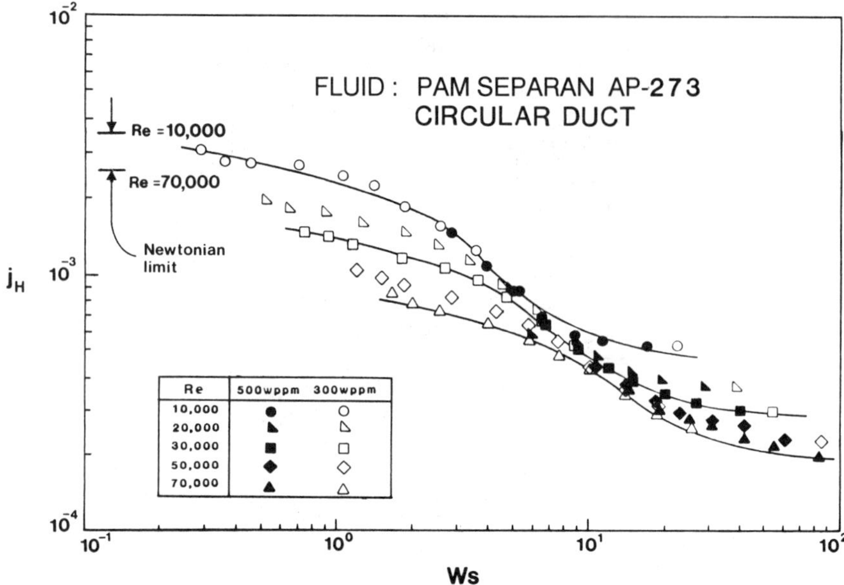

FIG. 53. Heat transfer factor, j_H, for fully established turbulent pipe flow of aqueous polyacrylamide solution as a function of the Weissenberg number with the Reynolds number as a parameter. Results are from Hartnett and Kwack [130].

the behavior of common Newtonian fluids such as air, water, oils, etc., can be predicted with sufficient accuracy for engineering design.

In the case of non-Newtonian fluids, the theoretical predictions yield low estimates of the heat transfer under laminar flow conditions. The fact that the available experimental heat transfer measurements lie above the predictions could reflect an inadequacy in the analytical model. For non-Newtonians in turbulent flow through rectangular channels, the situation is even more complicated. Some non-Newtonian fluids act as pseudoplastics, showing some reduction in friction and heat transfer as compared with a Newtonian fluid. Other non-Newtonian fluids experience large reductions in the friction factor and in heat transfer under turbulent flow conditions. It is difficult, if not impossible, to predict a priori which solutions will behave as pseudoplastic and which will behave as drag-reducing fluids.

Having stated that the behavior of ordinary Newtonian fluids is reasonably well understood, it should be noted that there are "extraordinary" Newtonian fluids, such as very dilute aqueous solutions of polyacrylamide or of polyethylene oxide. The viscosity of such fluids is invariant with shear rate and equal to that of water. These fluids exhibit the same behavior as water in a rotating viscometer or in a capillary tube viscometer. However, in turbulent flow in circular and noncircular channels, the dilute solutions of these polymers behave very differently than do Newtonian fluids, showing substantial reductions in pressure drop and heat transfer.

High concentrations of these same drag-reducing aqueous polymer solutions are viscoelastic, that is, they have measurable normal force differences and show a phase shift different from Newtonian fluids in an oscillatory viscometer. Furthermore, they are non-Newtonian with shear-dependent viscosity. They also reveal a substantial reduction in the dimensionless turbulent friction factor as a function of the Reynolds number. It appears that the presence of these particular drag-reducing polymers for both low and high concentrations modifies the turbulent structure of the flow, in effect laminarizing the flow. An interesting question arises with respect to very dilute concentrations of these drag-reducing aqueous polymer solutions: Are they viscoelastic Newtonian fluids? If so, what additional measurement is necessary to establish the viscoelastic nature of these fluids? It is clear that a knowledge of the viscosity alone is inadequate to predict their behavior in turbulent flow.

Drag reduction does not appear to be an adequate criterion to establish viscoelasticy. For example, a solution of 1000 wppm of Carbopol is viscoelastic but shows no drag reduction in turbulent flow, rather it acts as a pseudoplastic fluid. Furthermore, there are many types of additives (including soaps, different particle suspensions, etc.) that show drag reduction under turbulent flow conditions. It is highly unlikely that all of these fluids are viscoelastic. Rather, it is probable that different mechanisms are involved in modifying the turbulent structure of the flow field.

A related question surrounding these very dilute drag-reducing polymer solutions involves their behavior under laminar flow conditions. It is well established that the laminar pressure drop behavior agrees with the Newtonian prediction. However, there remains a question of whether such dilute polymer solutions in laminar flow through noncircular ducts generate secondary flows of the type found with high polymer concentrations. Measurements of the heat transfer performance of dilute concentrations of drag-reducing polymer additives are needed to answer this question. Perhaps the presence of nongravity-induced secondary flow as reflected in increased heat transfer performance could be an indication of viscoelasticity.

This brings up a related issue. Are the secondary flows associated with laminar flow in noncircular ducts only found in viscoelastic fluids? In other words, are there non-Newtonian fluids that show no secondary flows under such circumstances? Again, experimental laminar flow heat transfer studies of purely viscous non-Newtonian fluids in non-circular channels should be carried out to provide further insight into the behavior of non-Newtonian fluids. It appears that there are a sufficient number of unanswered questions about the behavior of non-Newtonian fluids in noncircular channels to keep researchers busy for many years.

NOMENCLATURE

a	half of the longer side of the rectangle, Fig. 1 and Table VI
a^*	geometric constant in Kozicki generalized Reynolds number, Eqs. (1.42) and (2.17) and Table VI
a_1, a_2	coefficients in velocity profile equations depending on aspect ratio and power law index, Eq. (2.10) and Table III
$A_1, A_2, \ldots, A_6$	coefficients in velocity profile equations depending on aspect ratio and power law index, Eq. (2.11) and Table V
A	area of the duct cross section, 4ab
b	half of the shorter side of the rectangle or half of the distance between parallel plates, Fig. 1
b^*	geometric constant in Kozicki generalized Reynold number, Eqs. (1.42) and (2.17) and Table VI
c	f Re product for Newtonian fluids, $16(a^* + b^*)$, Table VI
CMC	carboxymethylcellulose
C_p	specific heat of fluid
De	Deborah number, $\lambda/\bar{t}$, Eq. (1.52)
d_{ij}	shear rate tensor, component of, Eqs. (1.10) and (1.13)
D_h	hydraulic diameter, $4A/P$
$\frac{D}{Dt}$	material (substantial) derivative, Eq. (1.4)
e	energy per unit volume of the fluid element
f	Fanning friction factor, Eq. (1.32)
$\mathbf{F}_b$	body force (e.g. gravity force) per unit of fluid volume
g	acceleration due to Earth's gravity
g_x	component of g in x direction
g_y	component of g in y direction
g_z	component of g in z direction
Gr	Grashof number based on temperature difference, Eq. (1.47)
Gr_q	Grasof number based on heat flux, Eq. (1.48)
Gz	Graetz number, inversely proportional to x^+_{th} or x^*_{th}, Eq. (1.28)
$\mathrm{Gz}_{\dot{m}}$	Graetz number based on mass flow rate, Eq. (1.29)
h	convective heat transfer coefficient, $q''/(T_w - T_b)$
I	first invariant of shear rate tensor, Eq. (1.10)
II	second invariant of shear rate tensor, Eq. (1.10)
III	third invariant of shear rate tensor, Eq. (1.10)
j_H	Colburn heat transfer factor, $\mathrm{Nu}/(\mathrm{Re}\,\mathrm{Pr}^{1/3})$
j^*_H	Colburn heat transfer factor, based on Re* and Pr*, $\mathrm{Nu}/(\mathrm{Re}^*\mathrm{Pr}^{*1/3})$
K	consistency index in power law fluid model, $\tau = K\dot{\gamma}^n$
K^*	incremental pressure drop number, Eqs. (1.33), (2.20), and (2.21)
k	thermal conductivity of fluid
L_{hy}	hydrodynamic entrance length

L^+_{hy}	dimensionless hydrodynamic entrance length, $L_{hy}/(D_h \mathrm{Re}^+)$
L^*_{hy}	dimensionless hydrodynamic entrance length, $L_{hy}/(D_h \mathrm{Re}^*)$
L_{th}	thermal entrance length
L^+_{th}	dimensionless thermal entrance length, $L_{th}/(D_h \mathrm{Re}^+ \mathrm{Pr}^+)$
L^*_{th}	dimensionless thermal entrance length, $L_{th}/(D_h \mathrm{Re}^* \mathrm{Pr}^*)$, equal to L^+_{th}
M	coefficient in Metzner type equation, Eq. (3.7)
$\dot{m}$	mass flow rate
n	power law index in power law fluid model, $\tau = K\dot{\gamma}^n$
n^*	dimensionless coordinate normal to the perimeter P, either Z^* or Y^*
N	coefficient in Metzner type equation, Eq. (3.7)
Nu	Nusselt number, hD_h/k
P	perimeter of the duct cross section
P_h	heated part of the perimeter P
P^*_h	dimensionless P_h, P_h/D_h
p	pressure
Δp^*	dimensionless pressure drop, $\Delta p/(\rho U^2/2)$
PAM	polyacrylamide
Pe	Peclet number, $\rho C_p U D_h/k$, Eq. (1.43)
Pr	Prandtl number, Pe/Re, also see Eqs. (1.44)–(1.46)
Pr^+	Prandtl number, based on Re^+, $\mathrm{Pe}/\mathrm{Re}^+$
Pr^*	Prandtl number, based on Re^*, $\mathrm{Pe}/\mathrm{Re}^*$
$\mathbf{q}$	vector of heat flux per unit area
q'	heat flux per unit of duct length
q''	heat flux per unit of heating area
$\bar{q}$	ratio of the smaller heat flux to the larger heat flux in asymmetric case

Ra	Rayleigh number associated with Gr, Eq. (1.49)
Ra_q	Rayleigh number associated with Gr_q, Eq. (1.50)
Re	Reynolds number based on apparent or Newtonian viscosity, $\rho U D_h/\eta$
Re′	Metzner's generalized Reynold's number, Eq. (2.18)
Re^+	generalized Reynolds number, $\rho U^{2-n} D^n{}_h/K$
Re^*	Kozicki generalized Reynolds number, Eq. (1.42)
Re^*_N	generalized Reynolds number for Newtonian fluids, $\mathrm{Re}/(a^* + b^*)$
r	power constant in the approximate velocity profile equation, Eqs. (2.3)–(2.8)
s	power constant in approximate velocity profile equation, Eqs. (2.3)–(2.8)
S	heat source per unit volume of the fluid element
St	Stanton number, Nu/Pe
t	time
$\bar{t}$	characteristic time of flow
T	temperature
T^*	dimensionless temperature, Eqs. (1.35)–(1.37)
u	velocity component in axial direction
$u_{i,j,k}$	u, v, w, respectively
u^+	dimensionless velocity u/U^*
u^*	dimensionless velocity u/U
U^*	friction velocity, $\sqrt{\tau_w/\rho}$
U	mean velocity in axial direction
v	velocity component in transverse y direction
v^*	dimensionless velocity v/U
$\mathbf{V}$	velocity vector
w	velocity component in transverse z direction
w^*	dimensionless velocity w/U
wppm	weight parts per million
Ws	Weissenberg number, $\lambda U/D_h$

x	axial rectilinear coordinate
$x_{i,j,k}$	x, y, z, respectively
x_{hy}	dimensionless axial coordinate for the hydrodynamic entrance region $x/(D_h \mathrm{Re}^+)$
x^*_{hy}	dimensionless axial coordinate for the hydrodynamic entrance region based on Kozicki generalized Reynolds number, $x/(D_h \mathrm{Re}^*)$
x^+_{th}	dimensionless axial coordinate for the thermal entrance region, $x/(D_h \mathrm{Re}^+ \mathrm{Pr}^+)$
x^*_{th}	dimensionless axial coordinate for the thermal entrance region based on Kozicki generalized Reynolds number, $x/(D_h \mathrm{Re}^* \mathrm{Pr}^*)$, equal to x^+_{th}
y	transverse rectilinear coordinate, orthogonal to x
y^+	dimensionless distance from the wall, $\rho y_w U^*/\eta$
y_w	distance from the wall
Y^*	dimensionless peripheral coordinate, y/D_h
z	transverse rectilinear coordinate, orthogonal to x and y
Z^*	dimensionless peripheral coordinate, z/D_h

Greek Symbols

$\alpha_{1,2,\ldots,6}$	constants in velocity profile equation, Eq. (2.11) and Table IV
$\alpha^* = 2b/2a$	aspect ratio of duct cross section
β	volumetric coefficient of thermal expansion
$\beta_{1,2,\ldots,6}$	constants in velocity profile equation, Eq. (2.11) and Table IV
$\dot{\gamma}$	shear rate
Δ	difference of quantities
∇	vector-differential operator
ε_i	parameters in Ref. 26 used in calculation for Table VIII, $i = 0, 1, 2$
ζ_i	ith eigenvalue of the boundary value problem of Eqs. (2.25) and (2.26)
η	viscosity of Newtonian fluid or apparent viscosity of non-Newtonian fluid
η^*	dimensionless complex velocity gradient, Eq. (1.31)
λ	characteristic time of viscoelastic fluid
ν	kinematic fluid viscosity, η/ρ
ξ	dimensionless coordinate, y/b in Eq. (2.25)
ρ	density of fluid
τ_{ij}	shear stress component in j direction (i.e., x, y, z, direction) acting on the surface orthogonal to i direction (i.e., x, y, z direction)
$\bar{\bar{\tau}}$	stress tensor
ϕ	function of power law index, given by Eq. (2.28)
$\Phi(\alpha^*)$	geometric function, Eqs. (3.4) and (3.5)
ψ	function of power law index, given by Eq. (2.30)
Ψ_i	ith eigenfunction of the eigenvalue problem of Eqs. (2.25) and (2.26)

Subscripts

a	apparent value, based on apparent viscosity
b	bulk fluid property
fd	fully developed value

$H1$	constant wall heat flux axially and constant temperature peripherally, boundary condition, see also Table I and Eqs. (1.21) and (1.22)	m	mean, average value
		max	maximum value of a quantity
		p	predicted value
		$\bar{q}$	refers to a given ratio $\bar{q}$
		T	constant wall temperature boundary condition, see also Table I and Eqs. (1.21 and (1.22)
$H2$	constant wall heat flux axially and peripherally, boundary condition, see also Table I and Eqs. (1.21) and (1.22)	u	at upper wall
in	at duct inlet	w	at the wall
L	refers to wall with lower heat flux	x	at axial location x from the duct entrance
ms	measured value	0	for very small (zero) shear rate
M	refers to wall with higher heat flux	∞	for very large (infinite) shear rate

ACKNOWLEDGMENTS

The authors acknowledge the assistance of many of their colleagues in preparing this review paper. Special thanks are due Dr. E. Y. Kwack of the Jet Propulsion Laboratory, who began work on the review several years ago during the time he was associated with the Energy Resources Center of the University of Illinois at Chicago. He also provided one of the analyses cited in the text. The secretarial staff, especially Marion B. Deloney, who typed the text, and Dave Balderas, who was responsible for the preparation of the figures, performed above and beyond the usual call of duty, often working nights and weekends to complete the text. The assistance of Joseph Agostinelli in solving problems related to the use of the computer is also acknowledged. The authors are deeply grateful for all of their help. Last, but certainly not least, the authors acknowledge the financial support of the Engineering Division of the Office of Basic Energy Sciences of the U. S. Department of Energy under its Grant No. ER 13311.

REFERENCES

1. R. K. Shah and A. L. London, Laminar flow forced convection in ducts. *Adv. Heat Transfer (Suppl. 1)* (1978).
2. R. K. Shah and M. S. Bhatti, Laminar convective heat transfer in ducts. *In* "Handbook of Single Phase Convective Heat Transfer" (S. Kakac, R. K. Shah, and W. Aung, eds.), p. 3–1. Wiley (Interscience), New York, 1987.
3. M. S. Bhatti and R. K. Shah, Turbulent and transition flow convective heat transfer in ducts. "Handbook of Single Phase Convective Heat Transfer" (S. Kakac, R. K. Shah, and W. Aung, eds.), p. 4–1. Wiley (Interscience), New York, 1987.
4. T. F. Irvine, Jr. and J. Karni, Non-Newtonian fluid flow and heat transfer. "Handbook of Single-Phase Convective Heat Transfer" (S. Kakac, R. K. Shah, and W. Aung, eds.), p. 20–1. Wiley, New York, 1987.
5. A. Lawal and A. S. Majumdar, Laminar duct flow and heat transfer to purely viscous non-Newtonian fluids. *Adv. Transp. Processes* **5,** 352 (1987).
6. A. B. Metzner, Heat transfer in Non-Newtonian fluids. *Adv. Heat Transfer* **2,** 357 (1965).
7. A. V. Shenoy and R. A. Mashelkar, Thermal convection in Non-Newtonian fluids. *Adv. Heat Transfer* **15,** 143 (1982).
8. J. C. Slattery and R. B. Bird, Non-Newtonian flow past a sphere. *Chem. Eng. Sci.* **16,** 231 (1961).

9. J. S. Sutterby, Laminar converging flow of dilute polymer solutions in conical sections. *Trans. Soc. Rheol.* **9,** 227 (1965).
10. R. B. Bird, R. C. Armstrong, and O. Hassager, "Dynamics of polymeric liquids," Vol. 1. Wiley, New York, 1977.
11. H. L. Dryden, F. D. Murnaghan, and H. Bateman, *Comm. Hydrodyn. Bull.* No. 84, 197 (1932); reprinted by Dover, New York, 1956.
12. S. M. Marco and L. S. Han, A note on limiting laminar Nusselt number in ducts with constant temperature gradient by analogy to thin-plate theory. *Trans. ASME* **77,** 625 (1955).
13. D. B. Holmes and J. R. Vermeulen, Velocity profiles in ducts with rectangular cross sections. *Chem. Eng. Sci.* **23,** 717 (1968).
14. R. J. Goldstein and D. K. Kreid, Measurement of laminar flow development in a square duct using a Laser-Doppler flowmeter. *J. Appl. Mech.* **34,** 813 (1967).
15. E. M. Sparrow, C. W. Hixon, and G. Shavit, Experiments on laminar flow development in rectangular ducts. *J. Basic Eng.* **89,** 116 (1967).
16. G. F. Muchnik, S. D. Solomonov, and A. R. Gordon, Hydrodynamic development of a laminar velocity field in rectangular channels. *J. Eng. Phys. (USSR)* **25,** 1268 (1973).
17. H. F. P. Purday, "Streamline Flow." Constable, London, 1949; same as "An Introduction to the Mechanics of Viscous Flow." Dover, New York, 1949.
18. N. M. Natarajan and S. M. Lakshmanan, Laminar flow in rectangular ducts: Prediction of velocity profiles and friction factor. *Indian. J. Technol.* **10,** 435 (1972).
19. A. H. P. Skelland, "Non-Newtonian flow and heat transfer." Wiley, New York, 1967.
20. R. S. Schechter, On the steady flow of a non-Newtonian fluid in cylinder ducts. *AIChE J.* **7,** 445 (1961).
21. C. Tien, Laminar heat transfer of power law non-Newtonian fluid—the extension of Graetz-Nusselt problem. *Can. J. Chem. Eng.* **40,** 130 (1962).
22. J. A. Wheeler and E. H. Wissler, The friction factor-Reynolds number relation for the steady flow of pseudoplastic fluids through rectangular ducts, *AIChE J.* **11,** 207 (1966).
23. A. R. Chandrupatla, Analytical and experimental studies of flow and heat transfer characteristics of a non-Newtonian fluid in a square duct. Ph. D. thesis, Indian Institute of Technology, Madras, India, 1977.
24. F. S. Shih, Laminar flow in axisymmetric conduits by a rational approach. *Can. J. Chem. Eng.* **45,** 284–294 (1967).
25. W. Kozicki, C. H. Chou, and C. Tiu, Non-Newtonian flow in ducts of arbitrary cross-sectional shape. *Chem. Eng. Sci.* **21,** 665 (1966).
26. W. Kozicki and C. Tiu, Improved parametric characterization of flow geometries. *Can. J. Chem. Eng.* **49,** 562 (1971).
27. A. B. Metzner and J. C. Reed, Flow of non-Newtonian fluids-Correlation of the laminar, transition and turbulent-flow regions. *AIChE J.* **1,** 434 (1955).
28. J. P. Hartnett, E. Y. Kwack, and B. K. Rao, Hydrodynamic behavior of non-Newtonian fluids in a square duct. *J. Rheol.* **30**(S), S45 (1986).
29. J. P. Hartnett and M. Kostic, Heat Transfer to a viscoelastic fluid in laminar flow through a rectangular channel. *Int. J. Heat Mass Transfer* **28,** 1147 (1985).
30. A. E. Green and R. S. Rivlin, Steady flow of non-Newtonian fluids through tubes. *Q. Appl. Math.* **XV,** 257 (1956).
31. J. A. Wheeler and E. H. Wissler, Steady flow of non-Newtonian fluids in a square duct. *Trans. Soc. Rheol.* **10,** 353 (1966).
32. A. G. Dodson, P. Townsend, and K. Walters, Non-Newtonian flow in pipes of non-circular cross-section, *Comput. Fluids* **2,** 317 (1974).
33. L. Schiller, Die Entwicklung der laminaren Geschwindigkeitsverteilung und ihre Bedeutung für Zähigkeitmessungen. Z. Angew. Math. Mech. **2,** 96 (1922).

34. R. C. Gupta, Flow development in the hydrodynamic entrance region of a flat duct. *AIChE J.* **11,** 1149 (1965).
35. J. W. Williamson, Decay of symmetrical laminar distorted profiles between flat parallel plates. *J. Basic Eng.* **91,** 558 (1969).
36. W. D. Campbell and J. C. Slattery, Flow in the entrance of a tube. *J. Basic Eng.* **85,** 41 (1963).
37. M. S. Bhatti and C. W. Savery, Heat transfer in the entrance region of a straight channel: Laminar flow with uniform wall heat flux. *ASME Pap.* 76-HT-20 (1976); also in a condensed form in *J. Heat Transfer* **99,** 142 (1977).
38. E. Naito and M. Hishida, Laminar boundary layers in the entrance regions of two parallel planes and a circular tube (in Japanese). *Nagoya Kogyo Daigaku Gakuho* **24,** 143 (1972).
39. E. Naito, Laminar heat transfer in the entrance region between parallel plates—the case of uniform heat flux, *Heat Transfer Jpn. Res.* **4,** 63 (1975).
40. J. Yau and C. Tien, Simultaneous development of velocity and temperature profiles for laminar flow of a non-Newtonian fluid in the entrance region of flat ducts, *Can. J. Chem. Eng.* **41,** 139–145 (1963).
41. H. Schlichting, "Boundary Layer Theory," 6th Ed. McGraw-Hill, New York, 1966.
42. M. Collins and W. R. Schowalter, Laminar flow in the inlet region of a straight channel. *Phys. Fluids* **5,** 1122 (1962).
43. M. Roidt and R. D. Cess, An approximate analysis of laminar magnetohydrodynamic flow in the entrance region of a flat duct. *J. Appl. Mech.* **29,** 171 (1962).
44. M. Collins and W. R. Schowalter, Behavior of non-Newtonian fluids in the inlet region of a channel. *AIChE J.* **9,** 98 (1963).
45. L. S. Han, Simultaneous development of temperature and velocity profiles in flat ducts. *Proceedings Heat Transf. Conf. Boulder, Colorado* Part III, 591 (1961).
46. H. L. Langhaar, Steady flow in the transition lengths for incompressible laminar flow in rectangular ducts. *J. Appl. Mech.* **9,** A55 (1942).
47. L. S. Han, Hydrodynamic entrance lengths for incompressible laminar flow in rectangular ducts. *J. Appl. Mech.* **27,** 403 (1960).
48. E. M. Sparrow, S. H. Lin, and T. S. Lundgren, Flow development in the hydrodynamic entrance region of tubes and ducts. *Phys. Fluids* **7,** 338 (1964).
49. R. W. Miller and L. S. Han, Pressure losses for laminar flow in the entrance region of ducts of rectangular and equilateral triangular cross section. *J. Appl. Mech.* **38,** 1083 (1971).
50. D. P. Fleming and E. M. Sparrow, Flow in the hydrodynamic entrance region of ducts of arbitrary cross section. *J. Heat Transfer* **91,** 345 (1969).
51. C. I. Wiginton and C. Dalton, Incompressible laminar flow in the entrance region of a rectangular duct. *J. Appl. Mech.* **37,** 854 (1970).
52. J. R. Bodoia and J. F. Osterle, Finite difference analysis of plane Poiseuille and Couette flow developments. *Appl. Sci. Res.,* Sect. A10, 265 (1964).
53. V. L. Shah and K. Farnia, Flow in the entrance of annular tubes. *Comput. Fluids* **2,** 285 (1974).
54. S. V. Patankar and D. B. Spalding, "Heat and Mass Transfer in Boundary Layers," 2nd Ed. Intertext Books, London, 1970.
55. S. V. Patankar and D. B. Spalding, A calculation procedure for heat, mass and momentum transfer in three-dimensional parabolic flows. *Int. J. Heat Mass Transfer* **15,** 1787 (1972).
56. L. S. Caretto, R. M. Curr, and D. B. Spalding, Two numerical methods for three-dimensional boundary layers. *Comput. Methods Appl. Mech. Eng.* **1,** 39 (1972).
57. R. M. Curr, D. Sharma, and D. G. Tatchell, Numerical predictions of some three-dimensional boundary layers in ducts. *Comput. Methods Appl. Mech. Eng.* **1,** 143 (1972).
58. G. A. Carlson and R. W. Hornbeck, A numerical solution for laminar entrance flow in a square duct. *J. Appl. Mech.* **40,** 25 (1973).

59. G. S. Beavers, E. M. Sparrow, and R. A. Magnuson, Experiments on hydrodynamically developing flow in rectangular ducts of arbitrary aspect ratio. *Int. J. Heat Mass Transfer* **13,** 689 (1970).
60. T. S. Lundgren, E. M. Sparrow, and J. B. Starr, Pressure drop due to the entrance region in ducts of arbitrary cross section. *J. Basic Eng.* **86,** 620 (1964).
61. H. M. Cheng, Analytical investigation of fully developed laminar-flow forced convection heat transfer in rectangular ducts with uniform heat flux. M. S. thesis, Mech. Eng. Dept., MIT, Cambridge, 1957.
62. E. M. Sparrow and R. Siegel, A variational method for fully developed laminar heat transfer in ducts. *J. Heat Transfer* **81,** 157 (1959).
63. M. Iqbal, A. K. Khatry, and B. D. Aggarwala, On the second fundamental problem of combined free and forced convection through vertical non-circular ducts. *Appl. Sci. Res.* **26,** 183 (1972).
64. R. K. Shah, Laminar flow friction and forced convection heat transfer in ducts of arbitrary geometry. *Int. J. Heat Mass Transfer* **18,** 849 (1975).
65. F. W. Schmidt and M. E. Newell, Heat transfer in fully developed laminar flow through rectangular and isosceles triangular ducts. *Int. J. Heat Mass Transfer* **10,** 1121 (1967).
66. V. Javeri, Analyses of laminar thermal entrance region of elliptical and rectangular channels with Kontorovich method. *Wärme Stoffübertragung* **9,** 85 (1976).
67. V. Javeri, Magnetohydrodynamic channel flow heat transfer for temperature boundary conditions of the third kind. *Int. J. Heat Mass Transfer* **20,** 543 (1977).
68. V. Javeri, Heat transfer in laminar entrance region of a flat channel for the temperature boundary conditions of the third kind. *Wärme Stoffübertrag.* **10,** 127 (1977).
69. V. Javeri, Laminar heat transfer in a rectangular channel for the temperature boundary conditions of the third kind. *Int. J. Heat Mass Transfer* **21,** 1029 (1978).
70. C. Tien, Laminar heat transfer of power-law non-Newtonian fluid-Extension of Graetz-Nusselt Problem. *Can. J. Chem. Eng.* **June,** 130 (1962).
71. W. H. Suckow, P. Hrycak, and R. G. Griskey, Heat transfer to polymer solutions and melts flowing between parallel plates. *Polym. Eng. Sci.* **11,** 401 (1971).
72. J. Vlachopoulos and C. K. J. Keung, Heat transfer to a power-law fluid flowing between parallel plates. *AIChE J.* **18,** 1272 (1972).
73. P. B. Kwant and Th. N. M. Van Ravenstein, Non-isothermal laminar channel flow. *Chem. Eng. Sci.* **28,** 1935 (1973).
74. S. M. Richardson, Extended leveque solutions for flows of power law fluids in pipes and channels. *Int. J. Heat Mass Transfer* **22,** 1417 (1979).
75. R. M. Cotta and M. N. Ozisik, Laminar forced convection of power-law non-Newtonian fluids inside ducts. *Wärme Stoffübertrag.* **20,** 211 (1986).
76. J. C. Gottifredi and A. F. Flores, Extended Leveque solution for heat transfer to non-Newtonian fluids in pipes and flat ducts. *Int. J. Heat Mass Transfer* **28,** 903 (1985).
77. S. H. Lin and W. K. Hsu, Heat transfer to power-law non-Newtonian flow between parallel plates. *J. Heat Transfer* **102,** 382–384 (1980).
78. J. Yau and C. Tien, Document 7543, American Documentation Institute, Library of Congress, 1963.
79. T. Lin, Numerical solutions of heat transfer to yield power law fluids in the entrance region. M. S. thesis, Univ. of Wisconsin, Milwaukee, 1977.
80. T. Lin and V. L. Shah. Numerical solution of heat transfer to yield-power-law fluids flowing in the entrance region. *Int. Heat Transfer Conf., 6th, Toronto* **5,** 317 (1978).
81. A. R. Chandrupatla and V. M. K. Sastri, Laminar forced convection heat transfer of a non-Newtonian fluid in a square duct. *Int. J. Heat Mass Transfer* **20,** 1315 (1977).

82. A. R. Chandrupatla and V. M. K. Sastri, Constant wall temperature entry length laminar flow of and heat transfer to a non-Newtonian fluid in a square duct. *Int. Heat Transfer Conf., 6th, Toronto* **5,** 323 (1978).
83. A. Lawal and A. S. Majumdar, Laminar flow and heat transfer in power-law fluids flowing in arbitrary cross-sectional ducts. *Numer. Heat Transfer* **8,** 217 (1985).
84. P. Wibulswas, Laminar-flow heat-transfer in non-circular ducts. PhD dissertation, Department of Mechanical Engineering, University of London, 1966.
85. R. W. Lyczkowski, C. W. Solbrig, and D. Gidaspow, Forced convective heat transfer in rectangular ducts general case of wall resistances and peripheral conduction. Inst. Gas Technol., Tech. Inf. Center, File 3229, 3424 S. State Street, Chicago, Illinois, 1969.
86. R. K. Shah, Thermal entry length solutions for the circular tube and parallel plates. *Proc. Natl. Heat Transfer Conf., 3rd., Indian Inst. Technol., Bombay* **1,** HMI-11-75 (1975).
87. E. Y. Kwack, Personal communication, 1986.
88. F. P. Incropera and H. V. Mahaney, Personal communication, 1988.
89. W. M. Kays and H. C. Perkins, Forced convection, internal flow in ducts. *In* "Handbook of Heat Transfer Fundamentals" (W. M. Rohsenow, J. P. Hartnett, and E. N. Ganic, eds.). McGraw Hill, New York, 1985.
90. Y. I. Cho and J. P. Hartnett, Non-Newtonian fluids in circular pipe flow. *Adv. Heat Transfer* **15,** 59 (1982).
91. S. H. Clark and W. M. Kays, Laminar-flow forced convection in rectangular tubes. *Trans. ASME* **75,** 859 (1953).
92. W. E. Mercer, W. M. Pearce, and J. E. Hitchcock, Laminar forced convection in the entrance region between parallel flat plates. *J. Heat Transfer* **89,** 251 (1967).
93. G. Lombardi and E. M. Sparrow, Measurements of local transfer coefficients for developing laminar flow in flat rectangular ducts. *Int. J. Heat Mass Transfer* **17,** 1135 (1974).
94. H. Nakamura, A. Matsuura, J. Kiwaki, I. Yamada, and M. Hasatani, Laminar heat transfer of high viscosity Newtonian fluids in horizontal rectangular ducts. *Int. Chem. Eng.* **22,** 479 (1982).
95. M. Kostic, Heat transfer and hydrodynamics of water and viscoelastic fluid flow in a rectangular duct. Ph.D. thesis, University of Illinois at Chicago, 1984.
96. Y. Mori and Y. Uchida, Forced convective heat transfer between horizontal flat plates. *Int. J. Heat Mass Transfer* **9,** 803 (1966).
97. M. Akiyama, G. J. Hwang, and K. C. Cheng, Experiments on the onset of longitudinal vortices in laminar forced convection between horizontal plates. *J. Heat Transfer* **93,** 335 (1971).
98. S. Ostrach and Y. Kamotani, Heat transfer augmentation in laminar fully developed channel flow by means of heating from below. *J. Heat Transfer* **97,** 220 (1975).
99. Y. Kamotani and S. Ostrach, Effect on thermal instability on thermally developing laminar channel flow. *J. Heat Transfer* **98,** 62 (1976).
100. G. J. Hwang and C. L. Liu, An experimental study of convective instability in the thermal entrance region of a horizontal parallel-plate channel heated from below. *Can. J. Chem. Eng.* **54,** 521 (1976).
101. Y. Kamotani, S. Ostrach, and H. Miao, Convective heat transfer augmentation in thermal entrance regions by means of thermal instability. *J. Heat Transfer* **101,** 222 (1979).
102. D. G. Osborne and F. P. Incropera, Laminar, mixed convection heat transfer for flow between horizontal parallel plates with asymmetric heating. *Int. J. Heat Mass Transfer* **28,** 207 (1985).
103. B. K. Rao, Heat transfer to viscoelastic fluids in 5:1 rectangular duct. Ph.D. thesis, University of Illinois at Chicago, 1988.

104. J. P. Hartnett, C. Xie, and T. Zhong, Laminar heat transfer to aqueous carbopol solutions in a 2:1 rectangular duct. *Int. Conf. Heat Transfer Energy Conserv., Shenyang, Peoples Republic of China,* October (1988).
105. D. R. Oliver, Non-Newtonian heat transfer: An interesting effect observed in non-circular tubes. *Trans. Inst. Chem., Eng.* **47,** T18 (1969).
106. D. R. Oliver and R. B. Karim, Laminar flow non-Newtonian heat transfer in flattened tubes. *Can. J. Chem. Eng.* **49,** 236 (1971).
107. B. Mena, G. Best, P. Bautista, and T. Sanchez, Heat transfer in non-Newtonian flow through pipes. *Rheol. Acta* **17,** 455 (1978).
108. M. Kostic and J. P. Hartnett, The effects of fluid elasticity on laminar flow in rectangular duct. *ZAMM, Z. Angew. Math. Mech.* **66,** T239 (1986).
109. A. Lawal, Laminar flow and heat transfer to variable property power law fluids in arbitrary cross-sectional ducts. Ph.D. thesis, McGill University, Canada, 1985.
110. A. F. Emery, P. K. Neighbors, and F. B. Gessner, The numerical prediction of developing turbulent flow and heat transfer in a square duct. *J. Heat Transfer* **102,** 51 (1980).
111. O. C. Jones, Jr., An improvement in the calculation of turbulent friction in rectangular ducts. *J. Fluids Eng.* **98,** 173 (1981).
112. T. von Karman, The analogy between fluid friction and heat transfer. *Trans. ASME* **61,** 705 (1939).
113. R. G. Deissler and M. F. Taylor, Analysis of turbulent flow and heat transfer in non-circular passages. *NASA* **TR** R-31 (1959).
114. J. P. Hartnett, J. C. Koh, and S. T. McComas, A comparison of predicted and measured friction factor for turbulent flow through rectangular ducts. *J. Heat Transfer* **84,** 82 (1962).
115. J. Nikuradse, Untersuchungen über turbulente Strömungen in nicht kreisformigen Rohren. *Ing. Arch.* **1,** 306 (1930).
116. L. C. Hoagland, Fully developed turbulent flow in straight rectangular ducts: Secondary flow, its cause and effects on primary flow. Ph.D. thesis, MIT, 1960.
117. A. F. Emery, P. K. Neighbors, and F. B. Gessner, Computational procedure for developing turbulent flow and heat transfer in a square duct. *Numer. Heat Transfer* **2,** 339 (1979).
118. M. Kostic and J. P. Hartnett, Pressure drop and heat transfer in viscoelastic duct flow: A new look. *In* "Fundamentals of Convection in Non-Newtonian Fluids–HTD 79" (J. L. S. Chen, T. M. Ekmann, and G. P. Peterson, eds.). ASME, New York, 1987.
119. D. W. Dodge and A. B. Metzner, Turbulent flow of non-Newtonian fluids, *AIChE J.* **5,** 189 (1959).
120. C. B. Millikan, A critical discussion of turbulent flows in channels and circular tubes. *Appl. Mech. Proc. Int. Congr., 5th* 386 (1939).
121. M. Kostic and J. P. Hartnett, Predicting turbulent friction factors of non-Newtonian fluids in non-circular ducts. *Int. Commun. Heat Mass Transfer* **11,** 345 (1984).
122. J. P. Hartnett and B. K. Rao, Heat transfer and pressure drop for purely viscous non-Newtonian fluids in turbulent flow through rectangular passages. *Wärme Stoffübertrag.* **21,** 261 (1987).
123. J. P. Hartnett and M. Kostic, Turbulent friction factor correlations for purely viscous non-Newtonian fluids in rectangular channel. Submitted to *Can. J. Chem. Eng.* (1988).
124. S. S. Yoo, Heat transfer and friction factors for non-Newtonian fluids in circular tubes. Ph.D. thesis, University of Illinois, Chicago, 1974.
125. S. S. Yoo and J. P. Hartnett, Thermal entrance lengths for non-Newtonian fluids in turbulent pipe flow. *Lett. Heat Mass Transfer* **2,** 189 (1975).
126. P. S. Virk, H. S. Mickley, and K. A. Smith, The ultimate asymptote and mean flow structure in Toms' phenomenon, *Trans. ASME. J. Appl. Mech.* **37,** 488 (1970).

127. T. T. Tung, K. S. Ng, and J. P. Hartnett, Pipe friction factors for concentrated aqueous solutions of polyacrylamide. *Lett. Heat Mass Transfer* **5,** 59 (1978).
128. M. Kostic and J. P. Hartnett, Heat transfer performance of aqueous polyacrylamide solutions in turbulent flow through a rectangular channel. *Int. Commun. Heat Mass Transfer* **12,** 483 (1985).
129. E. Y. Kwack, Y. I. Cho, and J. P. Hartnett, Solvent effects on drag reduction of Polyox solutions in square and capillary tube flows. *J. Non-Newtonian Fluid Mech.* **9,** 79 (1981).
130. J. P. Hartnett and E. Y. Kwack, Empirical correlations of turbulent friction factors and heat transfer coefficients of aqueous polyacrylamide solutions. *Int. Symp. Heat Transfer Proc., Beijing* (1985).
131. A. J. Ghajar and M. Y. Azar, Empirical correlations for friction factor in drag-reducing turbulent pipe flows. *Int. Commun. Heat Mass Transfer,* **15**(6), 705 (1988).
132. L. Washington and W. M. Marks, Rectangular air passages. *Ind. Eng. Chem.* **29,** 337 (1937).
133. W. H. Lowdermilk, W. F. Weiland, and J. N. B. Livingood, Measurement of heat transfer and friction coefficient of air in noncircular ducts at high temperature surfaces. *NACA* RM E53J07 (1954).
134. R. T. Lancet, The effect of surface roughness on the convection of heat transfer coefficient for full developed turbulent flow in ducts with uniform heat flux. *J. Heat Transfer* **81,** 129 (1959).
135. S. Levy, R. A. Fuller, and R. O. Niemi, Heat transfer to water in thin rectangular channels. *J. Heat Transfer* **81,** 129 (1959).
136. J. B. Heineman, An experimental investigation of heat transfer to superheated steam in round and rectangular channels. ANL-6213, Argonne National Laboratory, Argonne, Illinois, (1960).
137. W. R. Gambill and R. D. Bundy, Heat transfer studies of water flow in thin rectangular channels; Part I. Heat transfer, burnout and friction; Part II. Boiling burnout heat flux. *Nucl. Sci. Eng.* **18,** 69, 80 (1964).
138. E. Battista and H. C. Perkins, Turbulent heat and momentum transfer in a square duct with moderate property variations. *Int. J. Heat Mass Transfer* **13,** 1063 (1970).
139. J. L. Novotny, S. T. McComas, E. M. Sparrow, and E. R. G. Eckert, Heat transfer in rectangular ducts with two heated and two unheated walls. *Univ. Minn. Heat Transfer Lab. Tech. Rep.* No. 52 (1963).
140. J. L. Novotny, S. T. McComas, E. M. Sparrow, and E. R. G. Eckert, Heat transfer for turbulent flow in rectangular ducts with two heated and two unheated walls. *AIChE J.* **10,** 466 (1964).
141. H. Barrow, An analytical and experimental study of turbulent gas flow between two smooth parallel walls with unequal heat fluxes. *Int. J. Heat Mass Transfer* **5,** 469 (1962).
142. E. M. Sparrow, J. R. Lloyd, and C. W. Hixon, Experiments on turbulent heat transfer in an asymetrically heated rectangular duct. *J. Heat Transfer* **88,** 170 (1966).
143. D. D. James, B. W. Martin, and D. G. Martin, Forced convection heat transfer in asymmetrically heated ducts of rectangular cross-section. *Proc. 3rd Int. Heat Transfer Conf. 3,* 85–98 (1966).
144. E. Brundrett and P. R. Burroughs, The temperature inner-law and heat transfer for turbulent air flow in a vertical square duct. *Int. J. Heat Mass Transfer* **10,** 1133 (1967).
145. H. M. Tan and W. M. S. Charters, An experimental investigation of forced convection heat transfer for fully-developed turbulent flow in a rectangular duct with asymmetric heating. *Sol. Energy* **13,** 1221 (1970).
146. F. D. Haynes and G. D. Ashton, Turbulent heat transfer in large aspect channels. *J. Heat Transfer* **102,** 384 (1980).
147. E. M. Sparrow, A. Garcia, and W. Chuck, Numerical and experimental turbulent heat transfer results for a one-sided heated rectangular duct. *Numer. Heat Transfer* **9,** 301 (1986).

148. M. Kostic and J. P. Hartnett, Heat transfer to water flowing turbulently through a rectangular duct with asymmetric heating. *Int. J. Heat Mass Transfer* **29,** 1283 (1986).
149. R. K. Shah and R. S. Johnson, Correlations for fully developed turbulent flow through circular and non-circular channels. *Proc. Natl. Heat Mass Transfer Conf., Madras* (1981).
150. H. Barrow, Convection heat transfer coefficients for turbulent flow between parallel plates with unequal heat fluxes. *Int. J. Heat Mass Transfer* **1,** 306 (1961).
151. A. P. Hatton and A. Quarmby, The effect of axially varying and unsymmetrical boundary conditions on heat transfer with turbulent flow between parallel plates. *Int. J. Heat Mass Transfer* **6,** 903 (1963).
152. A. P. Hatton, A. Quarmby, and I. Grundy, Further calculations on the heat transfer with turbulent flow between parallel plates. *Int. J. Heat Mass Transfer* **7,** 817 (1964).
153. A. B. Metzner and P. S. Friend, Heat transfer to turbulent non-Newtonian fluids. *Ind. Eng. Chem. J.* **51,** 879 (1959).

Subject Index

C

D

E

J

K

L

M

O

P

S

T

U

V

W